Lecture Notes in Computer Science 14460

Founding Editors

Gerhard Goos
Juris Hartmanis

Editorial Board Members

The series Lecture Notes in Computer Science (LNCS), including its subseries Lecture Notes in Artificial Intelligence (LNAI) and Lecture Notes in Bioinformatics (LNBI), has established itself as a medium for the publication of new developments in computer science and information technology research, teaching, and education.

LNCS enjoys close cooperation with the computer science R & D community, the series counts many renowned academics among its volume editors and paper authors, and collaborates with prestigious societies. Its mission is to serve this international community by providing an invaluable service, mainly focused on the publication of conference and workshop proceedings and postproceedings. LNCS commenced publication in 1973.

Anupam Chattopadhyay · Shivam Bhasin ·
Stjepan Picek · Chester Rebeiro
Editors

Progress in Cryptology – INDOCRYPT 2023

24th International Conference on Cryptology in India
Goa, India, December 10–13, 2023
Proceedings, Part II

 Springer

Editors
Anupam Chattopadhyay (iD)
Nanyang Technological University
Singapore, Singapore

Shivam Bhasin (iD)
Nanyang Technological University
Singapore, Singapore

Stjepan Picek (iD)
Radboud University
Nijmegen, The Netherlands

Chester Rebeiro (iD)
Indian Institute of Technology Madras
Chennai, India

ISSN 0302-9743   ISSN 1611-3349 (electronic)
Lecture Notes in Computer Science
ISBN 978-3-031-56234-1   ISBN 978-3-031-56235-8 (eBook)
https://doi.org/10.1007/978-3-031-56235-8

This Springer imprint is published by the registered company Springer Nature Switzerland AG
The registered company address is: Gewerbestrasse 11, 6330 Cham, Switzerland

Paper in this product is recyclable.

Foreword

We, at BITS Pilani, K K Birla Goa Campus, were immensely pleased to organize and host INDOCRYPT 2023, the 24th International Conference on Cryptology in India, jointly with Indian Statistical Institute, Kolkata, under the aegis of the Cryptology Research Society of India (CRSI). INDOCRYPT began in 2000 under the leadership of Prof. Bimal Roy at the Indian Statistical Institute, Kolkata, with an intention to target researchers and academicians in the domain of cryptology. Since its inception, this annual conference has not only been considered as the leading Indian venue on cryptology but also has gained recognition among the prestigious cryptology conferences in the world. This was the first time the conference has been held on one of the Indian campuses of BITS Pilani. I gather that INDOCRYPT 2023 received diverse and significant submissions from different countries. The conference also had exciting workshops lined up! I'm grateful to the General Co-chairs, Bimal Kumar Roy, Subhamoy Maitra (ISI Kolkata) and Indivar Gupta (DRDO), for coordinating all the issues related to the organization of the event. We would also like to take this opportunity to thank the Organizing Chair, Organizing Co-chairs, and all members of BITS Goa without whom this conference would not have taken shape.

INDOCRYPT targets researchers from Industry and academia in areas which include but are not limited to foundations, new primitives, cryptanalysis, provable security, cryptographic protocols, and (post)-quantum cryptography. Needless to say, in this era of fast-moving Information Technology, the focus areas of Indocrypt are relevant and impactful. At BITS Goa we are trying our best to take tiny steps in these emerging domains of IT to augment our own strength in related fields. We were excited to host INDOCRYPT while we were celebrating 20 years of our sojourn!

We welcome all delegates, guests and participants, approximately 250 in total, to the post-proceedings of INDOCRYPT 2023 and hope that they immersed themselves in thoughtful discussions and interaction during December 10–13, 2023, the dates of the conference, while enjoying our beautiful campus in the scenic state of Goa.

December 2023

Suman Kundu
Indivar Gupta
Bimal Kumar Roy

Preface

It is our utmost pleasure to welcome you to go through the following pages, which contain the proceedings of the 24th International Conference on Cryptology in India, commonly known as Indocrypt. The conference took place during 10th–13th December, 2023, at the beautiful campus of BITS Pilani, Goa.

We gathered a brilliant group of 62 researchers, who graciously agreed to serve on the Technical Program Committee. Based on a well-publicized Call for Papers, the conference received 86 full paper submissions. Out of these, 27 papers were finally accepted based on double-blind reviews and extensive discussions among the TPC members. Each paper received at least 3 reviews. For several papers, shepherding was done after the review process to ensure the quality of the final manuscript. The acceptance rate was 31.39%, which is slightly lower than that of the previous edition of this conference - 35.23%. The general theme of this edition of the conference was to look into quantum-resilient security. Matching this theme, we had a complete session organized by Bosch India Private Limited, who contributed to 2 invited papers. As part of the technical program, we also solicited tutorial speakers (from CDAC, Bengaluru, India), panel discussions and keynote speeches, which are not included in these proceedings.

We are very thankful to several people and organizations who played a huge supporting role behind the successful organization of this conference. The following list is our humble attempt to acknowledge their service and support. First and foremost, we would like to thank our sponsors, listed in no order of priority - CRSI, CDAC, DRDO, Bosch India Private Limited, Google India, Microsoft Research Lab - India, and BITS Pilani, Goa campus for being gracious hosts. We are very thankful for the service of the entire organization committee for their hard work through the last few months of event organization. Last but not least, we would like to thank our General Chairs, Bimal Roy, Indivar Gupta (SAG, DRDO) and Suman Kundu, for guidance and motivation.

To the scientifically curious reader, we thank you for your engagement in the Indocrypt conference, and hope to see you at future editions!

December 2023

Shivam Bhasin
Anupam Chattopadhyay
Stjepan Picek
Chester Rebeiro

Organization

General Chairs

Bimal Kumar Roy   Indian Statistical Institute, India
Suman Kundu   BITS Pilani, India
Indivar Gupta   DRDO, India

Technical Program Co-chairs

Anupam Chattopadhyay   NTU, Singapore
Shivam Bhasin   Temasek Labs, NTU, Singapore
Stjepan Picek   Radboud University, The Netherlands
Chester Rebeiro   IIT Madras, India

Organizing Chairs

Snehanshu Saha   BITS Pilani, India
Santonu Sarkar   BITS Pilani, India

Organizing Co-chairs

Hemant Rathore   BITS Pilani, India
Sravan Danda   BITS Pilani, India
Diptendu Chatterjee   BITS Pilani, India
Gargi Alavani   BITS Pilani, India
Prabal Paul   BITS Pilani, India

Publicity Chair

Shilpa Gondhali   BITS Pilani, India

Industry Chair

Kunal Korgaonkar   BITS Pilani, India

Web Co-chairs

Nilanjan Datta   IAI TCG CREST, India
Hemant Rathore   BITS Pilani, India
Shreenivas A. Naik   BITS Pilani, India

Publication Co-chairs

Debolina Ghatak   BITS Pilani, India
Saranya G. Nair   BITS Pilani, India

Program Committee

Avishek Adhakari   Presidency University, Kolkata, India
Subidh Ali   IIT Bhilai, India
Nalla Anandakumar   Continental Automotive, Singapore
Shi Bai   Florida Atlantic University, USA
Anubhab Baksi   Nanyang Technological University, Singapore
Shivam Bhasin   Temasek Lab, Nanyang Technological University,
                            Singapore
Rishiraj Bhattacharyya   University of Birmingham, UK
Christina Boura   University of Versailles, France
Suvradip Chakraborty   VISA Inc, USA
Anupam Chattopadhyay   Nanyang Technological University, Singapore
Sherman S. M. Chow   Chinese University of Hong Kong, China
Debayan Das   Purdue University, USA
Prem Laxman Das   SETS Chennai, India
Nilanjan Datta   IAI, TCG CREST, Kolkata, India
Avijit Dutta   IAI, TCG CREST, Kolkata, India
Ratna Dutta   IIT Kharagpur, India
Keita Emura   Kanazawa University, Japan
Andre Esser   Technology Innovation Institute, Abu Dhabi, UAE
Satrajit Ghosh   Indian Institute of Technology Kharagpur, India
Indivar Gupta   SAG DRDO, India
Takanori Isobe   University of Hyogo, Japan
Dirmanto Jap   Nanyang Technological University, Singapore
Mahavir Jhanwar   Ashoka University, India
Selçuk Kavut   Balikesir University, Turkey
Mustafa Khairallah   Seagate Technologies, Singapore

Invited Papers

Secure Boot in Post-Quantum Era

Megha Agrawal ⓘ, Kumar Duraisamy ⓘ, Karthikeyan Sabari Ganesan ⓘ,
Shivam Gupta ⓘ, Suyash Kandele ⓘ, Sai Sandilya Konduru ⓘ,
Harika Chowdary Maddipati ⓘ, K. Raghavendra ⓘ, Rajeev Anand Sahu ⓘ,
and Vishal Saraswat ⓘ

Bosch Global Software Technologies, Bangalore, India
{megha.agrawal,kumar.d,karthikeyansabari.ganesan,
shivam.gupta,suyash.kandele,
fixed-term.konduru.saisandilya,harikachowdary.maddipati,
k.raghavendra3,rajeevanand.sahu,
vishal.saraswat}@in.bosch.com

Abstract. Secure boot is a standard feature for ensuring the authentication and integrity of software. For this purpose, secure boot leverages the advantage of Public Key Cryptography (PKC). However, the fast-developing quantum computers have posed serious threats to the existing PKC. The cryptography community is already preparing to thwart the expected quantum attacks. Moreover, the standardization of post-quantum cryptographic algorithms by NIST has advanced to 4th round, after selecting and announcing the post-quantum encryption and signature schemes for standardization. Hence, considering the recent developments, it is high time to realize a smooth transition from conventional PKC to post-quantum PKC. In this paper, we have implemented the PQ algorithms recently selected by NIST for standardization– CRYSTALS-Dilithium, FALCON and SPHINCS+ as candidate schemes in the secure boot process. Furthermore, we have also proposed an idea of double signing the boot stages, for enhanced security, with signing a classical signature by a post-quantum signature. We have also provided efficiency analysis for various combinations of these double signatures.

Keywords: Secure Boot · Post-Quantum Cryptography · Dilithium · FALCON · SPHINCS+

Patent Landscape in the field of Hash-Based Post-Quantum Signatures

Megha Agrawal⬤, Kumar Duraisamy⬤, Karthikeyan Sabari Ganesan⬤,
Shivam Gupta⬤, Suyash Kandele⬤, Sai Sandilya Konduru⬤,
Harika Chowdary Maddipati⬤, K. Raghavendra⬤, Rajeev Anand Sahu⬤,
and Vishal Saraswat⬤

Bosch Global Software Technologies, Bangalore, India
{megha.agrawal,kumar.d,karthikeyansabari.ganesan,
shivam.gupta,suyash.kandele,
fixed-term.konduru.saisandilya,harikachowdary.maddipati,
k.raghavendra3,rajeevanand.sahu,
vishal.saraswat}@in.bosch.com

Abstract. Post-Quantum Cryptography (PQC) is one of the most fascinating topics of recent developments in cryptography. Following the ongoing standardization process of PQC by NIST, industry and academia both have been engaged in PQC research with great interest. One of the candidate algorithms finalized by NIST for the standardization of post-quantum digital signatures belongs to the family of Hash-based Signatures (HBS). In this paper, we thoroughly explore and analyze the state-of-the-art patents filed in the domain of post-quantum cryptography, with special attention to HBS. We present country-wise statistics of the patents filed on the topics of PQC. Further, we categorize and discuss the patents on HBS based on the special features of their construction and different objectives. This paper will provide scrutinized information and a ready reference in the area of patents on hash-based post-quantum signatures.

Keywords: Post-Quantum Cryptography · Patents · Hash-based signatures · XMSS · LMS · SPHINCS$^+$

Contents – Part II

Contents – Part I

Attacks

Secure Computation, Algorithm
Hardness, Privacy

Threshold-Optimal MPC with Friends and Foes

Nikolas Melissaris[1]([✉]), Divya Ravi[2], and Sophia Yakoubov[1]

[1] Aarhus University, Aarhus, Denmark
{nikolas,sophia.yakoubov}@cs.au.dk
[2] University of Amsterdam, Amsterdam, The Netherlands
d.ravi@uva.nl

Abstract. Alon *et al.* (Crypto 2020) initiated the study of MPC with *Friends and Foes* (FaF) security, which captures the desirable property that even up to h^* honest parties should learn nothing additional about other honest parties' inputs, even if the t corrupt parties send them extra information. Alon *et al.* describe two flavors of FaF security: *weak* FaF, where the simulated view of up to h^* honest parties should be indistinguishable from their real view, and *strong* FaF, where the simulated view of the honest parties should be indistinguishable from their real view *even in conjunction with the simulated/real view of the corrupt parties*. They give several initial FaF constructions with guaranteed output delivery (GOD); however, they leave some open problems. Their only construction which supports the optimal corruption bounds of $2t + h^* < n$ (where n denotes the number of parties) only offers weak FaF security and takes much more than the optimal three rounds of communication. In this paper, we describe two new constructions with GOD, both of which support $2t + h^* < n$. Our first construction, based on threshold FHE, is the first three-round construction that matches this optimal corruption bound (though it only offers weak FaF security). Our second construction, based on a variant of BGW, is the first such construction that offers strong FaF security (though it requires more than three rounds, as well as correlated randomness). Our final contribution is further exploration of the relationship between FaF security and similar security notions. In particular, we show that FaF security does not imply mixed adversary security (where the adversary can make t active and h^* passive corruptions), and that Best of Both Worlds security (where the adversary can make t active or $t + h^*$ passive corruptions, but not both) is orthogonal to both FaF and mixed adversary security.

Keywords: Secure Computation · Friends and Foes · Guaranteed Output Delivery · Round Complexity

N. Melissaris, D. Ravi and S. Yakoubov—Funded by the European Research Council (ERC) under the European Unions's Horizon 2020 research and innovation programme under grant agreement No 803096 (SPEC).
S. Yakoubov—Funded by the Danish Independent Research Council under Grant-ID DFF-2064-00016B (YOSO).

1 Introduction

A set of n mutually distrusting parties who have secrets $x_1, \ldots, x_n$ can use *secure multi-party computation* (MPC) [4,22] to compute a joint function $f(x_1, \ldots, x_n)$ of their secrets, without revealing anything more about those secrets to one another. MPC is typically parametrized by a threshold t such that as long as t or fewer participants collude, they cannot subvert the privacy and correctness guarantees of the computation. However, if t parties deviate from the protocol, no guarantees are made about what the remaining $n-t$ parties learn. Many MPC protocols (such as [13,15,21]) make use of this by relying on fall-back protocols where, in the event of cheating, if one or more parties are identified as definitely *not* being one of the t cheaters, they are entrusted with the others' secrets.

Of course, this is not what we would like to use in practice. We would like even our honest peers — who do not collude with some central malicious adversary — not to learn our secrets. Alon *et al.* [1] introduce MPC with Friends and Foes (or MPC with FaF security), which captures exactly this guarantee. Informally, a protocol achieves (t, h^*)-FaF security if, as in the standard definition of MPC, for any (non-uniform) adversary $\mathcal{A}$ there exists a simulator $\mathcal{S}_\mathcal{A}$ which produces a view indistinguishable from that of the t corrupt parties without seeing the inputs of the honest parties. However, for FaF security, there must additionally exist a simulator $\mathcal{S}_{\mathcal{A}_{\mathcal{H}^*}}$ for every subset of up to h^* of the honest parties which produces a view indistinguishable from that of those honest parties, without seeing the inputs of the *remaining* honest parties. This implies that no matter what messages the corrupt parties send, they can not cause any h^* honest parties to learn more about their peers' inputs than they should.

Alon *et al.* define two degrees of FaF security:

Weak FaF. Here, though the output of $\mathcal{S}_\mathcal{A}$ must be indistinguishable from the real view of the t corrupt parties and the output of $\mathcal{S}_{\mathcal{A}_{\mathcal{H}^*}}$ must be indistinguishable from the real view of the h^* honest parties, taken together, those views may be distinguishable from the set of real views. (That is, the simulated views may not be mutually consistent.)

Strong FaF. Here, the outputs of $\mathcal{S}_\mathcal{A}$ and $\mathcal{S}_{\mathcal{A}_{\mathcal{H}^*}}$ must be *jointly* indistinguishable from the real views of the t corrupt parties and h^* honest parties.

One can think of strong FaF as modeling the case where the adversary receives some feedback about what the honest parties learned, and weak FaF as modeling the case where there is no such feedback.

1.1 Prior Work

Alon *et al.* showed some inherent limitations on MPC with FaF security (Sect. 1.1), and gave some initial constructions (Sect. 1.1). Their results primarily focus on the notion of guaranteed output delivery (GOD) where corrupt parties cannot prevent the honest parties from obtaining the output.

Limitations. Alon *et al.* consider two parameters of MPC protocols: number of rounds, and thresholds. They showed that two-round MPC with weak FaF security (and thus also strong FaF security) and GOD is impossible even for the lowest possible thresholds ($t = h^* = 1$); so, three rounds is the best we could hope to achieve. They then showed that even weak FaF security (and thus also strong FaF security) is unachievable for certain thresholds irrespective of the number of rounds. In particular, let n be the number of participants, t the bound on the number of corrupt parties, and h^* the bound on the number of honest parties who should learn nothing about other honest parties' secrets. Alon *et al.* show the following:

- Weak FaF secure MPC with GOD is impossible when $2t + h^* \geq n$.
- Information theoretic (statistical) weak FaF secure MPC with GOD is impossible if:
  - $2t + 2h^* \geq n$ (even if broadcast is available).
  - $2t + 2h^* \geq n$ or $3t \geq n$ (when broadcast is *not* available).
- Information theoretic (perfect) weak FaF secure MPC with GOD is impossible (even when broadcast is available) when $3t + 2h^* \geq n$.

Constructions. Alon *et al.* give several initial constructions of FaF secure MPC with GOD. They describe a round-optimal (three-round) construction that achieves strong FaF security, but only for $5t + 3h^* < n$. They also describe a threshold-optimal ($2t + h^* < n$) construction that only achieves *weak* FaF security. Finally, they show several information theoretic constructions. We summarize all of the constructions in Fig. 1.

1.2   Related Work

Exploring the potential of FaF security belongs in the general area of exploring the robustness of different security models. Robustness is a highly desirable feature because it removes a potential denial of service threat and supports user participation. Towards this goal there is work that has been done by Koti *et al.* [18] which proposes a robust Privacy Preserving Machine Learning (PPML) framework for a variety of machine learning tasks, Dalskov *et al.* [7] in which the authors introduce a novel four-party honest-majority MPC protocol with active security which has guaranteed output delivery (with some extensions to their main protocol), and [19] where the authors introduce a robust actively secure 4-party protocol for secure training and inference.

Exploring FaF security is a relatively new endeavour and the following works concurrent to ours have shown some promising results. In [17], Koti *et al.* show a concretely efficient (1,1)-FaF secure 5PC protocol and in [12] Hedge *et al.* prove the necessity of semi-honest oblivious transfer for FaF-secure protocols with optimal resiliency and they show a ring-based 4PC protocol, which achieves fairness and GOD in the case of optimal corruptions: 1 semi-honest and 1 malicious adversaries.

Construction	FaF Level	Security	Threshold	Rounds	Assumptions	Preprocessing?
From GMW [1]	Weak	Comp	$2t + h^* < n$	Dependent on κ	OT, OWP	no
From DI [1]	Strong	Comp	$5t + 3h^* < n$	3	PRG	no
BGW emulation [1]	Strong	Statistical IT	$2t + 2h^* < n$	Dependent on κ	Broadcast	no
BGW emulation [1]	Strong	Perfect IT	$3t + 2h^* < n$	Dependent on κ	None	no
Ours						
TFHE-FaF	Weak	Comp	$2t + h^* < n$	3	Lattices, Broadcast	no
BGW-BT-Comp	Strong	Comp	$2t + h^* < n$	$O(\kappa)$	Enhanced Trapdoor Permutations	Beaver triples

Fig. 1. Constructions. n denotes the number of participants, t denotes the bound on the number of corruptions, h^* denotes the bound on the number of honest parties against whom we want privacy, and κ denotes the multiplicative depth of the circuit being evaluated.

1.3   Our Contributions

In this paper, we close two of the gaps left open by the constructions of Alon *et al.*. We also extend the study of how FaF security relates to other security notions. The focus of our work is FaF security with GOD.

First, we give a three-round construction that achieves weak FaF security for $2t + h^* < n$ in the CRS (common reference string) model. This is the first construction that is optimal both in terms of the number of rounds and in terms of the threshold (even though it does not achieve the stronger notion of FaF). Second, we give a construction that achieves strong FaF security for $2t + h^* < n$. This is the first strong FaF construction to achieve the optimal threshold (though the number of rounds depends on the multiplicative depth of the function being computed). A caveat of our second construction is that it relies on correlated randomness.

Finally, we further the study of the relationship between FaF security and other related notions of security. Recalling the standard notions, while actively corrupt parties are completely controlled by the adversary and may deviate arbitrarily from the protocol; passively corrupt parties follow the protocol steps but leak their internal states to the adversary. Alon *et al.* showed that *Mixed Adversary* security (where the adversary can make t active corruptions and h^* passive ones) does not imply FaF security in the computational setting. We show the other direction; that FaF security does not imply mixed adversary security. We additionally consider *Best of Both Worlds* (BoBW) security [14,16] (where the adversary can either make t active corruptions or $t + h^*$ passive ones, but not both). We show that FaF security does not imply BoBW security, and vice versa. These results are summarized in Fig. 2.

Technical Overview

Three-Round Weak FaF Construction. Our three-round construction is based on decentralized threshold FHE (described in Sect. 4), and follows the blueprint of Gordon *et al.* [11]. In the first round, the participants exchange public keys. They then encrypt their inputs *to the set of all participants' public keys*, and broadcast the resulting ciphertexts in the second round. Once they receive one another's ciphetexts, they perform the homomorphic computation of the function locally, and broadcast their individual partial decryptions in the third round. Everyone is then able to locally combine these partial decryptions and obtain the output. Gordon *et al.* show that this construction achieves guaranteed output delivery in the presence of a dishonest minority. We show that it has weak FaF security as long as $2t + h^* < n$.

Strong FaF Construction. Our strong FaF construction is based on BGW [4]. We proceed in three steps; first, we show that BGW with Beaver triple pre-processing [3] achieves guaranteed output delivery in the presence of an adaptive mixed adversary making t fail-stop corruptions (where fail-stop corruptions are similar to passive corruptions, except that the parties may additionally choose to abort at any step) and h^* passive corruptions, as long as $2t + h^* < n$. We then apply the compiler of Canetti *et al.* [6], which relies on adaptively secure commitments and zero knowledge proofs, to instead allow our mixed adversary to make t *active* corruptions and h^* passive corruptions. Finally, we rely on the observation of Alon *et al.* that adaptive security implies strong FaF security to obtain our result.

Relation of FaF to Other Notions. We consider FaF security, BoBW security and mixed adversary security. We describe several protocols that achieve some of these notions but not others, which, taken together with the results of Alon *et al.*, shows that all three notions are incomparable. In Fig. 2 we summarize what we know about the relationship of FaF, BoBW and mixed adversaries.

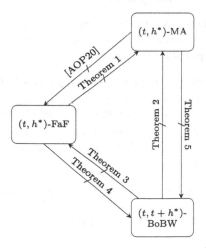

Fig. 2. Relationships of FaF to other notions. MA denotes security against mixed adversaries; BoBW denotes (active/passive) best of both worlds security.

1.4 Organization

In Sect. 2, we recall the definitions of FaF security. In Sect. 3, we describe our results about the relationship of FaF, BoBW and mixed adversary security. In Sect. 4, we describe decentralized threshold fully homomorphic encryption (dTFHE), which we use in one of our constructions. In Sect. 5 and Sect. 6, we describe our round optimal weak FaF and strong FaF constructions, respectively.

1.5 Notation

We use λ to denote the security parameter. By $\mathsf{poly}(\lambda)$ we denote a polynomial function in λ. By $\mathsf{negl}(\lambda)$ we denote a negligible function; that is, a function f such that $f(\lambda) < \frac{1}{p(\lambda)}$ holds for any polynomial $p(\cdot)$ and sufficiently large λ. We denote the set $\{1, \ldots, k\}$ by $[k]$ (or, equivalently, by $[1, \ldots, k]$).

2 Definitions

In this section, we recall the definitions given by Alon *et al.* [1] of Friends and Foes security. They consider a classical adversary $\mathcal{A}$ corrupting t of parties, who would like to leak unauthorized information to some honest parties. So, we really have *two separate adversaries* in this setting:

1. The adversary $\mathcal{A}$ who *actively* corrupts a subset $\mathcal{I} \subseteq [n]$ of the parties, meaning that she can instruct the parties in $\mathcal{I}$ to arbitrarily deviate from the protocol. ($\mathcal{A}$ is given auxiliary input $y_{\mathcal{A}}$.)
2. An adversary $\mathcal{A}_{\mathcal{H}^*}$ who *passively* corrupts a subset $\mathcal{H}^* \subseteq [n] \setminus \mathcal{I}$ of the honest parties. ($\mathcal{A}_{\mathcal{H}^*}$ is given auxiliary input $y_{\mathcal{H}^*}$.)

For security parameter λ, inputs $\boldsymbol{x} = (x_1, \ldots x_n)$ and auxiliary inputs $y_{\mathcal{A}}, y_{\mathcal{H}^*}$, we define the following random variables for a real-world execution of protocol Π that computes a function f:

$\mathsf{OUT}_{\mathcal{A},\Pi}^{\mathsf{REAL}}(1^\lambda, \boldsymbol{x})$ is the output of the non-active parties (where, non-active refers to the honest and passively corrupt parties) $\mathcal{H} = [n] \setminus \mathcal{I}$.
$\mathsf{VIEW}_{\mathcal{A},\Pi}^{\mathsf{REAL}}(1^\lambda, \boldsymbol{x})$ is $\mathcal{A}$'s view during an execution of the protocol.
$\mathsf{VIEW}_{\mathcal{A},\mathcal{A}_{\mathcal{H}^*},\Pi}^{\mathsf{REAL}}(1^\lambda, \boldsymbol{x})$ is $\mathcal{A}_{\mathcal{H}^*}$'s view during an execution of the protocol. Since $\mathcal{A}$'s best strategy in order to leak information is to send her entire view, we assume that $\mathcal{A}_{\mathcal{H}^*}$'s view includes $\mathcal{A}$'s view.

We can now formalize the global view of the real world execution of Π:

$$\mathsf{REAL}_{1^\lambda, \boldsymbol{x}, y_{\mathcal{A}}, y_{\mathcal{H}^*}}^{\Pi, \mathcal{A}, \mathcal{A}_{\mathcal{H}^*}} = \left(\mathsf{VIEW}_{\mathcal{A},\Pi}^{\mathsf{REAL}}(1^\lambda, \boldsymbol{x}), \mathsf{VIEW}_{\mathcal{A},\mathcal{A}_{\mathcal{H}^*},\Pi}^{\mathsf{REAL}}(1^\lambda, \boldsymbol{x}), \mathsf{OUT}_{\mathcal{A},\Pi}^{\mathsf{REAL}}(1^\lambda, \boldsymbol{x}) \right)$$

It is useful to define the following projection of the global view to the view of each of the adversaries and the non-active parties' output:

$$\mathsf{REAL}_{1^\lambda, \boldsymbol{x}, y_{\mathcal{A}}, y_{\mathcal{H}^*}}^{\Pi, \mathcal{A}, \mathcal{A}_{\mathcal{H}^*}}(\mathcal{A}) = \left(\mathsf{VIEW}_{\mathcal{A},\Pi}^{\mathsf{REAL}}(1^\lambda, \boldsymbol{x}), \mathsf{OUT}_{\mathcal{A},\Pi}^{\mathsf{REAL}}(1^\lambda, \boldsymbol{x}) \right)$$

and

$$\text{REAL}_{1^\lambda,x,y_\mathcal{A},y_{\mathcal{H}^*}}^{\Pi,\mathcal{A},\mathcal{A}_{\mathcal{H}^*}}(\mathcal{A}_{\mathcal{H}^*}) = \left(\text{VIEW}_{\mathcal{A},\mathcal{A}_{\mathcal{H}^*},\Pi}^{\text{REAL}}(1^\lambda,x), \text{OUT}_{\mathcal{A},\Pi}^{\text{REAL}}(1^\lambda,x)\right).$$

Similarly we can define the following random variables for an ideal-world execution:

$\text{OUT}_{\mathcal{A},f}^{\text{IDEAL}}(1^\lambda,x)$ is the output of the non-active parties in $\mathcal{H}$.
$\text{VIEW}_{\mathcal{A},f}^{\text{IDEAL}}(1^\lambda,x)$ is $\mathcal{A}$'s simulated view.
$\text{VIEW}_{\mathcal{A},\mathcal{A}_{\mathcal{H}^*},f}^{\text{IDEAL}}(1^\lambda,x)$ is $\mathcal{A}_{\mathcal{H}^*}$'s simulated view.

As before we can formalize the global view of the ideal world execution:

$$\text{IDEAL}_{1^\lambda,x,y_\mathcal{A},y_{\mathcal{H}^*}}^{f,\mathcal{A},\mathcal{A}_{\mathcal{H}^*}} = \left(\text{VIEW}_{\mathcal{A},f}^{\text{IDEAL}}(1^\lambda,x), \text{VIEW}_{\mathcal{A},\mathcal{A}_{\mathcal{H}^*},f}^{\text{IDEAL}}(1^\lambda,x), \text{OUT}_{\mathcal{A},f}^{\text{IDEAL}}(1^\lambda,x)\right)$$

We define the following projections:

$$\text{IDEAL}_{1^\lambda,x,y_\mathcal{A},y_{\mathcal{H}^*}}^{f,\mathcal{A},\mathcal{A}_{\mathcal{H}^*}}(\mathcal{A}) = \left(\text{VIEW}_{\mathcal{A},f}^{\text{IDEAL}}(1^\lambda,x), \text{OUT}_{\mathcal{A},f}^{\text{IDEAL}}(1^\lambda,x)\right)$$

and

$$\text{IDEAL}_{1^\lambda,x,y_\mathcal{A},y_{\mathcal{H}^*}}^{f,\mathcal{A},\mathcal{A}_{\mathcal{H}^*}}(\mathcal{A}_{\mathcal{H}^*}) = \left(\text{VIEW}_{\mathcal{A},\mathcal{A}_{\mathcal{H}^*},f}^{\text{IDEAL}}(1^\lambda,x), \text{OUT}_{\mathcal{A},f}^{\text{IDEAL}}(1^\lambda,x)\right).$$

2.1 FaF Security

We say that a protocol Π computes a functionality f with (t,h^*)-FaF security if the two following simulators exist for any adversary $\mathcal{A}$ that statically corrupts at most t parties:

- A simulator $\mathcal{S}_\mathcal{A}$ which simulates $\mathcal{A}$'s view in the real world, and
- A simulator $\mathcal{S}_{\mathcal{A}_{\mathcal{H}^*}}$ which simulates the view of any subset $\mathcal{H}^*$ of size at most h^* of the honest parties, such that when given $\mathcal{S}_\mathcal{A}$'s entire state, $\mathcal{S}_{\mathcal{A}_{\mathcal{H}^*}}$ can generate a view that is indistinguishable from the real world view of $\mathcal{H}^*$.

We say that $\mathcal{S}_{\mathcal{A}_{\mathcal{H}^*}}$ is given the entire state of $\mathcal{S}_\mathcal{A}$ because in the real world, nothing stops an adversary from sending her entire view to one (or more) honest parties.

FaF Functionality with GOD. In an ideal evaluation of the function f, the parties interact with the functionality as follows:

Inputs. Each party P_i is given input x_i. Adversary $\mathcal{A}$ is given auxiliary input $y_\mathcal{A} \in \{0,1\}^*$ and x_i for all $i \in \mathcal{I}$. Adversary $\mathcal{A}_{\mathcal{H}^*}$ is given auxiliary input $y_{\mathcal{H}^*} \in \{0,1\}^*$ and x_i for all $i \in \mathcal{H}^*$.

Parties Send Input. All non-active parties (i.e. the honest and passive parties) $i \in [n] \setminus \mathcal{I}$ send their inputs x_i to the functionality. $\mathcal{A}$ chooses inputs x_i' for $i \in \mathcal{I}$ as the input of each corrupt party and sends it to the functionality. For non-active parties i, we define $x_i' := x_i$.

Computation. The functionality computes $z = (z_1, \ldots, z_n) = f(x_1', \ldots, x_n')$ and sends z_i to each party i.

$\mathcal{A}_{\mathcal{H}^*}$ **receives** $\mathcal{A}$**'s state.** $\mathcal{A}_{\mathcal{H}^*}$ receives $\mathcal{A}$'s randomness, inputs, auxiliary input, and z_i for $i \in \mathcal{I}$.

Output. Each non-active party i outputs z_i, while the corrupted parties output nothing. $\mathcal{A}_{\mathcal{H}^*}$ and $\mathcal{A}$ output some function of their view.

Weak and Strong FaF Definitions. In the following, we use $\equiv$ to denote computational indistinguishability.

Definition 1 (Weak FaF). *Let Π be a protocol for computing f. We say that Π computes f with computational weak (t, h^*)-FaF security (with GOD), if the following holds. For every non-uniform PPT adversary $\mathcal{A}$ controlling a set $\mathcal{I} \subset [n]$ of size at most t in the real world, there exists a non-uniform PPT simulator $\mathcal{S}_{\mathcal{A}}$ controlling $\mathcal{I}$ in the ideal world; and for every subset of the remaining parties $\mathcal{H}^* \subset [n] \setminus \mathcal{I}$ of size at most h^* controlled by a non-uniform passive PPT adversary $\mathcal{A}_{\mathcal{H}^*}$ there exists a non uniform PPT simulator $\mathcal{S}_{\mathcal{A}_{\mathcal{H}^*}}$, controlling $\mathcal{H}^*$ in the ideal world such that*

$$\mathsf{IDEAL}_{1^\lambda, x, y_{\mathcal{A}}, y_{\mathcal{H}}}^{\mathcal{S}_{\mathcal{A}}, \mathcal{S}_{\mathcal{A}_{\mathcal{H}^*}}}(\mathcal{S}_{\mathcal{A}}) \equiv \mathsf{REAL}_{1^\lambda, x, y_{\mathcal{A}}, y_{\mathcal{H}}}^{\mathcal{A}, \mathcal{A}_{\mathcal{H}}}(\mathcal{A})$$

and

$$\mathsf{IDEAL}_{1^\lambda, x, y_{\mathcal{A}}, y_{\mathcal{H}}}^{\mathcal{S}_{\mathcal{A}}, \mathcal{S}_{\mathcal{A}_{\mathcal{H}^*}}}(\mathcal{S}_{\mathcal{A}_{\mathcal{H}^*}}) \equiv \mathsf{REAL}_{1^\lambda, x, y_{\mathcal{A}}, y_{\mathcal{H}}}^{\mathcal{A}, \mathcal{A}_{\mathcal{H}}}(\mathcal{A}_{\mathcal{H}^*})$$

for any set of inputs $x \in (\{0,1\}^*)^n$, any auxiliary inputs $(y_{\mathcal{A}}, y_{\mathcal{H}^*}) \in (\{0,1\}^*)^2$, and any large enough security parameter $\lambda \in \mathbb{N}$.

Definition 2 (Strong FaF). *For $\mathcal{A}, \mathcal{S}_{\mathcal{A}}, \mathcal{S}_{\mathcal{A}_{\mathcal{H}^*}}$ defined as in Definition 1, we say that Π computes f with computational strong (t, h^*)-FaF security (with GOD), if*

$$\mathsf{IDEAL}_{1^\lambda, x, y_{\mathcal{A}}, y_{\mathcal{H}}}^{\mathcal{S}_{\mathcal{A}}, \mathcal{S}_{\mathcal{A}_{\mathcal{H}^*}}} \equiv \mathsf{REAL}_{1^\lambda, x, y_{\mathcal{A}}, y_{\mathcal{H}}}^{\mathcal{A}, \mathcal{A}_{\mathcal{H}}}$$

for any set of inputs $x \in (\{0,1\}^*)^n$, any auxiliary inputs $(y_{\mathcal{A}}, y_{\mathcal{H}^*}) \in (\{0,1\}^*)^2$, and any large enough security parameter $\lambda \in \mathbb{N}$.

The main difference is that in the strong notion of FaF security we want the simulated views of $\mathcal{A}$ and $\mathcal{A}_{\mathcal{H}^*}$ to be indistinguishable from the real views *even when taken together.*

3   Relation of FaF to Other Notions

Somewhat surprisingly, Alon *et al.* show that standard security against a static adversary making $t + h^*$ active corruptions does not imply (t, h^*) FaF security. Informally, this is because the simulator $\mathcal{S}_{\mathcal{A}_{\mathcal{H}^*}}$ for the honest parties is not allowed to choose which input to send to the ideal functionality, so it does not have as much power as the standard security simulator $\mathcal{S}$. Security against a (t, h^*) mixed adversary making t active corruptions and h^* passive corruptions also does not imply (t, h^*) FaF security. This is because the mixed adversary simulator can decide the active parties' inputs based on the passive parties' inputs; however, the FaF simulator $\mathcal{S}_{\mathcal{A}}$ does not know any of the honest parties' inputs when simulating.

On the other hand, Alon *et al.* show that security against an *adaptive* adversary making $t + h^*$ active corruptions *does* imply (t, h^*) FaF security. This is because a simulator for an adaptive adversary needs to be able to handle corruptions which occur after the end of the protocol execution, at which point it cannot choose the input even for actively corrupt parties. We observe that the proof given by Alon *et al.* also shows that security against an adaptive *mixed* (t, h^*) adversary making t active corruptions and h^* passive corruptions implies (t, h^*) FaF security, and we use this in our strong FaF construction.

Here, we further explore the relationship between FaF security and other security notions. First, we show the other direction: that (t, h^*) FaF security does not imply security against a (t, h^*) mixed adversary making t active corruptions and h^* passive corruptions, making the FaF and mixed adversary models incomparable. We do so by giving an example (Example 1) of a protocol that achieves (t, h^*) FaF security but not (t, h^*) mixed adversary security. We also consider $(t, t + h^*)$ Best of Both Worlds (BoBW) security, where the same protocol must tolerate either t active corruptions *or* $t + h^*$ passive corruptions (but not both). We show that BoBW is incomparable to both FaF and mixed adversaries.

Example 1 ($\Pi_{\neg MA}$). Consider a function $f(x_1, \ldots, x_n)$ such that the output of f does not reveal the set of inputs $(x_1, \ldots, x_n)$ such as the XOR function. So we have that $f(x_1, \ldots, x_n) = x_1 \oplus \cdots \oplus x_n$ Suppose a protocol Π_{FaF} computes any f with (t, h^*) FaF security[1]. Now, consider a function $g((x_1, \rho_1), \ldots, (x_n, \rho_n))$, where each party P_i has an additional input ρ_i. g returns $(x_1, \ldots, x_n)$ to everyone if at least $t + 1$ of the ρ_i's are equal, and returns $f(x_1, \ldots, x_n)$ to everyone otherwise. The following protocol $\Pi_{\neg MA}$ computes $f(x_1, \ldots, x_n)$ with (t, h^*) FaF security but not with security against a (t, h^*) mixed adversary.

1. Each party P_i chooses ρ_i uniformly at random from a large space.
2. The parties use Π_{FaF} to compute $g((x_1, \rho_1), \ldots, (x_n, \rho_n))$.
3. Each party outputs the value returned by Π_{FaF}.

---

[1] Here, it is implicitly assumed that the values of (t, h^*) are such that they admit FaF security.

Theorem 1. *Protocol* $\Pi_{\neg MA}$ *(Example 1) computes* $f(x_1, \ldots, x_n)$ *with* (t, h^*) *FaF security but not with security against a* (t, h^*) *mixed adversary making* t *active corruptions and* h^* *passive corruptions.*

Proof. $\Pi_{\neg MA}$ achieves FaF security, because the adversary cannot possibly guess the honest parties' randomly chosen values ρ_i, and so cannot exploit the additional leakage given by the output of g.

However, the mixed adversary knows h^* passive party inputs and randomly chosen ρ_i's, and can choose one of those to set corrupt parties' ρ_i's to. This allows the mixed adversary to learn all parties' inputs. This is clearly insecure due to our assumption that the set of inputs are not revealed by the output of f.

Remark 1. We observe that since the above reduction (Example 1) is information-theoretic, plugging in the statistically-secure FaF protocol of [1] to instantiate Π_{FaF} would yield a statistically-secure protocol $\Pi_{\neg MA}$ that satisfies (t, h^*) FaF security but not (t, h^*) mixed security. This shows a separation between statistical FaF and mixed security at the protocol level, which was left as an open question in [1] (the only known separation at the protocol level was for computational security).

Theorem 2. *If* Π_{FaF} *from Example 1 is replaced with a protocol* Π_{BoBW} *which has* $(t, t + h^*)$ *BoBW security, then protocol* $\Pi_{\neg MA}$ *(Example 1) computes* $f(x_1, \ldots, x_n)$ *with* $(t, t + h^*)$ *BoBW security, but not with security against a* (t, h^*) *mixed adversary making* t *active corruptions and* h^* *passive corruptions.*

Proof. $\Pi_{\neg MA}$ achieves BoBW security, since an adversary making just t corruptions cannot guess honest parties' ρ_i values and thus cannot exploit the additional leakage, and an adversary making $t + h^*$ corruptions cannot dishonestly choose passive parties' values ρ_i to be equal.

However, as in the proof of Theorem 1, the mixed adversary knows h^* passive party inputs and randomly chosen ρ_i's, and can choose one of those to set corrupt parties' ρ_i's to.

We next show that BoBW security does not imply FaF security, by giving an example (Example 2) of a protocol that achieves $(t, t + h^*)$ BoBW security but not (t, h^*) FaF security.

Example 2 ($\Pi_{\neg FaF}$). Consider a function $f(x_1, \ldots, x_n)$ as in Example 1. Suppose a protocol Π_{BoBW} that computes any function with $(t, t+h^*)$ BoBW security. The following protocol $\Pi_{\neg FaF}$ computes $f(x_1, \ldots, x_n)$ with $(t, t + h^*)$ BoBW security, but not with (t, h^*) FaF security.

1. The parties use Π_{BoBW} to compute $f(x_1, \ldots, x_n)$.
2. If a party P_i receives a special "attack" message from t parties, it sends its input x_i to the *other* $n - t$ parties. (Note that *sending* an "attack" message is not part of the instructions.)
3. Each party outputs the value returned by Π_{BoBW}.

Theorem 3. *Protocol* $\Pi_{\neg \mathsf{FaF}}$ *(Example 2) computes* $f(x_1, \ldots, x_n)$ *with* $(t, t + h^*)$ *BoBW security, but not with* (t, h^*) *FaF security.*

Proof. $\Pi_{\neg \mathsf{FaF}}$ achieves $(t, t + h^*)$ BoBW security: if there are only passive corruptions, no party will send an "attack" message, and thus the second step of $\Pi_{\neg \mathsf{FaF}}$ will never come into play. If there are only t active corruptions, they are able to trigger the attack, but will not learn the honest parties' inputs because those inputs are only sent to parties who didn't send attack messages. $\Pi_{\neg \mathsf{FaF}}$ does not achieve (t, h^*) FaF security: the t corrupt parties can easily trigger an attack, causing all the honest parties to learn one another's inputs. This violates security due to our assumption that the output of f does not reveal any party's input.

It remains to show that neither FaF or mixed adversary security imply BoBW security. This follows from the fact that in both FaF and mixed adversary security, the simulator can choose the inputs of the actively corrupt parties. However, in BoBW security, in the case where the adversary only makes passive corruptions, the simulator is unable to choose the inputs of any parties.

Example 3 ($\Pi_{\neg \mathsf{BoBW}}$). Consider a function $f(x_1, \ldots, x_n)$ as in the previous examples, and a protocol Π_{FaF} that computes f with (t, h^*) FaF security. Now, consider a function $g((x_1, y_1), \ldots, (x_n, y_n))$, where each party P_i has an additional input y_i. g returns $y_i \wedge f(x_1, \ldots, x_n)$ to each party P_i. It returns $\perp$ to everyone else. The following protocol $\Pi_{\neg \mathsf{BoBW}}$ computes $g((x_1, y_1), \ldots, (x_n, y_n))$ with (t, h^*) FaF security but not with $(t, t + h^*)$ BoBW security.

1. The parties use Π_{FaF} to compute $z = f(x_1, \ldots, x_n)$.
2. Each party P_i outputs $y_i \wedge z$.

Theorem 4. *Protocol* $\Pi_{\neg \mathsf{BoBW}}$ *(Example 3) computes* $g((x_1, y_1), \ldots, (x_n, y_n))$ *with* (t, h^*) *FaF security but not with* $(t, t + h^*)$ *BoBW security.*

Proof. $\Pi_{\neg \mathsf{BoBW}}$ achieves (t, h^*) FaF security: a simulator can always set one of the actively corrupt parties' auxiliary inputs y_i to be 1 to learn the output of f.

However, it does *not* achieve BoBW security: in the case where the adversary can only make passive corruptions, if all parties' auxiliary inputs y_i are 0, the simulator does not learn the output of f^2, which it needs in order to simulate successfully.

Theorem 5. *If* Π_{FaF} *from Example 3 is replaced with a protocol* Π_{MA} *which has* (t, h^*) *mixed adversary security, then protocol* $\Pi_{\neg \mathsf{BoBW}}$ *(Example 3) computes* $g((x_1, y_1), \ldots, (x_n, y_n))$ *with* (t, h^*) *mixed adversary security but not with* $(t, t + h^*)$ *BoBW security.*

The proof is the same as the proof of Theorem 4.

---

[2] We assume that the function f is such that the output of f depends on the inputs of all parties.

Remark 2. We design the function g in such a way that *any* party can learn the output of f by setting their auxiliary input y_i to 1. This gives us FaF and mixed adversary security; no matter whom the adversary actively corrupts, the simulator can use that party to learn the output of f, and simulate for the rest.

4   Building Block: Decentralized Threshold FHE

We recap the definitions of d-out-of-n decentralized threshold fully homomorphic encryption (dTFHE) as presented by Boneh et al. [5].

Syntax. A dTFHE scheme is a tuple of PPT algorithms (DistGen, Enc, Eval, PDec, Combine, SimPDec) with the following syntax:

DistGen$(1^\lambda, 1^\kappa, i; \rho_i) \to (\text{pk}_i, \text{sk}_i)$: On input the security parameter λ, a depth bound κ, party index i and randomness ρ_i, the distributed setup outputs a public-secret key pair $(\text{pk}_i, \text{sk}_i)$ for party i. The public key of the scheme is denoted by $\text{pk} = (\text{pk}_1 \parallel \text{pk}_2 \parallel \ldots \parallel \text{pk}_n)$.

Enc$(\text{pk}, m; \rho) \to \text{c}$: On input a public key pk and a plaintext m in the message space $\mathcal{M}$, the randomized algorithm outputs a ciphertext c.

Eval$(\text{pk}, C, \text{c}_1, \ldots, \text{c}_k) \to \text{c}$: On input a public key pk, a circuit $C : \mathcal{M}^k \to \mathcal{M}$ of depth at most κ, and a set of k ciphertexts $\text{c}_1, \ldots, \text{c}_k$ (where $k = \text{poly}(\lambda)$), the evaluation algorithm outputs an encrypted evaluation c.

PDec$(\text{pk}, \text{sk}_i, \text{c}) \to \text{d}_i$: On input the public key pk, a ciphertext c and a secret key sk_i the algorithm outputs a partial decryption d_i.

Combine$(\text{pk}, \{\text{d}_i\}_{i \in S}) \to m \backslash \bot$: On input a public key pk and a set partial decryptions $\{\text{d}_i\}_{i \in S}$ where $S \subseteq [n]$, the combination algorithm outputs a plaintext m or the symbol $\bot$.

SimPDec$(\text{c}, \text{pk}, \{\text{sk}_i\}_{i \in \mathcal{I}}, z) \to \{\text{d}_i\}_{i \in [n] \backslash \mathcal{I}}$: On input a ciphertext c, the public key pk, the secret keys of at most d parties, and the target plaintext z, the simulated decryption algorithm outputs partial decryptions on behalf of the rest of the parties which are consistent with c decrypting to z.

Properties. As in a standard homomorphic encryption scheme, we require that a dTFHE scheme satisfies correctness and security, which we describe informally below. We give the formal definitions in the full version [20].

Correctness. Informally, a dTFHE scheme is said to be *correct* if combining at least $d+1$ partial decryptions of any honestly generated ciphertext output by the evaluation algorithm returns the correct evaluation of the corresponding circuit on the underlying plaintexts.

Semantic Security. Informally, a dTFHE scheme satisfies *semantic security* if no PPT adversary can distinguish between encryptions of a pair of (adversarially chosen) plaintext messages m_0 and m_1 of the same length, even given the secret keys corresponding to a subset $\mathcal{I}$ of the parties for any set $\mathcal{I}$ of size at most d. Since we use the dTFHE scheme as a tool in our semi-malicious

MPC construction[3], we define the notion with respect to a semi-malicious adversary $\mathcal{A}$.

Simulation Security. Informally, a dTFHE scheme satisfies *simulation security* if there exists an efficient algorithm SimPDec that takes as input a ciphertext c, the public key pk, the secret keys of at most d parties and the target plaintext z, and outputs a set of partial decryptions on behalf of the rest of the parties such that its output is computationally indistinguishable from the output of the real algorithm PDec that outputs partial decryptions of the ciphertext c using the corresponding secret keys for the same subset of parties. Similar to semantic security, we define this notion with respect to a semi-malicious adversary $\mathcal{A}$.

5   Three-Round MPC with Weak FaF and Guaranteed Output Delivery

In this section, we present a three-round MPC construction in the CRS model (where it is assumed that parties have access to a common reference string at the beginning of the protocol execution) that achieves weak FaF security and GOD when $n > 2t + h^*$. This is round-optimal, following the impossibility of two-round MPC with weak FaF security and GOD shown in [1] (which holds even in the CRS model). Specifically, their result shows that there are functionalities that cannot be computed with $(1, 1)$ weak FaF security and GOD in less than three rounds, for any $n \geq 3$[4].

Our construction is based on the three-round construction of Gordon *et al.* [11] that achieves standard security with GOD against $t < n/2$ active corruptions. At a high-level, this construction uses the tool of distributed threshold fully homomorphic encryption scheme (dTFHE) with threshold t and proceeds as follows. First, the distributed setup allows the parties to obtain their individual public/secret key pairs. Each of them broadcasts their public key. Next, the parties broadcast encryptions of their input, which can be homomorphically evaluated to compute an encryption of the output. In the last round, the parties compute and broadcast partial decryptions of the output ciphertext, which can be combined to obtain the output.

We observe that the above construction admits (t, h^*) weak FaF security and GOD when $n > 2t + h^*$, if a dTFHE scheme with threshold $(t + h^*)$ is used instead. Intuitively, security of such a dTFHE scheme ensures that the joint view of the active and passively corrupt parties comprising of $(t + h^*)$ secret keys does not reveal any information about the inputs of the honest parties (beyond the output of computation). Correctness of such a scheme ensures that even if up

---

[3] where semi-malicious security [2] refers to security against an adversary who needs to follow the protocol specification, but has the liberty to decide the input and random coins in each round.

[4] The proof in [1] is a simple observation regarding the impossibility of computing the AND functionality with GOD in two rounds against 2 corrupted parties, proved by Gennaro et al. [9]; which holds in the common reference string (CRS) model.

to t parties abort, guaranteed output delivery is achieved because the partial decryptions sent by the remaining $n - t > t + h^*$ parties suffice to compute the output.

Similar to the work of Gordon et al. [11], we present a three-round construction $\Pi_{\mathsf{wFaF}}^{\mathsf{sm}}$ that is secure against semi-malicious adversaries. Semi-malicious security was introduced by Aashrov et al. [2] and subsequently used in many works as a stepping-stone on the way to achieving active security. Recall that a semi-malicious adversary needs to follow the protocol specification, but has the liberty to decide the input and random coins in each round. Additionally, the parties controlled by the semi-malicious adversary may choose to abort at any time. To upgrade a semi-malicious construction to achieve active security, the general round-preserving compiler of Asharov et al. uses UC NIZKs (non-interactive zero-knowledge proofs) in the CRS model. We should point out that since FaF security becomes relevant only in settings with more than 3 parties, the standard security notion for 2-party NIZK used as a building block would not affect FaF security of the overall protocol.

We give a formal description of the protocol $\Pi_{\mathsf{wFaF}}^{\mathsf{sm}}$ below.

Inputs: Each party P_i has an input $x_i \in \{0,1\}^\lambda$.
Output: $f(x_1, \ldots, x_n)$, where the function f is represented by a circuit C.
Tools: A $(t + h^*)$-out-of-n dTFHE scheme (DistGen, Enc, Eval, PDec, Combine, SimPDec). Such a scheme can be built based on LWE [5,11].

Round 1: Each party P_i does the following:
  – Computes $(\mathsf{pk}_i, \mathsf{sk}_i) \leftarrow \mathsf{DistGen}(1^\lambda, 1^\kappa, i; \rho_i)$ using randomness ρ_i.
  – Broadcasts pk_i.
Round 2: All parties set $\mathsf{pk} := (\mathsf{pk}_1 \| \ldots \| \mathsf{pk}_n)$ (where a default public key is used corresponding to parties who have aborted in the first round).

  Each party P_i does the following:
  – Computes the encryption of its input as $c_i \leftarrow \mathsf{Enc}(\mathsf{pk}, x_i)$.
  – Broadcasts c_i.
Round 3: Each party P_i does the following:
  – Computes the homomorphic evaluation of the circuit C on the ciphertexts as $c \leftarrow \mathsf{Eval}(\mathsf{pk}, C, c_1, \ldots, c_n)$, where c_j is computed using default input and randomness if P_j aborted during the previous rounds.
  – Computes her own partial decryption as $d_i \leftarrow \mathsf{PDec}(\mathsf{pk}, \mathsf{sk}_i, c)$.
  – Broadcasts the partial decryption d_i.
Output Computation: Let $S \subseteq [n]$ denote the set of parties who have not yet aborted. Each party combines the partial decryptions broadcast by parties in S to obtain the output as $z \leftarrow \mathsf{Combine}(\mathsf{pk}, \{d_j\}_{j \in S})$.

Protocol $\Pi_{\mathsf{wFaF}}^{\mathsf{mal}}$. Let $\Pi_{\mathsf{wFaF}}^{\mathsf{mal}}$ denote the three-round construction obtained by applying the compiler of Asharov et al. [2] to the three-round protocol $\Pi_{\mathsf{wFaF}}^{\mathsf{sm}}$ in the semi-malicious setting. In particular, in every round of $\Pi_{\mathsf{wFaF}}^{\mathsf{mal}}$, each party executes the actions of the corresponding round of $\Pi_{\mathsf{wFaF}}^{\mathsf{sm}}$ along with a non-interactive zero-knowledge proving that she is following the protocol consistently with respect to certain random coins. This compiler is round-preserving and preserves security of the underlying construction (i.e. if the underlying protocol achieves GOD, so does the compiled protocol).

We state the formal theorem below.

Theorem 6. *Let f be an efficiently computable n-party function and let $n > 2t + h^*$. Assuming a setup with CRS and the existence of a $(t + h^*)$-out-of-n decentralized threshold fully homomorphic encryption scheme, the three-round protocol $\Pi_{\mathsf{wFaF}}^{\mathsf{mal}}$ achieves (t, h^*) weak FaF security with guaranteed output delivery.*

Proof. To prove the theorem, we construct the simulators $\mathcal{S}_{\mathcal{A}}$ and $\mathcal{S}_{\mathcal{A}_{\mathcal{H}^*}}$ for $\Pi_{\mathsf{wFaF}}^{\mathsf{sm}}$ in the semi-malicious setting. This suffices to complete the proof, since active security would directly follow from the result of [2].

We begin with the description of $\mathcal{S}_{\mathcal{A}}$. Let $\mathcal{I}$ and $\mathcal{H}^*$ denote the set of indices of the t semi-malicious corrupt parties and the remaining parties respectively.

First Round: $\mathcal{S}_{\mathcal{A}}$ simulates public keys of honest parties by honestly generating key pairs.

Second Round: $\mathcal{S}_{\mathcal{A}}$ simulates input ciphertexts by broadcasting $c_j \leftarrow \mathsf{Enc}(\mathsf{pk}, 0)$ on behalf of P_j, where $j \in [n] \setminus \mathcal{I}$.

Third Round: $\mathcal{S}_{\mathcal{A}}$ does the following:

- Computes $c \leftarrow \mathsf{Eval}(\mathsf{pk}, C, c_1, \ldots, c_n)$ honestly, where c_i is computed using default input and randomness if P_i ($i \in \mathcal{I}$) aborted during the previous rounds.
- Reads the witness tape of the semi-malicious adversaries to learn the inputs x_i and secret keys sk_i for each $i \in \mathcal{I}$. If P_i has aborted, x_i is set as the default input.
- Invokes the ideal functionality $\mathcal{F}$ on inputs x_i for $i \in \mathcal{I}$ and receives the output z.
- Runs the simulated decryption algorithm to obtain partial decryptions as $\{d_j\}_{j \in [n] \setminus \mathcal{I}} \leftarrow \mathsf{SimPDec}(c, \mathsf{pk}, \{\mathsf{sk}_i\}_{i \in \mathcal{I}}, z)$.
- Broadcasts the partial decryption d_j on behalf of party P_j ($j \in [n] \setminus \mathcal{I}$).

Consider the following sequence of hybrids:

Hyb_0: Same as the real world execution of $\Pi_{\mathsf{wFaF}}^{\mathsf{sm}}$.

Hyb_1: Same as Hyb_0, except that the partial decryptions for honest parties are computed as $\mathsf{SimPDec}(c, \mathsf{pk}, \{\mathsf{sk}_i\}_{i \in \mathcal{I}}, z)$ instead of $\mathsf{PDec}(\mathsf{pk}, \mathsf{sk}_i, c)$. It follows from simulation security of the dTFHE scheme that this hybrid is indistinguishable from the previous hybrid.

Hyb_2: Same as Hyb_1, except that the input ciphertexts of honest parties are computed as $c_j \leftarrow \mathsf{Enc}(\mathsf{pk}, 0)$ using a dummy input 0 for $j \in [n] \setminus \mathcal{I}$, instead of using the actual input x_j. It follows from the semantic security of the dTFHE scheme that this hybrid is indistinguishable from the previous hybrid.

Since Hyb_2 corresponds to the ideal execution and every pair of consecutive hybrids are indistinguishable, this completes the proof that the view of $\mathcal{A}$ in the real world is indistinguishable from her view in the ideal world execution.

Next, suppose we fix the adversary $\mathcal{A}$ corrupting the parties in $\mathcal{I}$ where $\mathcal{I}$ is of size at most t, and let $\mathcal{H}^* \subseteq [n] \setminus \mathcal{I}$ of size at most h^*. The passive simulator $\mathcal{S}_{\mathcal{A}_{\mathcal{H}^*}}$ works very similarly to $\mathcal{S}_{\mathcal{A}}$. The only difference is that the messages the simulator sends to the adversary on behalf of the parties in $\mathcal{H}^*$ are the actual messages computed as per protocol specifications. For instance, the input ciphertexts will be computed as encryptions of the real inputs of parties in $\mathcal{H}^*$ (unlike in $\mathcal{S}_{\mathcal{A}}$ where it was computed as encryptions of dummy input 0). Similarly, the partial decryptions of the parties in $\mathcal{H}^*$ would be computed honestly using their secret keys (not as output of SimPDec). Lastly, the sequence of hybrids and indistinguishability can be argued as above; it follows from the semantic and simulation security of the dTFHE scheme having threshold $(t + h^*)$.

Remark 3. Note that the protocol described in this section is not strongly FaF secure; this is because the simulator $\mathcal{S}_{\mathcal{A}}$ does not know the honest parties' inputs, and instead encrypts a default value on their behalf. However, this simulation cannot then be consistent with an honest party's simulated view, since their view must include their real input and randomness that maps that input to the ciphertext they broadcast.

6   Optimal-Threshold MPC with Strong FaF and Guaranteed Output Delivery

Alon *et al.* [1] showed that even weak FaF with GOD is impossible if $2t + h^* \geq n$. Their proof holds even if arbitrary correlated randomness is available to the parties. So, the best that we could possible hope for is strong FaF with GOD and $2t + h^* + 1 = n$.

In this section we prove that BGW with Beaver triple preprocessing and augmented with adaptive zero knowledge proofs achieves exactly this. We do this in three steps, as described in Fig. 3. First, in Sect. 6.1, we prove that BGW with Beaver triple preprocessing achieves security with GOD against adaptive mixed (fail-stop/passive) adversaries. Second, in Sect. 6.2, we use the compiler of Canetti *et al.* [6] to show that our protocol against adaptive mixed (fail-stop/passive) adversaries can be augmented with adaptive commitments and zero knowledge proofs of correct behavior in order to achieve security against adaptive mixed (*active*/passive) adversaries. Third, we invoke a theorem from Alon *et al.* [1] to argue that any such protocol achieves GOD with strong FaF security.

We restate the theorem of Alon *et al.* [1] which connects adaptive security to strong FaF below. Alon *et al.* prove that $(t + h^*)$ adaptive security implies (t, h^*) FaF; however, their proof can be strengthened (with no modifications necessary) to show that *adaptive mixed security with a corruption budget of t active corruptions and h^* passive corruptions* implies (t, h^*) FaF. As mentioned in Sect. 3, security against a (t, h^*) mixed adversary making t active corruptions and h^* passive corruptions does not imply (t, h^*) FaF security because the mixed adversary simulator can decide the active parties' inputs based on the passive

Fig. 3. Getting (t, h^*)-FaF Security from BGW

parties' inputs but the FaF simulator $\mathcal{S}_{\mathcal{A}}$ does not know the honest parties' inputs. However, in the case of an adaptive mixed adversary, any party that is corrupted by an adaptive malicious adversary after the protocol has terminated can essentially be viewed as a passive corruption since the adversary cannot control their input.

Theorem 7 ([1], Theorem 5.3). *Let* type $\in$ {computational, statistical, perfect}, *and let* Π *be an n-party protocol computing some n-party functionality f with* type *adaptive mixed security with a corruption budget of t active corruptions and* h^* *passive corruptions. Then* Π *computes f with* type *strong* (t, h^*)-*FaF-security.*

6.1 Adaptive BGW Against Mixed (Fail-Stop/Passive) Adversaries

We first recall BGW (Sect. 6.1), and how Beaver triples can be used to improve the corruption threshold (Sect. 6.1).

Brief Overview of BGW Without Preprocessing. Let $\mathbb{F}$ be a finite field. A secret value s is secret shared using a polynomial $f_s(x) \in \mathbb{F}[x]$ of degree d such that $f_s(0) = s$ and each party P_i holds $f_s(i)$. The evaluation of the circuit then proceeds gate by gate. In order to add two secret shared values x and y, each party P_i can locally add the shares $f_x(i)$ and $f_y(i)$ that they are holding to get $f_{x+y}(i) = f_x(i) + f_y(i)$. This is a valid sharing of $x + y$, because f_{x+y} is of the same degree as f_x and f_y and $f_{x+y}(0) = f_x(0) + f_y(0) = x + y$. To reconstruct an output z, all parties broadcast their share $f_z(i)$, and everyone interpolates the polynomial.

Multiplication gates pose more of a challenge. If a party P_i computes $f'_{xy}(i) = f_x(i)f_y(i)$, she gets a point on a polynomial f'_{xy} such that $f'_{xy}(0) = xy$, which is what we wanted. The caveat is that f'_{xy} is of degree $2d$; if the degree keeps growing in this way, it will be too high to admit interpolation given only $n - t$ points (which is necessary if we would like to withstand fail-stop corruptions). Additional work needs to be done to reduce the degree of this polynomial: $2d+1$ parties P_i need to reshare their points $f'_{xy}(i)$ using a new d-degree polynomial (that is, party P_i will pick a random polynomial r_i of degree t such that $r_i(0) = f'_{xy}(i)$ and send $r_i(j)$ to all other parties P_j). Each party P_j can locally compute $f_{xy}(j) = \sum_{i \in \mathcal{R}} \lambda_i r_i(j)$ where the λ_i's are the appropriate Lagrange coefficients.

Threshold Requirements. Now, consider the (t, h^*)-FaF setting. In order to withstand fail-stop corruptions, even during degree reduction, we need $2d < n - t$. In order to have FaF security, we need $t + h^* \leq d$. We thus need

$$t + h^* < \frac{n - t}{2}$$

$$\Rightarrow 3t + 2h^* < n,$$

which is not optimal.

Brief Overview of BGW with Preprocessing. In order to approach the optimal threshold, we change the way we do multiplication to rely on *Beaver triple pre-processing* [3]. Since this is work that explores the feasibility of FaF we can assume that the Beaver triples are generated by a trusted third party. We leave open the question of how such triples can be generated with security against a FaF adversary, eliminating the need for setup by a trusted third party. We give each party shares $f_a(i), f_b(i), f_c(i)$ of a randomly chosen a and b, and of their product $c = ab$. We use these shares in order to multiply x and y as follows. Each party P_i computes $f_\delta(i) = f_x(i) - f_a(i)$ and $f_\epsilon(i) = f_y(i) - f_b(i)$ and broadcasts these values. All parties reconstruct $\delta = x - a$ and $\epsilon = y - b$, and compute $f_{xy}(i) = f_c(i) + \epsilon f_x(i) + \delta f_y(i) - \delta\epsilon$. We can see that f_{xy} has the same degree d, and that $f_{xy}(0) = c + \epsilon x + \delta y - \delta\epsilon = c + (y-b)x + (x-a)y - (x-a)(y-b) = c + xy - xb + yx - ya - xy + xb + ya - ab = xy$, because $c = ab$.

Threshold Requirements. As in Sect. 6.1, in order to have FaF security, we need $t + h^* \leq d$. However, in order to withstand fail-stop corruptions, now we only need $d < n - t$. Putting these together, we need $t + h^* < n - t \Rightarrow 2t + h^* < n$, which is optimal.

Adaptive Security of BGW with Preprocessing. We now prove that BGW with Preprocessing with $2t + h^* < n$ and with $d = t + h^*$ is adaptively secure.

Theorem 8. *The construction summarized in Sect. 6.1 with $2t + h^* < n$ and with $d = t + h^*$ achieves security with guaranteed output delivery against an adaptive adversary with a budget of t fail-stop corruptions and h^* passive corruptions.*

We follow the blueprint of Damgård and Nielsen [8] for proving the adaptive security of BGW. We start by describing a simulator $\mathcal{S}_{static}$ for a static adversary, who is given the inputs of the passive and the fail-stop corrupted parties (which are similar to passive corruptions except that the adversary can choose to abort them at any step in the real world and substitute their input with a default input in the ideal world). $\mathcal{S}_{static}$ interacts with the adversary on behalf of the set $\mathcal{H}$ of the $n - t - h^*$ honest parties. She shares the input 0 on behalf of honest parties $P_i \in \mathcal{H}$. If a fail-stop party P_i fails to share an input, that party implicitly gives the degree 0 sharing of 0, where every share is 0 (thereby, its default input can be considered as 0). $\mathcal{S}_{static}$ forwards these inputs to the ideal functionality to obtain the output. $\mathcal{S}_{static}$ follows the protocol on behalf of the honest parties up until it's time to reconstruct the output. When it's time to reconstruct the output, $\mathcal{S}_{static}$ computes the difference $\delta = z - z'$, where z' is the computed output (shared on the polynomial $f_{z'}$), and z is the output dictated by the ideal functionality. $\mathcal{S}_{static}$ chooses a random polynomial Δ of degree d such that $\Delta(0) = \delta$ and $\Delta(i) = 0$ for all $i \in \mathcal{I}$. Notice that $f_{z'} + \Delta$ is a sharing f_z of the desired output $z = z' + \delta$. $\mathcal{S}_{static}$ uses shares $f_z(i)$ on behalf of honest parties $P_i \in \mathcal{H}$. (We make use of an assumption that the output z is produced by a multiplication gate. This forces the polynomial f_z used for the output to be a random degree d polynomial with the only constraint being that $f_z(0) = z$.)

We now describe the simulator $\mathcal{S}$ for an adaptive adversary. $\mathcal{S}$ starts out much like $\mathcal{S}_{static}$, by interacting with the adversary on behalf of the initial set $\mathcal{H}$ of honest parties, and maintaining a record of their views (including their shares of all intermediate values). However, unlike $\mathcal{S}_{static}$, at any point $\mathcal{S}$ may be asked to explain the view of one of these honest parties P_i, in the event that P_i becomes corrupt. It is insufficient for $\mathcal{S}$ to hand over the current simulated view of P_i, since $\mathcal{S}$ used the input 0 on behalf of P_i; this makes the simulated view clearly distinguishable from the real view, where the real input would have been used. So, $\mathcal{S}$ must adjusts the shares she has stored to account for the use of the real input. Upon the corruption of party P_i, the simulator makes the following adjustments:

Shares of Input x_i: When party P_i is corrupted, $\mathcal{S}$ learns the real input x_i. Let $\mathcal{H}$ be the set of parties who were honest before the corruption of party P_i, and $\mathcal{I}$ be the set of parties who were corrupt. Note that $|\mathcal{I}| < t + h^* = d$; so, the views of the parties in $\mathcal{I}$ contain no information about $\{f_{x_i}(j)\}_{j \in \mathcal{H}}$, even if she knew x_i. $\mathcal{S}$ cannot change the shares of parties in $\mathcal{I}$, so she picks a random difference polynomial Δ s.t. $\Delta(0) = x_i$ and $\Delta(i) = 0$ for $i \in \mathcal{I}$. $\mathcal{S}$ updates the sharing polynomial as $f_{x_i} := f_{x_i} + \Delta$.

Addition or Multiplication by a Constant: In this case we only have local computation, so $\mathcal{S}$ simply recomputes the shares of the honest parties that were affected by the change to f_{x_i}.

Multiplication: Recall that for the multiplication of x by y each party j has published her share of δ, which is $f_\delta(j) = f_x(j) - f_a(j)$, and her share ϵ, which is $f_\epsilon(j) = f_y(j) - f_b(j)$ (where a, b and c are the Beaver triple s.t. $c = ab$). We will consider only x, a and δ; the case for y, b and ϵ is analogous. Since the adversary already saw $\delta = x - a$, and x might have changed, $\mathcal{S}$ must adjust a accordingly. Let x_{old} be the old value of x (shared on $f_{x_{old}}$), and a_{old} be the old value of a (shared on $f_{a_{old}}$). $\mathcal{S}$ defines $a := a_{old} + (x - x_{old})$, and lets $f_a := f_{a_{old}} + f_x - f_{x_{old}}$. Observe that $\delta = x_{old} - a_{old} = x_{old} - (a - (x - x_{old})) = x_{old} - a + x - x_{old} = x - a$, as desired. The shares on f_a are now consistent with the shares of δ previously published.

Note that we have changed the values of a and b, but not the value of c; so, it may no longer be the case that $c = ab$. However, this is not a problem, since the view of the adversary has no information about c (as she has insufficient shares).

Output Reconstruction: For an output z, the adversary has already seen all of the points on f_z. So, now we need to fix P_i's view to be consistent with $f_z(i)$. Recall that we assume that the output is produced by a multiplication gate, which uses a Beaver triple a, b, c. Let c_{old} be the old value of c (shared on $f_{c_{old}}$). Let a (shared on f_a) and b (shared on f_b) be the rest of that Beaver triple, and let x (shared on f_x) and y (shared on f_y) be the two inputs to the multiplication gate. Recall that $\delta = x - a$ and $\epsilon = y - b$ are fixed. $\mathcal{S}$ defines $c := z - \left(\epsilon x + \delta y - \delta \epsilon\right)$, and $f_c := f_z - (\epsilon f_x + \delta f_y - \delta \epsilon)$. This is consistent with what the adversary has seen.

As before, it may no longer be the case that $c = ab$; however, this is not a problem, since the adversary will have seen too few shares to be able to tell.

6.2 Adaptive BGW Against Mixed (Active/Passive) Adversaries

We now discuss how the security of the construction in Sect. 6.1 can be boosted to achieve guaranteed output delivery against a mixed adaptive adversary controlling t parties *actively* and h^* parties passively. This can be done by using the generic compiler of Canetti *et al.* [6] that transforms a protocol secure against adaptive fail-stop corruptions to a protocol secure against adaptive active corruptions. At a high-level, this compiler follows the GMW compiler paradigm [10] where the parties (a) run an augmented coin-tossing protocol to obtain their respective uniformly distributed random tapes and commitments to other parties' random tapes, (b) commit to their inputs, (c) run the underlying fail-stop adaptively secure MPC protocol, while proving in each round using zero-knowledge that the computations have been done correctly. We can use the same compiler to upgrade security of the construction in Sect. 6.1 (achieving adaptive security against t fail-stop corruptions and h^* passive corruptions) to adaptive security against t active corruptions and h^* passive corruptions with a minor simplification. Since our underlying protocol satisfies perfect correctness (i.e. the protocol results in the correct output when everyone executes the protocol steps honestly, irrespective of the choice of random tapes of the parties), the augmented coin-tossing protocol used to determine the random tapes of the parties in the compiler of Canetti *et al.* can be avoided. The rest of the compiler remains the same; it relies on adaptively secure commitment and zero-knowledge tools (which can be based on enhanced trapdoor permutations). Since FaF security becomes relevant only in settings with more than 3 parties, the standard security notion for 2-party NIZK used as a building block would not affect FaF security of the overall protocol.

We argue that this simplified compiler preserves guaranteed output delivery. This is because, whenever an actively corrupt party misbehaves in the compiled protocol (for instance, a party aborts or the zero-knowledge proof showing the correctness of her actions in round r of the underlying protocol fails), such a scenario can be translated to an analogous scenario in the underlying protocol where the same party is fail-stop corrupt and stops communicating in round r. It is now easy to see that since the underlying protocol achieved GOD against t fail-stop and h^* passive corruptions, the same guarantees must hold against t active and h^* passive corruptions in the compiled protocol as well.

We state the formal theorem below.

Theorem 9. *The construction summarized in Sect. 6.2 with $2t + h^* < n$ and with $d = t + h^*$ achieves security with guaranteed output delivery against an adaptive adversary with a budget of t active corruptions and h^* passive corruptions.*

References

1. Alon, B., Omri, E., Paskin-Cherniavsky, A.: MPC with friends and foes. In: Micciancio, D., Ristenpart, T. (eds.) Advances in Cryptology – CRYPTO 2020. CRYPTO 2020. LNCS, vol. 12171, pp. 677–706. Springer, Cham (2020). https://doi.org/10.1007/978-3-030-56880-1_24
2. Asharov, G., Jain, A., López-Alt, A., Tromer, E., Vaikuntanathan, V., Wichs, D.: Multiparty computation with low communication, computation and interaction via threshold FHE. In: Pointcheval, D., Johansson, T. (eds.) Advances in Cryptology – EUROCRYPT 2012. EUROCRYPT 2012. LNCS, vol. 7237, pp. 483–501. Springer, Berlin, Heidelberg (2012). https://doi.org/10.1007/978-3-642-29011-4_29
3. Beaver, D.: Efficient multiparty protocols using circuit randomization. In: Feigenbaum, J. (ed.) Advances in Cryptology – CRYPTO '91. CRYPTO 1991. LNCS, vol. 576, pp. 420–432. Springer, Berlin, Heidelberg (1992). https://doi.org/10.1007/3-540-46766-1_34
4. Ben-Or, M., Goldwasser, S., Wigderson, A.: Completeness theorems for non-cryptographic fault-tolerant distributed computation (extended abstract). In: 20th ACM STOC, pp. 1–10. ACM Press, May 1988. https://doi.org/10.1145/62212.62213
5. Boneh, D., et al.: Threshold cryptosystems from threshold fully homomorphic encryption. In: Shacham, H., Boldyreva, A. (eds.) Advances in Cryptology – CRYPTO 2018. CRYPTO 2018. LNCS, vol. 10991, pp. 565–596. Springer, Cham (2018). https://doi.org/10.1007/978-3-319-96884-1_19
6. Canetti, R., Lindell, Y., Ostrovsky, R., Sahai, A.: Universally composable two-party and multi-party secure computation. In: 34th ACM STOC, pp. 494–503. ACM Press, May 2002. https://doi.org/10.1145/509907.509980
7. Dalskov, A.P.K., Escudero, D., Keller, M.: Fantastic four: Honest-majority four-party secure computation with malicious security. In: Bailey, M., Greenstadt, R. (eds.) USENIX Security 2021, pp. 2183–2200. USENIX Association, August 2021
8. Damgård, I., Nielsen, J.B.: Adaptive versus static security in the UC model. In: Chow, S.S.M., Liu, J.K., Hui, L.C.K., Yiu, S.M. (eds.) Provable Security. ProvSec 2014. LNCS, vol. 8782, pp. 10–28. Springer, Cham (2014). https://doi.org/10.1007/978-3-319-12475-9_2
9. Gennaro, R., Ishai, Y., Kushilevitz, E., Rabin, T.: On 2-round secure multiparty computation. In: Yung, M. (ed.) Advances in Cryptology – CRYPTO 2002. CRYPTO 2002. LNCS, vol. 2442, pp. 178–193. Springer, Berlin, Heidelberg (2002). https://doi.org/10.1007/3-540-45708-9_12

10. Goldreich, O., Micali, S., Wigderson, A.: How to play any mental game or a completeness theorem for protocols with honest majority. In: Aho, A. (ed.) 19th ACM STOC, pp. 218–229. ACM Press, May 1987. https://doi.org/10.1145/28395.28420

11. Gordon, S.D., Liu, F.H., Shi, E.: Constant-round MPC with fairness and guarantee of output delivery. In: Gennaro, R., Robshaw, M. (eds.) Advances in Cryptology – CRYPTO 2015. CRYPTO 2015. LNCS, vol. 9216, pp. 63–82. Springer, Berlin, Heidelberg (2015). https://doi.org/10.1007/978-3-662-48000-7_4

12. Hegde, A., Koti, N., Kukkala, V.B., Patil, S., Patra, A., Paul, P.: Attaining GOD beyond honest majority with friends and foes. In: Agrawal, S., Lin, D. (eds.) Advances in Cryptology – ASIACRYPT 2022. ASIACRYPT 2022. LNCS, vol. 13791, pp. 556–587. Springer, Cham (2022). https://doi.org/10.1007/978-3-031-22963-3_19

13. Ishai, Y., Kumaresan, R., Kushilevitz, E., Paskin-Cherniavsky, A.: Secure computation with minimal interaction, revisited. In: Gennaro, R., Robshaw, M. (eds.) Advances in Cryptology – CRYPTO 2015. CRYPTO 2015. LNCS, vol. 9216, pp. 359–378. Springer, Berlin, Heidelberg (2015). https://doi.org/10.1007/978-3-662-48000-7_18

14. Ishai, Y., Kushilevitz, E., Lindell, Y., Petrank, E.: On combining privacy with guaranteed output delivery in secure multiparty computation. In: Dwork, C. (ed.) Advances in Cryptology – CRYPTO 2006. CRYPTO 2006. LNCS, vol. 4117, pp. 483–500. Springer, Berlin, Heidelberg (2006). https://doi.org/10.1007/11818175_29

15. Ishai, Y., Kushilevitz, E., Paskin, A.: Secure multiparty computation with minimal interaction. In: Rabin, T. (ed.) Advances in Cryptology – CRYPTO 2010. CRYPTO 2010. LNCS, vol. 6223, pp. 577–594. Springer, Berlin, Heidelberg (2010). https://doi.org/10.1007/978-3-642-14623-7_31

16. Katz, J.: On achieving the best of both worlds in secure multiparty computation. In: Johnson, D.S., Feige, U. (eds.) 39th ACM STOC, pp. 11–20. ACM Press, June 2007. https://doi.org/10.1145/1250790.1250793

17. Koti, N., Kukkala, V.B., Patra, A., Gopal, B.R.: PentaGOD: stepping beyond traditional GOD with five parties. In: Yin, H., Stavrou, A., Cremers, C., Shi, E. (eds.) ACM CCS 2022, pp. 1843–1856. ACM Press, November 2022. https://doi.org/10.1145/3548606.3559369

18. Koti, N., Pancholi, M., Patra, A., Suresh, A.: SWIFT: super-fast and robust privacy-preserving machine learning. In: Bailey, M., Greenstadt, R. (eds.) USENIX Security 2021, pp. 2651–2668. USENIX Association, August 2021

19. Koti, N., Patra, A., Rachuri, R., Suresh, A.: Tetrad: Actively secure 4PC for secure training and inference. Cryptology ePrint Archive, Report 2021/755 (2021). https://eprint.iacr.org/2021/755

20. Melissaris, N., Ravi, D., Yakoubov, S.: Threshold-optimal MPC with friends and foes. Cryptology ePrint Archive, Paper 2022/1526 (2022). https://eprint.iacr.org/2022/1526

21. Patra, A., Ravi, D.: On the exact round complexity of secure three-party computation. In: Shacham, H., Boldyreva, A. (eds.) Advances in Cryptology – CRYPTO 2018. CRYPTO 2018. LNCS, vol. 10992, pp. 425–458. Springer, Cham (2018). https://doi.org/10.1007/978-3-319-96881-0_15

22. Yao, A.C.C.: How to generate and exchange secrets (extended abstract). In: 27th FOCS, pp. 162–167. IEEE Computer Society Press, October 1986. https://doi.org/10.1109/SFCS.1986.25

Network-Agnostic Perfectly Secure Message Transmission Revisited

Nidhish Bhimrajka◉, Ashish Choudhury$^{(\boxtimes)}$◉, and Supreeth Varadarajan◉

International Institute of Information Technology Bangalore, Bengaluru, India
{nidhish.bhimrajka,ashish.choudhury,supreeth.varadarajan}@iiitb.ac.in

Abstract. *Secure Message Transmission* (SMT) is one of the fundamental primitives in secure distributed computing. Consider two nodes **A** and **B**, connected by n *node-disjoint* channels, a subset of which may be under the control of a *computationally-unbounded* Byzantine (malicious) adversary. An SMT protocol enables **A** to securely send any message to **B** by interacting over the channels. Known SMT protocols are secure in the presence of up to $\frac{n}{2}$ and $\frac{n}{3}$ faulty channels in a *synchronous* and *asynchronous* network, respectively. A *network-agnostic* SMT protocol can tolerate up to $t_s < \frac{n}{2}$ and $t_a < \frac{n}{3}$ faulty channels in synchronous and asynchronous networks respectively. Deligios and Liu-Zhang (FC 2023) have shown that network-agnostic SMT protocol is possible if and only if $2t_a + t_s < n$ holds. Their protocol with the *optimal* threshold conditions require $t_s - t_a$ communications rounds in a *synchronous* network, which is *not* optimal. It has been left as an open problem to design a network-agnostic SMT protocol with the *optimal* threshold conditions and the *optimal* number of communication rounds in a synchronous network. We resolve this open problem by presenting the *first* such protocol.

Keywords: secure message transmission · perfect security · synchronous and asynchronous network · secret sharing

1 Introduction

Secure Message Transmission (SMT) [6] is one of the fundamental primitives in cryptographic protocols. Consider a distributed *connected* network and two designated nodes **A** and **B**, who are part of the network. A subset of the nodes in the network, *excluding* **A** and **B**, may be under the control of a Byzantine (malicious) adversary Adv. The goal of any SMT protocol is to enable **A**, which has some message $m^{\mathbf{A}}$, to securely send $m^{\mathbf{A}}$ to **B**, even if the nodes under the control of Adv behave arbitrarily during the protocol execution. Following [6], we abstract the underlying network and assume that **A** and **B** are connected by n *node-disjoint* channels $c_1, \ldots, c_n$, such that a subset of these channels can

This research is an outcome of the R&D work undertaken in the project under the Visvesvaraya PhD Scheme of the Ministry of Electronics & Information Technology, Government of India, being implemented by Digital India Corporation.

be under the control of Adv. SMT protocol allows one to emulate the effect of a virtual *complete* network over an *incomplete*, but *connected* network, following which one can run more complex distributed protocols over such a virtual complete network, such as secure MPC protocols [3,10].

Traditionally, SMT protocols are classified with respect to the behaviour of the underlying network. *Synchronous* SMT (SSMT) protocols are designed assuming the network behaves synchronously. In this communication model, it is *assumed* that any message sent over a channel will be delivered to the intended recipient within some *known* time Δ. As a result, SSMT protocols are executed as a sequence of communication *rounds*, where the duration of each round is Δ. On the other hand, *asynchronous* SMT (ASMT) protocols are designed *without* making any such timing assumption, and messages sent over any channel can be *arbitrarily*, yet *finitely* delayed. The only guarantee in this model is that any message sent over a channel will be delivered *eventually*. In any ASMT protocol, the adversary is given the power to control the schedule of message delivery. Consequently, neither **A** nor **B** will know how long to wait for an intended message since there is no way to distinguish between a *corrupt* channel and a slow but *non-faulty* channel over which the messages are arbitrarily delayed.

We are interested in SMT with *perfect security*, where the adversary is assumed to be *computationally-unbounded*. SSMT protocols with perfect security can be designed iff Adv can corrupt up to $t_s < \frac{n}{2}$ channels [6]. Moreover, it is also known that the *optimal* number of communication rounds required by any SSMT protocol is *constant* and *independent* of t_s and n. On the other hand, ASMT protocols with perfect security can tolerate up to $t_a < \frac{n}{3}$ corrupt channels [4]. SSMT and ASMT protocols are designed assuming that the protocol participants will know the underlying network's exact behaviour *beforehand*. We envision a scenario where this is not the case and aim to design an SMT protocol that achieves (simultaneous) security guarantees in both networks, namely a protocol that remains secure if the network behaves synchronously even if up to $t_s < \frac{n}{2}$ channels are corrupt and which also remains secure if the network behaves asynchronously even if up to $t_a < \frac{n}{3}$ channels are corrupt. Such a protocol is called a *network-agnostic* SMT protocol. The main advantage of such a protocol is that the protocol participants *need not* have to worry about the behaviour of the underlying network and will get the best possible security guarantees, *irrespective* of the behaviour of the network. There are several challenges in designing network-agnostic SMT protocols. Known SSMT protocols become insecure when the network behaves asynchronously if even a single message over a *non-faulty* channel is delayed arbitrarily. On the other hand, ASMT protocols (designed to tolerate up to $\frac{n}{3}$ faulty channels) become insecure when the network behaves synchronously since the adversary can corrupt *more* than $\frac{n}{3}$ channels.

Our Motivation and Results: The work of [5] studies network-agnostic SMT protocols with perfect security. They show that such protocols can be designed iff $2t_a + t_s < n$ holds. To show the *optimality* of this threshold, they present a protocol that requires $t_s - t_a$ rounds if the network behaves *synchronously*. This is *not* optimal since the optimal number of rounds for any SMT protocol in a *synchronous* network is *constant*. Designing a network-agnostic SMT protocol with perfect security and condition $2t_a + t_s < n$, which requires a *constant* number of rounds in a synchronous network, is left as an *open problem* in [5]. In this work, we resolve this open problem by presenting the *first* such protocol. Our results, compared with relevant existing results, are presented in Table 1.

Table 1. Best known SMT protocols with perfect security. Here, r denotes the number of rounds if the network behaves synchronously. The communication complexity is measured in bits, where $\mathbb{F}$ and $\mathbb{G}$ denote a finite field and group, respectively, over which computations are performed in the underlying protocols. Moreover, ℓ denotes the size of $m^{\mathbf{A}}$ in terms of field/group elements.

Network Setting	Resilience	Reference	#r	Communication Complexity	Computation Complexity
Synchronous	$t_s < n/2$	[1]	2	$\mathcal{O}\left(\frac{n\ell}{n-2t_s}\log|\mathbb{F}|\right)$	$\text{Exp}(n, t_s)$
		[9]	3	$\mathcal{O}\left(\frac{n\ell}{n-2t_s}\log|\mathbb{F}|\right)$	$\text{Poly}(n, t_s)$
		[8]	2	$\mathcal{O}\left(\frac{n\ell}{n-2t_s}\log|\mathbb{F}|\right)$	$\text{Poly}(n, t_s)$
Asynchronous	$t_a < n/3$	[4]		$\mathcal{O}\left(\frac{n\ell}{n-3t_a}\log|\mathbb{F}|\right)$	$\text{Poly}(n, t_a)$
Network Agnostic	$t_s < n/2,$	[5]	$t_s - t_a$	$\mathcal{O}(n \cdot (t_s - t_a)^2 \cdot \ell \log|\mathbb{F}|)$	$\text{Poly}(n, t_s, t_a)$
	$t_a < n/3,$				
	$2t_a + t_s < n$	**This work**	6	$\mathcal{O}\left(\binom{n}{t_s} \cdot n^3 \cdot \ell \log|\mathbb{G}|\right)$	$\text{Exp}(n, t_s, t_a)$

1.1 Technical Overview

We follow the *generic compiler* of [5] for designing the network-agnostic SMT protocol. The key ingredient of the compiler is a generic SSMT protocol, called Π_{SSMT}, which provides *all* security guarantees if the network behaves *synchronously*. Additionally, it *also* provides certain guarantees if the network behaves *asynchronously*. Namely, the *privacy* is achieved, and it will be ensured that both **A** and **B** terminate after a "fixed" time. Additionally, a "weaker" form of *correctness* is also achieved, where **B** either outputs $m^{\mathbf{A}}$ or $\bot$.

The instantiation of Π_{SSMT} in [5] is based on the properties of error-correcting codes over a finite field $\mathbb{F}$ and requires $t_s - t_a$ rounds if the network behaves *synchronously*. We provide a *different* instantiation of Π_{SSMT}, which may require up to 6 rounds. Our instantiation exploits the properties of *replicated secret-sharing* (RSS) over a group $\mathbb{G}$ [7]. In more detail, **A** first tries to securely send $m^{\mathbf{A}}$ to **B** by computing RSS-shares of $m^{\mathbf{A}}$ and sending these shares to **B** over different channels in such a way that *privacy* is achieved even if up to t_s channels are under the control of the adversary. The next step is for **B** to identify how many channels have delayed the communication and try to "report" it back to **A** using an instance of *reliable message transmission* (RMT) protocol.[1] Now, based

---

[1] An RMT protocol is an SMT protocol with *no* privacy and its goal is to emulate a *public* channel between **A** and **B**, which *cannot* be tampered by the adversary.

on this information, **A** and **B** either interactively identify the *corrupt* channels, which delivered *incorrect* shares, followed by **B** using the correct shares to recover $m^\mathbf{A}$; otherwise, **A** and **B** interact, where **A** now computes "fresh" shares of $m^\mathbf{A}$ as per a *new* instance of RSS, with a reduced "degree" of replication and communicates these shares to **B**, followed by **A** and **B** interactively identifying the incorrectly delivered shares, after which **B** recovers **A**'s message.

The instantiation of Π_{SSMT} in [5] also follows a similar idea, where **A** and **B** interactively identify corrupt channels through *several* iterations of exchanging shares and applying error-detection. However, the number of iterations required in their instantiation is *not* constant. In contrast, our instantiation of Π_{SSMT} may require at most *two* iterations, which will take at most 6 rounds. Since our protocol is slightly technical, we defer to Sect. 3 for the complete details.

2   Preliminaries and Definitions

We consider a distributed *connected* network, abstracted as an *undirected* graph $G = (V, E)$, with **A** and **B** being two special nodes from V, such that G is an *incomplete* graph. Following [6], we abstract the underlying paths between **A** and **B** as *vertex-disjoint* paths, called *channels*, and assume that there exists n channels $c_1, \ldots, c_n$ between **A** and **B**. The distrust in the network is modelled by a *computationally unbounded adversary* Adv, who can corrupt a subset of the nodes from V *apart* from **A** and **B** in a *malicious* (Byzantine) fashion and force them to behave in any *arbitrary* fashion during the execution of any protocol.

We consider a communication model where **A** and **B** have access to *local* clocks and are *not aware* apriori about the network conditions when executing any protocol, where the underlying network could behave either in a *synchronous* fashion or in an *asynchronous* fashion. If the network behaves *synchronously*, then the local clocks of **A** and **B** are synchronized, and every message being sent is guaranteed to be delivered within some *known* time bound Δ. The protocols in this model can be conveniently described as a sequence of communication *rounds*, where for every $r \in \mathbb{N}$ with $r \geq 1$, any message received in the time slot $[r\Delta, (r+1)\Delta]$ as per the local clock of the receiving node is regarded as a round-r message. Moreover, in this model, it is assumed that Adv can corrupt at most t_s nodes in the network, which amounts to corrupting at most t_s out of the n channels in the channel-abstraction. Furthermore, we assume that $t_s < \frac{n}{2}$, which is necessary for any SMT protocol in the *synchronous* model [6].

If the network behaves *asynchronously*, then the local clocks of **A** and **B** are *not* synchronized, and there is *no known* upper bound on message delays. To model the worst-case scenario, the adversary is allowed to schedule the delivery of messages *arbitrarily*, with the restriction that every message being sent is *eventually* delivered. The protocols in this model are described in an *event-based* fashion. That is, upon receiving a message, the receiving node adds the message to a pool of received messages and checks whether a list of conditions specified in the underlying protocol is satisfied to decide its next set of actions. In the asynchronous model, it is assumed that Adv can corrupt at most t_a nodes, which amounts to corrupting at most t_a channels. Furthermore, we assume that $t_a < \frac{n}{3}$, which is necessary for any SMT protocol in the *asynchronous* model [4].

Following [5], we describe protocols both in a round-based fashion as well as an event-based fashion, depending upon whether a given protocol is designed for a synchronous network or an asynchronous network. Moreover, in a protocol described in a round-based fashion, if a message is received *outside* of the time allocated for a certain round (as per the local clock of the receiving party), then it is *ignored* by the receiving party. We assume that $2t_a + t_s < n$ holds, which is necessary for any network-agnostic SMT protocol [5]. In our protocols, all computations are performed over a finite group $(\mathbb{G}, +)$.

2.1 Definitions

A *message transmission* (MT) protocol allows **A** and **B**, connected by multiple channels, to communicate messages, even when a subset of the channels is under the control of Adv. We consider two types of MT protocols. A *reliable message transmission protocol* (RMT) allows **A** and **B** to *robustly* exchange messages in the presence of Adv. However, it provides *no* privacy guarantees. A *secure message transmission* (SMT) protocol allows to exchange messages robustly and *also* maintains the *privacy* of the messages. We next formally define MT. As done in [5], we slightly deviate from the standard definition of [6] by requiring that the "sender" outputs a Boolean value as an *indicator* of whether the protocol has succeeded or not. Similarly, the "receiver" is allowed to output $\perp$, indicating that it could not produce a "valid" output. We also define MT in terms of a pair of abstract *sender* **S** and *receiver* **R**, where any distinct node from $\{\mathbf{A}, \mathbf{B}\}$ can play the role of **S** and **R**. Looking ahead, this is done because in our instantiation of Π_{SSMT} (recall the compiler of [5] from Sect. 1.1 and also see Sect. 2.2), there will be multiple instances of RMT, where both **A** and **B** will interchangeably play the role of **S** and **R**.

Definition 1 (Message Transmission (MT) [5]). *Let* **S** *(the sender) and* **R** *(the receiver) be two distinct nodes from* $\{\mathbf{A}, \mathbf{B}\}$. *Let* Π *be a protocol executed between* **S** *with input* $m^{\mathbf{S}} \in \mathbb{G}$ *and output* $b \in \{0, 1\}$ *and* **R** *with output* $m^{\mathbf{R}} \in \mathbb{G} \cup \{\perp\}$, *with* Adv *corrupting up to* t *channels. We say that* Π *is a protocol for MT achieving:*

- t-**correctness**: *if* $m^{\mathbf{R}} = m^{\mathbf{S}}$ *and* $b = 1$ *holds;*
- t-**termination**: *if both* **S** *and* **R** *terminate;*
- t-**weak correctness**: *if one of the following hold.*
    - $m^{\mathbf{R}} = m^{\mathbf{S}}$ *and* $b = 1$;
    - $m^{\mathbf{R}} = m^{\mathbf{S}}$ *and* $b = 0$;
    - $m^{\mathbf{R}} = \perp$ *and* $b = 0$.
- t-**privacy**: *if for all* $m_1^{\mathbf{S}}, m_2^{\mathbf{S}} \in \mathbb{G}$, *for all* $k \geq 1$ *and for all* $\mathcal{I} \subseteq \{1, \dots, n\}$ *such that* $|\mathcal{I}| \leq t$, *the probability distributions of* $T_{\mathcal{I}, m_1^{\mathbf{S}}}^k$ *and* $T_{\mathcal{I}, m_2^{\mathbf{S}}}^k$ *are identical. Here* $T_{\mathcal{I}, m_1^{\mathbf{S}}}^k$ *and* $T_{\mathcal{I}, m_2^{\mathbf{S}}}^k$ *denotes the random variables whose values are the k-th messages communicated on the channels* $\{c_i\}_{i \in \mathcal{I}}$ *during the execution of* Π *when* **S** *has input* $m_1^{\mathbf{S}}$ *and* $m_2^{\mathbf{S}}$ *respectively.*

2.2   Existing Building Blocks

Synchronous RMT with Asynchronous Detection. Protocol Π_{RMT} presented in [5] is an RMT protocol that achieves t_s-correctness and t_s-termination in a *synchronous* network. The protocol also achieves t_a-weak correctness and t_a-termination in an *asynchronous* network. The protocol is standard: **S** sends its message over all the channels, and **R** takes a "majority" decision over the received messages. The proof of Lemma 1 is available in [5].

Lemma 1 ([5]). *If $t_a \leq t_s < \frac{n}{2}$, then protocol Π_{RMT} achieves the following.*

- **Synchronous Network**: *If the network behaves synchronously, then the protocol achieves t_s-correctness and t_s-termination, such that **R** and **S** obtain their output at the end of the first and second rounds, respectively.*
- **Asynchronous Network**: *If the network behaves asynchronously, then the protocol achieves t_a-weak correctness and t_a-termination, where **S** and **R** have an output at (local) time Δ and 2Δ respectively.*
- *The protocol incurs a communication of $\mathcal{O}(n \log |\mathbb{G}|)$ bits.*

A Generic Compiler for Network-Agnostic SMT. The work of [5] presents a generic compiler for designing a network-agnostic SMT protocol, given black-box access to a *synchronous* SMT protocol with certain asynchronous guarantees and a completely asynchronous SMT protocol. Namely, let Π_{SSMT} be an SMT protocol with the following guarantees:

- If the network behaves *synchronously*, then Π_{SSMT} provides t_s-correctness, t_s-privacy, and t_s-termination.
- If the network behaves *asynchronously*, then Π_{SSMT} provides t_a-weak correctness, t_a-privacy, and t_a-termination.
- *Irrespective* of the network type, **S** and **R** gets an output at local time $r^{\mathbf{S}}\Delta$ and $r^{\mathbf{R}}\Delta$ respectively.[2]

Moreover, let Π_{ASMT} be a *non-interactive* SMT protocol, which provides t_a-correctness, t_a-privacy, and t_a-termination if the network behaves *asynchronously*.[3] Then [5] shows how to "stitch" together protocols Π_{SSMT} and Π_{ASMT} generically and get a *network-agnostic* protocol Π_{naSMT}, which provides all the security guarantees, irrespective of the behaviour of the network.

The idea behind Π_{naSMT} is as follows: **A** and **B** first run an instance of Π_{SSMT}, playing the role of **S** and **R**, respectively, where the input of **A** is the message $m^{\mathbf{A}}$ which it wants to securely send to **B**. If the network behaves *synchronously*, then we are done, and the asynchronous protocol *need not* be executed. If the network behaves *asynchronously*, then the t_a-weak correctness of Π_{SSMT} guarantees that if **B** obtains any output other than $\perp$, then it is bound to be $m^{\mathbf{A}}$. However, it

---

[2] This implies that if the network behaves *synchronously*, then **S** and **R** get their respective output after round $r^{\mathbf{S}}$ and $r^{\mathbf{R}}$, respectively.

[3] By *non-interactive* we mean that in the protocol only **S** communicates messages to **R** and there is no need for any "feedback" from **R**.

might also be possible that $\mathbf{B}$ obtains $\bot$ as an output. In the latter case, the t_a-weak correctness of Π_{SSMT} guarantees that $\mathbf{A}$ detects this and consequently $\mathbf{A}$ *re-attempts* to send its message, but via an instance of Π_{ASMT}, by playing the role of $\mathbf{S}$. It might be possible that $\mathbf{B}$ has already obtained $m^{\mathbf{A}}$ through the instance of Π_{SSMT} (but $\mathbf{A}$ is *not* aware of this), in which case $\mathbf{B}$ *ignores* all the communication from $\mathbf{A}$, done as part of Π_{ASMT}. This is fine since Π_{ASMT} is *non-interactive* and *does not* require feedback from $\mathbf{B}$.

3   Synchronous SMT with Asynchronous Detection

In this section, we provide our instantiation of protocol Π_{SSMT} (see Fig. 1), which is the first ingredient for the generic compiler of [5]. The protocol provides t_s-correctness, t_s-privacy, and t_s-termination in a *synchronous* network and, at the same time, provides t_a-weak correctness, t_a-privacy, and t_a-termination in an *asynchronous* network. Moreover, the number of rounds required in the protocol is *constant* if the network behaves *synchronously*.

The high-level idea of the protocol is as follows: for simplicity, imagine that the network will behave *synchronously*. Then $\mathbf{A}$ can securely send $m^{\mathbf{A}}$ to $\mathbf{B}$ in a *constant* number of rounds as follows. Node $\mathbf{A}$ computes *shares* for $m^{\mathbf{A}}$, as per an instance of *replicated secret-sharing* (RSS) [7], and sends these shares to $\mathbf{B}$. The "degree of replication" is set to $n - t_s$, where each share $m_k^{\mathbf{A}}$ is sent (*replicated*) over a *distinct* subset S_k of $n - t_s$ channels. The privacy is guaranteed since there will be at least one share of $m^{\mathbf{A}}$, which will be sent over a subset of $n - t_s$ channels, *none* of which is under the control of Adv. Moreover, since $t_s < \frac{n}{2}$, it will be guaranteed that each subset of channels S_k consists of at least one *non-faulty* channel, which correctly delivers the share $m_k^{\mathbf{A}}$. Node $\mathbf{B}$, upon receiving the shares, can *detect* if any channel within S_k has delivered an *incorrect* $m_k^{\mathbf{A}}$, by comparing all the "versions" of the received $m_k^{\mathbf{A}}$. If any "inconsistency" is detected, then $\mathbf{B}$ reports it back to $\mathbf{A}$ (through an instance of Π_{RMT}). Upon receiving these inconsistencies, $\mathbf{A}$ can identify the *corrupt* channels within S_k which had delivered incorrect shares to $\mathbf{B}$ and reliably sends this information to $\mathbf{B}$ (through an instance of Π_{RMT}). Node $\mathbf{B}$, upon receiving the identity of corrupt shares discards them and uses the remaining shares to reconstruct $m^{\mathbf{A}}$.

Next, consider the case when the network behaves *asynchronously*, where the adversary can arbitrarily *delay* communication over *non-faulty* channels and *additionally* corrupt shares sent over the *faulty* channels. A *trivial* case will be if the adversary delays communication over *more* than t_s channels, in which case $\mathbf{B}$ is guaranteed that the network is *asynchronous* and hence it can safely terminate the protocol with $\bot$ as the output, following which $\mathbf{A}$ will also terminate with output 0. The other *easier* case to deal with is when the adversary delays communication over *at most* t_a channels. Since $2t_a + t_s < n$ holds, in this case, it will still be guaranteed that at least one *non-faulty* channel within each S_k has correctly delivered $m_k^{\mathbf{A}}$, as the adversary can corrupt *only* up to t_a copies of $m_k^{\mathbf{A}}$. Consequently, $\mathbf{B}$ and $\mathbf{A}$ can interactively try to find out the identity of incorrectly delivered shares, as they would have done in a *synchronous* network.

The challenging case is when the network behaves *asynchronously* and if the adversary delays communication over d channels, where $t_a < d < t_s$, such that all the delayed channels are *non-faulty*. In this case, it will *not* be guaranteed that at least one *non-faulty* channel within S_k has *correctly* delivered $m_k^{\mathbf{A}}$, and all the versions of the delivered $m_k^{\mathbf{A}}$ may be *incorrect* but the *same*. Moreover, **B** will *not* be able to distinguish whether the network is behaving *synchronously* and the *correctness* holds or the network is behaving *asynchronously*, and the adversary has tampered with the *correctness*.

To overcome the above challenge, **B** first finds out the set of channels $\mathcal{D}$ over which communication is delayed and reliably sends it to **A** through an instance of Π_{RMT} so that both **A** and **B** learn d, the cardinality of the set $\mathcal{D}$. If $d \leq t_a$, then as discussed above, *irrespective* of the network type, both **A** and **B** can go ahead and interactively identify the faulty channels which delivered incorrect shares to **B**, followed by **B** recovering $m^{\mathbf{A}}$ using the remaining shares. For the "problematic" case when $t_a < d < t_s$, node **A** computes "fresh" shares of $m^{\mathbf{A}}$ through a fresh instance of RSS, and the shares are sent to **B** over the channels, where the channels in $\mathcal{D}$ are *not* considered. Moreover, the degree of replication for the new instance of RSS is carefully selected so that the *privacy* is maintained, *irrespective* of the network type. For this, **A** and **B** check over how many channels the adversary can *further* delay the communication if the network behaves *asynchronously* and still conceal this fact from **B**. Namely, if the network behaves *asynchronously* and adversary delays communication over *more* than $t_s - d$ channels when **A** sends *new* RSS-shares of $m^{\mathbf{A}}$, then **B** will be *sure* that the network is behaving *asynchronously* (since the *total* number of channels over which communication is delayed for **B** across *all* the rounds will cross t_s, which can happen *only* in an *asynchronous* network); consequently, **B** can terminate the protocol there itself. Hence, the adversary now *cannot* afford to delay communication over more than $t_s - d$ channels. So, depending upon the value of $t_s - d$, the degree of replication of the new RSS is decided.

If $(t_s - d) > t_a$, then the degree of replication of the new RSS is set to be $(n - d) - (t_s - d) = n - t_s$, where after *excluding* all the channels in $\mathcal{D}$, each fresh share of $m^{\mathbf{A}}$ is sent (replicated) over a distinct subset of $n - t_s$ channels. This will ensure that even if the network behaves *synchronously* and d is in the range $t_a < d < t_s$, the privacy of $m^{\mathbf{A}}$ is maintained when **A** sends fresh shares of $m^{\mathbf{A}}$ to **B**. This is because the adversary can now control at most $t_s - d$ channels out of $n - d$ channels, and there will be at least one new share of $m^{\mathbf{A}}$, which will be sent over a set of $(n - d) - (t_s - d)$ channels, *none* of which will be under the control of the adversary. On the other hand, if $(t_s - d) \leq t_a$, then the degree of replication of new RSS can be safely set to $(n - d) - t_a$, where after *excluding* all the channels in $\mathcal{D}$, each fresh share of $m^{\mathbf{A}}$ is sent (replicated) over a distinct subset of $n - d - t_a$ channels. This will ensure *privacy* in either network since the number of channels that will still be under the control of the adversary in either network will be *at most* t_a, and hence there will be at least one new share of $m^{\mathbf{A}}$ not learnt by the adversary.

Protocol $\Pi_{\mathsf{SSMT}}(\mathbf{A}, \mathbf{B})$

Let $K_s \overset{def}{=} \binom{n}{t_s}$, let $Z_1, \ldots, Z_{K_s}$ be different subsets of the channels $\{c_1, \ldots, c_n\}$ of size t_s and for $i = 1, \ldots, K_s$, let $S_i = \{c_1, \ldots, c_n\} \setminus Z_i$.

Implicit Termination Conditions: If during any round, $\mathbf{A}$ finds that the *total* number of channels over which the communication has been delayed so far exceeds t_s, or if any instance of Π_{RMT} outputs $\perp$ for $\mathbf{A}$ (as $\mathbf{R}$) or outputs $b = 0$ (as $\mathbf{S}$), then $\mathbf{A}$ outputs $b = 0$ for Π_{SSMT} and terminates. A similar assumption is made for $\mathbf{B}$.

- **Round 1:**
  - **(RSS of the Message with Degree of Replication $n - t_s$)**: $\mathbf{A}$ on having the input $m^{\mathbf{A}}$, randomly picks $m_1^{\mathbf{A}}, \ldots, m_{K_s}^{\mathbf{A}} \in \mathbb{G}$ such that $m^{\mathbf{A}} = m_1^{\mathbf{A}} + \ldots + m_{K_s}^{\mathbf{A}}$ and sends $m_i^{\mathbf{A}}$ to $\mathbf{B}$ over all the channels in S_i, for $i = 1, \ldots, K_s$.
  - Let $\mathcal{D} \subseteq \{c_1, \ldots, c_n\}$ be the set of channels over which communication is delayed for $\mathbf{B}$ during this round.
- **Round 2:**
  - **(Communicating the Identity of Delayed Channels)**: $\mathbf{B}$ acts as $\mathbf{S}$ and invokes an instance $\Pi_{\mathsf{RMT}}^{(\mathbf{B},1)}$ of Π_{RMT} with input $\mathcal{D}$. Let $|\mathcal{D}| = d$.
  - **(Receiving the Identity of Delayed Channels)**: $\mathbf{A}$ upon obtaining $\mathcal{D}$ as the output (as $\mathbf{R}$) from the instance $\Pi_{\mathsf{RMT}}^{(\mathbf{B},1)}$, computes $d = |\mathcal{D}|$.
  - $\mathbf{A}$ and $\mathbf{B}$ check the following and execute further steps accordingly.
    - **Case 1**: $d \leq t_a$
    - **Case 2**: $(t_a < d < t_s$ and $(t_s - d) \leq t_a)$ or $d = t_s$
    - **Case 3**: $t_a < d < l_s$ and $(l_s - d) > t_a$

Case 1

- **Round 2:**
  - **(Reliably Sending Back the Conflicting Shares)**: $\mathbf{B}$ does the following.
    - For $k = 1, \ldots, K_s$, add the tuple $(k, i, j, m_{k,i}^{\mathbf{B}}, m_{k,j}^{\mathbf{B}})$ to a *conflict list* L_1 (initialized to $\emptyset$), provided $c_i, c_j \in S_k$ and corresponding to the set S_k, $\mathbf{B}$ has received the shares $m_{k,i}^{\mathbf{B}}$ and $m_{k,j}^{\mathbf{B}}$ over c_i and c_j respectively during **Round 1**, such that $m_{k,i}^{\mathbf{B}} \neq m_{k,j}^{\mathbf{B}}$.
    - Act as $\mathbf{S}$ and invoke an instance $\Pi_{\mathsf{RMT}}^{(\mathbf{B},2)}$ of Π_{RMT} with input L_1.
  - **(Resolving Conflicts)**: $\mathbf{A}$ does the following.
    - Upon computing L_1 as the output from the instance $\Pi_{\mathsf{RMT}}^{(\mathbf{B},2)}$, initialize a list of faulty channels $\mathcal{F}_1$.
    - Corresponding to each $(k, i, j, m_{k,i}^{\mathbf{B}}, m_{k,j}^{\mathbf{B}}) \in L_1$, add c_i or c_j to $\mathcal{F}_1$, depending upon whether $m_k^{\mathbf{A}} \neq m_{k,i}^{\mathbf{B}}$ or $m_k^{\mathbf{A}} \neq m_{k,j}^{\mathbf{B}}$ holds.
- **Round 3:**
  - **(Sending the List of Faulty Channels)**: $\mathbf{A}$ does the following.
    - Act as $\mathbf{S}$ and invoke an instance $\Pi_{\mathsf{RMT}}^{(\mathbf{A},1)}$ of Π_{RMT} with input $\mathcal{F}_1$.
  - **(Recovering the message)**: $\mathbf{B}$ upon computing $\mathcal{F}_1$ as the output from the instance $\Pi_{\mathsf{RMT}}^{(\mathbf{A},1)}$, proceed to recover $\mathbf{A}$'s message as follows
    - For $k = 1, \ldots, K_s$, set $m_k^{\mathbf{B}}$ to $m_{k,i}^{\mathbf{B}}$, provided $c_i \notin \mathcal{F}_1$ and if corresponding to the set S_k, $\mathbf{B}$ has received the share $m_{k,i}^{\mathbf{B}}$ during **Round 1** over c_i.[a]
    - If $m_1^{\mathbf{B}}, \ldots, m_{K_s}^{\mathbf{B}}$ are set, then set $m^{\mathbf{B}} = m_1^{\mathbf{B}} + \ldots + m_{K_s}^{\mathbf{B}}$ else set $m^{\mathbf{B}} = \perp$.
- **Round 4:**
  - $\mathbf{B}$ outputs $m^{\mathbf{B}}$ and terminates.
  - $\mathbf{A}$ upon computing the output $b_{\mathsf{RMT}}^{(\mathbf{A},1)} \in \{0, 1\}$ during the instance $\Pi_{\mathsf{RMT}}^{(\mathbf{A},1)}$, outputs $b = b_{\mathsf{RMT}}^{(\mathbf{A},1)}$ and terminates.

Fig. 1. Synchronous SMT with asynchronous detection

Case 2

For simplicity and without loss of generality, let $\mathcal{D} = \{c_1, \ldots, c_d\}$ be the set of channels over which the communication is delayed for **B** during Round 1, where $|\mathcal{D}| = d$. Let $K_u \overset{def}{=} \binom{n-d}{t_a}$, let $U_1, \ldots, U_{K_u}$ be different subsets of the channels $\{c_{d+1}, \ldots, c_n\}$ of size t_a and for $i = 1, \ldots, K_u$, let $T_i = \{c_{d+1}, \ldots, c_n\} \setminus U_i$.

- **Round 3:**
  - (**Fresh RSS-Sharing of the Message with Degree of Replication** $n - d - t_a$): **A** does the following.
    - Randomly pick shares $\mathsf{m}_1^{\mathbf{A}}, \ldots, \mathsf{m}_{K_u}^{\mathbf{A}}$ such that $m^{\mathbf{A}} = \mathsf{m}_1^{\mathbf{A}} + \ldots + \mathsf{m}_{K_u}^{\mathbf{A}}$ and send $\mathsf{m}_i^{\mathbf{A}}$ to **R** over all the channels in T_i, for $i = 1, \ldots, K_u$.
- **Round 4:**
  - (**Reliably Sending Back the Conflicting Shares for New Sharing**): **B** does the following.
    - For $k = 1, \ldots, K_u$, add the tuple $(k, i, j, \mathsf{m}_{k,i}^{\mathbf{B}}, \mathsf{m}_{k,j}^{\mathbf{B}})$ to a *conflict list* L_2 (initialized to $\emptyset$), provided $c_i, c_j \in T_k$ and corresponding to the set T_k, **B** has received the shares $\mathsf{m}_{k,i}^{\mathbf{B}}$ and $\mathsf{m}_{k,j}^{\mathbf{B}}$ over c_i and c_j respectively during **Round 3**, such that $\mathsf{m}_{k,i}^{\mathbf{B}} \neq \mathsf{m}_{k,j}^{\mathbf{B}}$.
    - Act as **S** and invoke an instance $\Pi_{\mathsf{RMT}}^{(\mathbf{B},3)}$ of Π_{RMT} with input L_2.
  - (**Resolving Conflicts**): **A** does the following.
    - Upon computing L_2 as the output from the instance $\Pi_{\mathsf{RMT}}^{(\mathbf{B},3)}$, initialize a list of faulty channels $\mathcal{F}_2$.
    - Corresponding to each $(k, i, j, \mathsf{m}_{k,i}^{\mathbf{B}}, \mathsf{m}_{k,j}^{\mathbf{B}}) \in L_2$, add c_i or c_j to $\mathcal{F}_2$, depending upon whether $\mathsf{m}_k^{\mathbf{A}} \neq \mathsf{m}_{k,i}^{\mathbf{B}}$ or $\mathsf{m}_k^{\mathbf{A}} \neq \mathsf{m}_{k,j}^{\mathbf{B}}$ holds.
- **Round 5:**
  - (**Sending the List of Faulty Channels**): **A** does the following.
    - Act as **S** and invoke an instance $\Pi_{\mathsf{RMT}}^{(\mathbf{A},2)}$ of Π_{RMT} with input $\mathcal{F}_2$.
  - (**Recovering the message**): **B** upon computing $\mathcal{F}_2$ as the output from the instance $\Pi_{\mathsf{RMT}}^{(\mathbf{A},2)}$, proceed to recover **A**'s message as follows
    - For $k = 1, \ldots, K_u$, set $\mathsf{m}_k^{\mathbf{B}}$ to $\mathsf{m}_{k,i}^{\mathbf{B}}$, provided $c_i \notin \mathcal{F}_2$ and if corresponding to the set T_k, **B** has received the shares $\mathsf{m}_{k,i}^{\mathbf{B}}$ during **Round 3** over c_i.[b]
    - If $\mathsf{m}_1^{\mathbf{B}}, \ldots, \mathsf{m}_{K_u}^{\mathbf{B}}$ are set, then set $m^{\mathbf{B}} = \mathsf{m}_1^{\mathbf{B}} + \ldots + \mathsf{m}_{K_u}^{\mathbf{B}}$ else set $m^{\mathbf{B}} = \perp$.
- **Round 6:**
  - **B** outputs $m^{\mathbf{B}}$ and terminates.
  - **A** upon computing the output $b_{\mathsf{RMT}}^{(\mathbf{A},2)} \in \{0,1\}$ during the instance $\Pi_{\mathsf{RMT}}^{(\mathbf{A},2)}$, outputs $b = b_{\mathsf{RMT}}^{(\mathbf{A},2)}$ and terminates.

Case 3

For simplicity and without loss of generality, let $\mathcal{D} = \{c_1, \ldots, c_d\}$ be the set of channels over which the communication is delayed for **B** during Round 1, where $|\mathcal{D}| = d$. Let $K_v \overset{def}{=} \binom{n-d}{t_s-d}$, let $V_1, \ldots, V_{K_v}$ be different subsets of the channels $\{c_{d+1}, \ldots, c_n\}$ of size $t_s - d$ and for $i = 1, \ldots, K_v$, let $W_i = \{c_{d+1}, \ldots, c_n\} \setminus V_i$.

- **Round 3:**
  - (**Fresh RSS-Sharing of the Message with Degree of Replication** $n - d - (t_s - d)$): **A** does the following.
    - Randomly pick shares $\mathfrak{m}_1^{\mathbf{A}}, \ldots, \mathfrak{m}_{K_v}^{\mathbf{A}}$ such that $m^{\mathbf{A}} = \mathfrak{m}_1^{\mathbf{A}} + \ldots + \mathfrak{m}_{K_v}^{\mathbf{A}}$ and send $\mathfrak{m}_i^{\mathbf{A}}$ to **R** over all the channels in W_i, for $i = 1, \ldots, K_v$.
- **Round 4:**
  - (**Reliably Sending Back the Conflicting Shares for New Sharing**): **B** does the following.
    - For $k = 1, \ldots, K_v$, add the tuple $(k, i, j, \mathfrak{m}_{k,i}^{\mathbf{B}}, \mathfrak{m}_{k,j}^{\mathbf{B}})$ to a *conflict list* L_3

Fig. 1. (*continued*)

(initialized to $\emptyset$), provided $c_i, c_j \in W_k$ and corresponding to the set W_k, **B** has received the shares $\mathsf{m}_{k,i}^{\mathbf{B}}$ and $\mathsf{m}_{k,j}^{\mathbf{B}}$ over c_i and c_j respectively during **Round 3**, such that $\mathsf{m}_{k,i}^{\mathbf{B}} \neq \mathsf{m}_{k,j}^{\mathbf{B}}$.
- Act as **S** and invoke an instance $\Pi_{\mathsf{RMT}}^{(\mathbf{B},4)}$ of Π_{RMT} with input L_3.
- **(Resolving Conflicts):** **A** does the following.
  - Upon computing L_3 as the output from the instance $\Pi_{\mathsf{RMT}}^{(\mathbf{B},4)}$, initialize a list of faulty channels $\mathcal{F}_3$.
  - Corresponding to each $(k, i, j, \mathsf{m}_{k,i}^{\mathbf{B}}, \mathsf{m}_{k,j}^{\mathbf{B}}) \in L_3$, add c_i or c_j to $\mathcal{F}_3$, depending upon whether $\mathsf{m}_k^{\mathbf{A}} \neq \mathsf{m}_{k,i}^{\mathbf{B}}$ or $\mathsf{m}_k^{\mathbf{A}} \neq \mathsf{m}_{k,j}^{\mathbf{B}}$ holds.
- **Round 5:**
  - **(Sending the List of Faulty Channels):** **A** does the following.
    - Act as **S** and invoke an instance $\Pi_{\mathsf{RMT}}^{(\mathbf{A},3)}$ of Π_{RMT} with input $\mathcal{F}_3$.
  - **(Recovering the message):** **B** upon computing $\mathcal{F}_3$ as the output from the instance $\Pi_{\mathsf{RMT}}^{(\mathbf{A},3)}$, proceed to recover **A**'s message as follows
    - For $k = 1, \ldots, K_v$, set $\mathsf{m}_k^{\mathbf{B}}$ to $\mathsf{m}_{k,i}^{\mathbf{B}}$, provided $c_i \notin \mathcal{F}_3$ and if corresponding to the set W_k, **B** has received the shares $\mathsf{m}_{k,i}^{\mathbf{B}}$ during **Round 3** over c_i.[c]
    - If $\mathsf{m}_1^{\mathbf{B}}, \ldots, \mathsf{m}_{K_v}^{\mathbf{B}}$ are set, then set $m^{\mathbf{B}} = \mathsf{m}_1^{\mathbf{B}} + \ldots + \mathsf{m}_{K_v}^{\mathbf{B}}$, else set $m^{\mathbf{B}} = \bot$.
- **Round 6:**
  - **B** outputs $m^{\mathbf{B}}$ and terminates.
  - **A** upon computing the output $b_{\mathsf{RMT}}^{(\mathbf{A},3)} \in \{0,1\}$ during the instance $\Pi_{\mathsf{RMT}}^{(\mathbf{A},3)}$, outputs $b = b_{\mathsf{RMT}}^{(\mathbf{A},3)}$ and terminates.

---

[a] If there are several candidates for c_i, then consider the one with the least index.
[b] If there are several candidates for c_i, then consider the one with the least index.
[c] If there are several candidates for c_i, then consider the one with the least index.

Fig. 1. (*continued*)

Irrespective of the degree of replication in the new instance of RSS, it can be shown that each new share of $m^{\mathbf{A}}$ will be delivered over at least one non-faulty channel. Consequently, irrespective of the case, **B** can always detect any inconsistency among the received shares and, by interacting with **A**, can identify the corrupt shares, after which, using the remaining shares, **B** can recover $m^{\mathbf{A}}$. In the protocol, we use what we call as *implicit termination conditions*. These conditions are triggered when **A** (resp. **B**) is sure that the network is behaving *asynchronously*, for instance, if the *total* number of channels over which communication has been delayed so far *exceeds* t_s or if during any instance of Π_{RMT}, node **A** outputs 0 as **S** or $\bot$ as **R**. As and when these conditions are triggered, **A** (resp. **B**) terminates the protocol Π_{SSMT} with output 0 (resp. $\bot$). Protocol Π_{SSMT} is a *round*-based protocol, and as explained in Sect. 2, if a message is received *outside* of the time allocated for a certain round (as per the local clock of the receiving party), then it is ignored by the corresponding receiving party.

We next prove the properties of the protocol. We start with the *synchronous* network and show that **A** or **B** will *not* terminate the protocol "pre-maturely".

Lemma 2. *If the network behaves synchronously, then the implicit termination condition will never be triggered in the protocol Π_{SSMT}.*

Proof. The implicit termination conditions are triggered only if the communication is delayed over *more* than t_s channels or if any instance of Π_{RMT} outputs $b = 0$ for the underlying **S** or $\perp$ for the underlying **R**. Since the network is assumed to behave *synchronously*, *none* of these conditions will hold.

We next prove a helping lemma to prove the correctness in the synchronous network. Namely, we show that irrespective of the way **A** secret-shares its message, both **A** and **B** will identify all incorrect shares delivered to **B**.

Lemma 3. *If the network behaves synchronously, then in the protocol Π_{SSMT}, irrespective of the degree of replication used by **A**, both **A** and **B** will identify all the incorrect shares delivered to **B**. Moreover, any channel that is not under the control of* Adv *will never be included by **A** in the list of faulty channels.*

Proof. We first claim that **A** and **B** will agree on the value of d, and consequently, both will agree on which of the three cases to pursue from Round 2 onward. This trivially follows from the t_s-correctness of Π_{RMT}, which ensures that **A** obtains the set of delayed channels $\mathcal{D}$ as output during Round 2 from the instance $\Pi_{\mathsf{RMT}}^{(\mathbf{B},1)}$. Now there are three possible cases, depending upon the value of $d = |\mathcal{D}|$.

- **Case 1:** $d \leq t_a$: In this case, **A** and **B** decide to continue with the degree of replication $n - t_s$. Since $n > 2t_s$, it implies that each S_k contains at least $t_s + 1$ channels. Consequently, each share $m_k^{\mathbf{A}}$ is sent by **A** over at least $t_s + 1$ channels. Moreover, since Adv can corrupt at most t_s channels, it follows that S_k has at least one *non-faulty* channel, say c_i, and consequently, the condition $m_{k,i}^{\mathbf{B}} = m_k^{\mathbf{A}}$ holds, where $m_{k,i}^{\mathbf{B}}$ is the share received by **B** over c_i during Round 1. Now consider the case when some channel $c_j \in S_k$ is corrupted by the adversary and delivers the share $m_{k,j}^{\mathbf{B}} \neq m_k^{\mathbf{A}}$. In this case, **B** will find that $m_{k,i}^{\mathbf{B}} \neq m_{k,j}^{\mathbf{B}}$ and consequently add $(k, i, j, m_{k,i}^{\mathbf{B}}, m_{k,j}^{\mathbf{B}})$ to L_1. Since L_1 is communicated to **A** through Π_{RMT}, the t_s-correctness of Π_{RMT} ensures that **A** obtains the output L_1 through $\Pi_{\mathsf{RMT}}^{(\mathbf{B},2)}$ at the end of Round 2. As a result, **A** will identify that c_j is *corrupt* and include c_j to $\mathcal{F}_1$, since $m_{k,j}^{\mathbf{B}} \neq m_k^{\mathbf{A}}$ will hold. Moreover, **A** will *not* include c_i to $\mathcal{F}_1$ since $m_{k,i}^{\mathbf{B}} = m_k^{\mathbf{A}}$ will hold. Since $\mathcal{F}_1$ is communicated to **B** through Π_{RMT}, the t_s-correctness of Π_{RMT} ensures that **B** obtains $\mathcal{F}_1$ through $\Pi_{\mathsf{RMT}}^{(\mathbf{A},1)}$ at the end of Round 3. Hence, **B** will identify that c_j is a faulty channel at the end of Round 3.
- **Case 2:** $(t_a < d < t_s$ and $(t_s - d) \leq t_a)$ or $d = t_s$: In this case, **A** and **B** decide to use the degree of sharing t_a, where **A** computes the shares $\mathsf{m}_1^{\mathbf{A}}, \ldots, \mathsf{m}_{K_u}^{\mathbf{A}}$ for $m^{\mathbf{A}}$, and each share $\mathsf{m}_k^{\mathbf{A}}$ is communicated over the set of channels T_k of size $n - d - t_a$. Now, there are two possible sub-cases.
    - **If** $d = t_s$: In this case, each set of channels T_k consists of only *non-faulty* channels since Adv can corrupt at most t_s channels, all of which are included in $\mathcal{D}$. Moreover, T_k *excludes* all the channels in $\mathcal{D}$. Consequently, in this case, *every* share $\mathsf{m}_{k,i}^{\mathbf{B}}$ received by **B** over any channel $c_i \in T_k$ is bound to be correct and satisfy the condition $\mathsf{m}_{k,i}^{\mathbf{B}} = \mathsf{m}_k^{\mathbf{A}}$.

- **If** $(t_a < d < t_s$ and $(t_s - d) \leq t_a)$: In this case, each set T_k consists of at least one *non-faulty* channel, say c_i, since $n - d - t_a - (t_s - d) > 0$. Consequently, the condition $m_{k,i}^B = m_k^A$ holds, where $m_{k,i}^B$ is the share received by **B** over c_i during Round 3. Now consider the case when some channel $c_j \in T_k$ is corrupted by the adversary and delivers the share $m_{k,j}^B \neq m_k^A$. In this case, **B** will find that $m_{k,i}^B \neq m_{k,j}^B$ and consequently add the tuple $(k, i, j, m_{k,i}^B, m_{k,j}^B)$ to L_2. Since L_2 is communicated to **A** through Π_{RMT}, the t_s-correctness of Π_{RMT} ensures that **A** obtains the output L_2 through $\Pi_{\mathsf{RMT}}^{(\mathbf{B},3)}$ at the end of Round 4. As a result, **A** will identify that c_j is *corrupt* and include c_j to $\mathcal{F}_2$, since $m_{k,j}^B \neq m_k^A$ will hold. Moreover, **A** will *not* include c_i to $\mathcal{F}_2$ since $m_{k,i}^B = m_k^A$ will hold. Since $\mathcal{F}_2$ is communicated to **B** through Π_{RMT}, the t_s-correctness of Π_{RMT} ensures that **B** obtains $\mathcal{F}_2$ through $\Pi_{\mathsf{RMT}}^{(\mathbf{A},2)}$ at the end of Round 5. Consequently, **B** identifies c_j as a faulty channel at the end of Round 5.
- **Case 3:** $t_a < d < t_s$ and $(t_s - d) > t_a$: In this case, **A** and **B** decide to use the degree of sharing $t_s - d$, where **A** computes the shares $m_1^A, \ldots, m_{K_v}^A$ for m^A and each share m_k^A is communicated over the set of channels W_k of size $n - d - (t_s - d) = n - t_s$. Since $n > 2t_s$, it follows that each set W_k consists of at least one *non-faulty* channel, say c_i, over which **B** will receive the share $m_{k,i}^B = m_k^A$ during Round 3. Now, similar to case 2, it can be shown that if any channel $c_j \in W_k$ delivers an incorrect share $m_{k,j}^B \neq m_k^A$ to **B** during Round 3, then at the end of Round 4 and Round 5, respectively, **A** and **B** will identify that c_j is a faulty channel. We omit the proof due to repetition.

Lemma 4. *If the network is synchronous, then* Π_{SSMT} *achieves* t_s-*correctness.*

Proof. If the network behaves *synchronously*, then from the proof of Lemma 3, both **A** and **B** will agree on the value of $|\mathcal{D}| = d$ and, consequently, will know which of the three cases to pursue further from Round 2 onward. We consider each of the three cases and show that t_s-correctness holds in each of those cases.

- **Case 1:** $d \leq t_a$: From the proof of Lemma 3, it follows that during Round 3, after obtaining $\mathcal{F}_1$ through $\Pi_{\mathsf{RMT}}^{(\mathbf{A},1)}$, **B** will learn the identity of *all faulty* channels which delivered incorrect shares to **B** during Round 1. Moreover, corresponding to each S_k, there will be at least one *non-faulty* channel in S_k, through which **B** would have received the share m_k^A during Round 1. Consequently, after identifying all faulty channels through $\mathcal{F}_1$, **B** would compute the output m^B for Π_{SMT}, which will be the same as m^A. On the other hand, from the t_s-correctness of Π_{RMT}, it follows that **A** would compute the output $b_{\mathsf{RMT}}^{(\mathbf{A},1)} = 1$ during $\Pi_{\mathsf{RMT}}^{(\mathbf{A},1)}$ and hence would output $b = 1$ for Π_{SMT}.
- **Case 2:** $(t_a < d < t_s$ and $(t_s - d) \leq t_a)$ or $d = t_s$: In this case, from the proof of Lemma 3, it follows that during Round 5, after obtaining the output $\mathcal{F}_2$ through the instance $\Pi_{\mathsf{RMT}}^{(\mathbf{A},2)}$, **B** will learn the identity of *all* the *faulty* channels which delivered incorrect shares to **B** during Round 3. Moreover, corresponding to each set of channels T_k, there will be at least one *non-faulty* channel in T_k, through which **B** would have received the share m_k^A

during Round 3. Consequently, after identifying all faulty channels through $\mathcal{F}_2$, $\mathbf{B}$ would compute the output $m^{\mathbf{B}}$ for Π_{SMT}, which will be the same as $m^{\mathbf{A}}$. From the t_s-correctness of Π_{RMT}, it follows that $\mathbf{A}$ would compute the output $b_{\mathsf{RMT}}^{(\mathbf{A},2)} = 1$ during the instance $\Pi_{\mathsf{RMT}}^{(\mathbf{A},2)}$ and outputs $b = 1$ for Π_{SMT}.

- **Case 3:** $t_a < d < t_s$ and $(t_s - d) > t_a$: In this case, from the proof of Lemma 3, it follows that during Round 5, after obtaining the output $\mathcal{F}_3$ through the instance $\Pi_{\mathsf{RMT}}^{(\mathbf{A},3)}$, $\mathbf{B}$ will learn the identity of *all* the *faulty* channels which delivered incorrect shares to $\mathbf{B}$ during Round 3. Moreover, corresponding to each set of channels W_k, there will be at least one *non-faulty* channel in W_k, through which $\mathbf{B}$ would have received the share $\mathsf{m}_k^{\mathbf{A}}$ during Round 5. Consequently, after identifying all faulty channels through $\mathcal{F}_3$, $\mathbf{B}$ would compute the output $m^{\mathbf{B}}$ for Π_{SMT}, which will be the same as $m^{\mathbf{A}}$. From the t_s-correctness of Π_{RMT}, it follows that $\mathbf{A}$ would compute the output $b_{\mathsf{RMT}}^{(\mathbf{A},3)} = 1$ during the instance $\Pi_{\mathsf{RMT}}^{(\mathbf{A},3)}$ and outputs $b = 1$ for Π_{SMT}.

Lemma 5. *If the network behaves synchronously, then the protocol Π_{SSMT} achieves t_s-termination and takes at most 6 rounds.*

Proof. Follows from the fact that the protocol is round-based and from the proof of Lemma 3, both $\mathbf{A}$ and $\mathbf{B}$ will know which case they have to pursue.

Lemma 6. *If the network is synchronous, then Π_{SSMT} achieves t_s-privacy.*

Proof. From the proof of Lemma 3, both $\mathbf{A}$ and $\mathbf{B}$ will know which of the three cases they are in during the protocol. We show that irrespective of the case, the view of the adversary will be independent of $m^{\mathbf{A}}$. This is because, in each case, there will be at least one share of $m^{\mathbf{A}}$, which is sent only on the channels which are *not* under the control of Adv. Since the shares are picked randomly by $\mathbf{A}$, the privacy follows. In more detail, we have the following three cases.

- **A and B are in Case 1:** For simplicity and without loss of generality, let Z_1 be the set of t_s channels under the control of Adv. We show that throughout the protocol, the adversary does not learn anything about the share $m_1^{\mathbf{A}}$. During Round 1, the adversary *does not* learn anything about $m_1^{\mathbf{A}}$, since it is sent over the channels in S_1, none of which is under the control of Adv. During Round 2, $\mathbf{B}$ sends the set $\mathcal{D}$, which *does not* add anything about $m_1^{\mathbf{A}}$ to the view of Adv. Additionally, $\mathbf{B}$ also sends the list of conflicting shares through L_1, and these shares will be already known to Adv since shares that are delivered correctly will never conflict with each other. Hence, through L_1, the adversary does not learn anything additional about $m_1^{\mathbf{A}}$. During Round 3, $\mathbf{A}$ identifies and sends the list of faulty channels that delivered incorrect shares to $\mathbf{B}$ during Round 1. These channels will be already known to Adv and do not add anything additional about $m_1^{\mathbf{A}}$ to the view of Adv. Now, since the share $m_1^{\mathbf{A}}$ is picked randomly, it follows that the view of Adv remains independent of $m^{\mathbf{A}}$. Namely, for every candidate $m^{\mathbf{A}}$, there is a corresponding $m_1^{\mathbf{A}}$, consistent with the view of Adv.

- **A and B are in Case** 2: From the proof of the previous case, the shares learnt by **B** during Round 1 are independent of $m^{\mathbf{A}}$. Now, there are two possible sub-cases, depending upon the value of d.

  - If $d = t_s$: In this case, all the channels in the sets $U_1, \ldots, U_{K_u}$ are non-faulty and not under the control of Adv. This is because all the channels in $\mathcal{D}$ are under the control of Adv, which has the capability to corrupt up to t_s channels. Consequently, during Round 3, the shares $m_1^{\mathbf{A}}, \ldots, m_{K_u}^{\mathbf{A}}$ are communicated only over non-faulty channels, as no share is sent over the channels in $\mathcal{D}$ by **A**. As a result, the adversary does not learn anything about these shares. Hence, no conflicting shares will be delivered to **B**, so L_2 will be $\emptyset$. As a result, the set $\mathcal{F}_2$ will also be $\emptyset$, and hence Adv does not learn anything about the shares $m_1^{\mathbf{A}}, \ldots, m_{K_u}^{\mathbf{A}}$ through L_2 and $\mathcal{F}_2$. Since the shares $m_1^{\mathbf{A}}, \ldots, m_{K_u}^{\mathbf{A}}$ are picked randomly by **A**, it follows that the view of Adv remains independent of $m^{\mathbf{A}}$.

  - If $t_a < d < t_s$ and $(t_s - d) \leq t_a$: Since all the channels in $\mathcal{D}$ are under the control of Adv and $d < t_s$, it implies that Adv can further corrupt up to $t_s - d$ channels during Round 3, since during Round 3, **A** does not send shares over the channels in $\mathcal{D}$. For simplicity and without loss of generality, let U_1 be the set of $(t_s - d)$ channels which are under the control of Adv during Round 3. We claim that throughout the protocol, the adversary does not learn anything about the share $m_1^{\mathbf{A}}$. During Round 3, the adversary *does not* learn anything about the share $m_1^{\mathbf{A}}$ since it is sent over the channels in U_1, none of which is under the control of Adv. During Round 4, **B** sends the list of conflicting shares through L_2, and these shares will be already known to Adv since shares that are delivered correctly will never conflict with each other. Hence, through L_2, the adversary does not learn anything additional about $m_1^{\mathbf{A}}$. During Round 5, **A** identifies and sends the list of faulty channels that delivered incorrect shares to **B** during Round 3. These channels will be already known to Adv and do not add anything additional about $m_1^{\mathbf{A}}$ to the view of Adv. Now since $m_1^{\mathbf{A}}$ is picked randomly, it follows that the view of Adv remains independent of $m^{\mathbf{A}}$. Namely, for every candidate $m^{\mathbf{A}}$, there is a corresponding $m_1^{\mathbf{A}}$, consistent with the view of Adv.

- **A and B are in Case** 3: From the proof of Case 1, the shares learnt by **B** during Round 1 are independent of $m^{\mathbf{A}}$. Note that all the channels in $\mathcal{D}$ are under the control of Adv, and it has the capacity to corrupt up to $t_s - d$ channels during Round 3 since no share is communicated by **A** during Round 3, over the channels in $\mathcal{D}$. For simplicity and without loss of generality, let V_1 be the set of $t_s - d$ channels which are under the control of Adv during Round 3. We claim that throughout the protocol, the adversary does not learn anything about the share $m_1^{\mathbf{A}}$. The proof for this is similar to the previous case and is omitted.

We next proceed to prove the properties in the *asynchronous* network. The termination is easy to prove since the protocol is round-based, and hence by (local) time 6Δ, both $\mathbf{A}$ and $\mathbf{B}$ will terminate. Moreover, the privacy follows from the proof of t_s-privacy (Lemma 6) and the fact that $t_a \leq t_s$.

Lemma 7. *Protocol* Π_{SSMT} *achieves* t_a-*privacy and* t_a-*termination.*

We next proceed to prove the t_a-weak correctness. We first consider the easier case when $\mathbf{B}$ outputs $\perp$ and show that in this case, $\mathbf{A}$ outputs 0.

Lemma 8. *If the network behaves asynchronously and if* $\mathbf{B}$ *terminates* Π_{SSMT} *with output* $\perp$, *then the protocol achieves* t_a-*weak correctness.*

Proof. Let $\mathbf{B}$ terminate Π_{SSMT} with output $\perp$. Then, from the protocol steps, it follows that the implicit termination condition is triggered for $\mathbf{B}$, as otherwise $\mathbf{B}$ would output an element from $\mathbb{G}$. We consider all possible scenarios for $\mathbf{B}$ in which the implicit termination condition would have been triggered for $\mathbf{B}$ and show that in all cases, $\mathbf{A}$ will terminate Π_{SSMT} with output 0.

- **B terminates at the end of Round** 1: this happens because $\mathbf{B}$ does not receive messages over more than t_s channels during Round 1. Consequently, $\mathbf{B}$ does not invoke the instance $\Pi_{\mathsf{RMT}}^{(\mathbf{B},1)}$ of Π_{RMT} during Round 2 to send $\mathcal{D}$, and as a result, communication over more than t_s channels will be delayed for $\mathbf{A}$ during Round 2. Hence, from the t_a-weak correctness of Π_{RMT}, $\mathbf{A}$ will output 0 (as $\mathbf{R}$) for the instance $\Pi_{\mathsf{RMT}}^{(\mathbf{B},1)}$ and due to the implicit termination condition, will terminate Π_{SSMT} with output $b = 0$ at the end of Round 2.
- **B terminates at the end of Round** 3: this happens under the following conditions.[4]
    - $\mathbf{B}$ *as* $\mathbf{S}$ *outputs* $b = 0$ *during the instance* $\Pi_{\mathsf{RMT}}^{(\mathbf{B},1)}$. In this case, $\mathbf{B}$ stops participating in the protocol from Round 4 onwards. Now, there are two possibilities. Either $\mathbf{A}$ (as $\mathbf{R}$) outputs $\perp$ during the instance $\Pi_{\mathsf{RMT}}^{(\mathbf{B},1)}$, in which case $\mathbf{A}$ terminates Π_{SSMT} with output $b = 0$ at the end of Round 2 due to the implicit termination condition. Else, communication over more than t_s channels will be delayed for $\mathbf{A}$ during Round 4 since $\mathbf{B}$ will not send any messages, and due to the implicit termination condition, $\mathbf{A}$ will terminate Π_{SSMT} with output $b = 0$.
    - $\mathbf{B}$ *executes Case 1 and outputs* $b = 0$ *(as* $\mathbf{S}$) *during the instance* $\Pi_{\mathsf{RMT}}^{(\mathbf{B},2)}$. This case is similar to the previous case.
    - $\mathbf{B}$ *executes Case 1 and outputs* $\perp$ *(as* $\mathbf{R}$) *during the instance* $\Pi_{\mathsf{RMT}}^{(\mathbf{A},1)}$. In this case, the t_a-weak correctness of Π_{RMT} guarantees that $\mathbf{A}$ as $\mathbf{S}$ outputs 0 for the instance $\Pi_{\mathsf{RMT}}^{(\mathbf{A},1)}$ during Round 4, and hence $\mathbf{A}$ will terminate Π_{SSMT} with output $b = 0$.
    - $\mathbf{B}$ *executes Case 2 and the communication over more than* t_s *channels is delayed by the end of Round 3.* In this case, $\mathbf{B}$ stops participating in the protocol from Round 4 onward. So the communication over more than t_s channels will be delayed for $\mathbf{A}$ during Round 4 since $\mathbf{B}$ will not send any

---

[4] Note that the implicit termination condition *cannot* be triggered for $\mathbf{B}$ during Round 2 since Round 2 involves communication from $\mathbf{B}$ to $\mathbf{A}$, and no instance of Π_{RMT} is invoked by $\mathbf{B}$ (as $\mathbf{S}$) during Round 1.

messages, and due to the implicit termination condition, $\mathbf{A}$ will terminate Π_{SSMT} at the end of Round 4 with output $b = 0$.

- **B terminates at the end of Round** 5: this happens under the following conditions.[5]

  - $\mathbf{B}$ *executes Case 2 and outputs* 0 *(as* $\mathbf{S}$*) during the instance* $\Pi_{\mathsf{RMT}}^{(\mathrm{B},3)}$: This implies that $\mathbf{B}$ does not send any messages to $\mathbf{A}$ during Round 6. Now, there are two possibilities for $\mathbf{A}$. If $\mathbf{A}$ (as $\mathbf{R}$) outputs $\perp$ during the instance $\Pi_{\mathsf{RMT}}^{(\mathrm{B},3)}$, then due to the implicit termination condition, $\mathbf{A}$ would terminate Π_{SSMT} with output 0 at the end of Round 4. Otherwise, during Round 6, communication over more than t_s channels will be delayed for $\mathbf{A}$, and so due to the implicit termination condition, $\mathbf{A}$ would terminate Π_{SSMT} with output 0 at the end of Round 6.

  - $\mathbf{B}$ *executes Case 3 and outputs* 0 *(as* $\mathbf{S}$*) during the instance* $\Pi_{\mathsf{RMT}}^{(\mathrm{B},4)}$: This case is similar to the previous case.

We next consider the case when $\mathbf{B}$ outputs $m^{\mathbf{B}} \in \mathbb{G}$ and show that $m^{\mathbf{B}} = m^{\mathbf{A}}$. For this, we first prove an analogue of Lemma 3 for the *asynchronous* network.

Lemma 9. *If the network behaves asynchronously and if* $\mathbf{B}$ *outputs* $m^{\mathbf{B}} \in \mathbb{G}$, *then in the protocol* Π_{SSMT}, *irrespective of the degree of secret-sharing used by* $\mathbf{A}$, *both* $\mathbf{A}$ *and* $\mathbf{B}$ *will identify all the incorrect shares delivered to* $\mathbf{B}$. *Moreover, any channel that is not under the control of* Adv *will never be included by* $\mathbf{A}$ *in the list of faulty channels.*

Proof. Let $\mathbf{B}$ output $m^{\mathbf{B}} \in \mathbb{G}$, implying that the implicit termination condition is *not* triggered for $\mathbf{B}$. We first claim that $\mathbf{A}$ and $\mathbf{B}$ will agree on the value of d, and consequently, both will agree on which of the three cases to pursue further from Round 2 onward. This trivially follows from the t_a-weak correctness of Π_{RMT} and the fact that $\mathbf{B}$ (as $\mathbf{S}$) outputs 1 during the instance $\Pi_{\mathsf{RMT}}^{(\mathrm{B},1)}$ since the implicit termination condition is not triggered for $\mathbf{B}$ and hence $\mathbf{A}$ would have obtained the set of delayed channels $\mathcal{D}$ as output during Round 2 from the instance $\Pi_{\mathsf{RMT}}^{(\mathrm{B},1)}$. Now, there are three possible cases, depending upon $d = |\mathcal{D}|$.

- **Case 1:** $d \le t_a$: In this case, $\mathbf{A}$ and $\mathbf{B}$ decide to continue with the degree of sharing t_s. Since $n > 2t_a + t_s$, it implies that each S_k is of size at least $n - t_s$ and hence contains at least $2t_a + 1$ channels. Out of these $2t_a + 1$ channels, Adv can delay communication over at most $d \le t_a$ *non-faulty* channels and corrupt at most t_a channels. Hence, it follows that corresponding to the set S_k, there exists at least one *non-faulty* channel, say c_i, which delivers the share $m_{k,i}^{\mathbf{B}}$ to $\mathbf{B}$ during Round 1, such that $m_{k,i}^{\mathbf{B}} = m_k^{\mathbf{A}}$ holds. Now consider the case when some channel $c_j \in S_k$ is corrupted by the adversary and delivers the share $m_{k,j}^{\mathbf{B}} \neq m_k^{\mathbf{A}}$ to $\mathbf{B}$ during Round 1. In this case, $\mathbf{B}$ will find that $m_{k,i}^{\mathbf{B}} \neq m_{k,j}^{\mathbf{B}}$ and consequently add the tuple $(k, i, j, m_{k,i}^{\mathbf{B}}, m_{k,j}^{\mathbf{B}})$ to L_1. Since L_1

---

[5] Note that the implicit termination condition *cannot* be triggered for $\mathbf{B}$ during Round 4 since Round 4 involves communication from $\mathbf{B}$ to $\mathbf{A}$, and no instance of Π_{RMT} is invoked by $\mathbf{B}$ (as $\mathbf{S}$) during Round 3.

is communicated to $\mathbf{A}$ through Π_{RMT} and the implicit termination condition is not triggered for $\mathbf{B}$, it implies that $\mathbf{B}$ (as $\mathbf{S}$) outputs 1 during the instance $\Pi_{\mathsf{RMT}}^{(\mathbf{B},2)}$. Consequently, the t_a-weak correctness of Π_{RMT} ensures that $\mathbf{A}$ obtains the output L_1 through the instance $\Pi_{\mathsf{RMT}}^{(\mathbf{B},2)}$ at the end of Round 2. As a result, $\mathbf{A}$ will identify that c_j is *corrupt* and include c_j to $\mathcal{F}_1$ since $m_{k,j}^{\mathbf{B}} \neq m_k^{\mathbf{A}}$ will hold. Moreover, $\mathbf{A}$ will *not* include c_i to $\mathcal{F}_1$, since $m_{k,i}^{\mathbf{B}} = m_k^{\mathbf{A}}$ will hold. Since $\mathcal{F}_1$ is communicated to $\mathbf{B}$ through Π_{RMT} and the implicit termination condition is not triggered for $\mathbf{B}$, it implies that $\mathbf{B}$ (as $\mathbf{R}$) outputs $\mathcal{F}_1$ through the instance $\Pi_{\mathsf{RMT}}^{(\mathbf{A},1)}$ at the end of Round 3. Consequently, $\mathbf{B}$ will identify that c_j is a faulty channel at the end of Round 3.

- **Case 2:** $(t_a < d < t_s$ **and** $(t_s - d) \leq t_a)$ **or** $d = t_s$: In this case, $\mathbf{A}$ and $\mathbf{B}$ decide to use the degree of sharing t_a, where $\mathbf{A}$ computes the shares $m_1^{\mathbf{A}}, \ldots, m_{K_u}^{\mathbf{A}}$ for $m^{\mathbf{A}}$ and each share $m_k^{\mathbf{A}}$ is communicated over the set of channels T_k of size $n - d - t_a$. Now among these $n - d - t_a$ channels, Adv can delay communication over at most $t_s - d$ channels, as otherwise, it would trigger the implicit termination condition for $\mathbf{B}$ (since we are considering the case when the communication over d channels has already been delayed and if additionally now communication over more than $t_s - d$ channels is delayed, then in *total* communication over more than t_s channels is delayed), which is a contradiction. Hence each share $m_k^{\mathbf{A}}$ will be delivered over at least $n - d - t_a - (t_s - d) = n - t_a - t_s$ channels. Out of these, the adversary can corrupt at most t_a channels, and hence at least $n - 2t_a - t_s$ channels will correctly deliver the share $m_k^{\mathbf{A}}$. Since $n > 2t_a + t_s$, it follows that corresponding to T_k, there exists at least one *non-faulty* channel, say c_i, which delivers the share $m_{k,i}^{\mathbf{B}}$ to $\mathbf{B}$ during Round 3, such that $m_{k,i}^{\mathbf{B}} = m_k^{\mathbf{A}}$ holds. Now consider the case when some channel $c_j \in T_k$ is corrupted by the adversary and delivers the share $m_{k,j}^{\mathbf{B}} \neq m_k^{\mathbf{A}}$ to $\mathbf{B}$ during Round 3. In this case, $\mathbf{B}$ will find that $m_{k,i}^{\mathbf{B}} \neq m_{k,j}^{\mathbf{B}}$ and consequently add the tuple $(k, i, j, m_{k,i}^{\mathbf{B}}, m_{k,j}^{\mathbf{B}})$ to L_2. Since L_2 is communicated to $\mathbf{A}$ through Π_{RMT} and the implicit termination condition is not triggered for $\mathbf{B}$, it implies that $\mathbf{B}$ (as $\mathbf{S}$) outputs 1 during the instance $\Pi_{\mathsf{RMT}}^{(\mathbf{B},3)}$. Consequently, the t_a-weak correctness of Π_{RMT} ensures that $\mathbf{A}$ obtains the output L_2 through the instance $\Pi_{\mathsf{RMT}}^{(\mathbf{B},3)}$ at the end of Round 4. As a result, $\mathbf{A}$ will identify that c_j is *corrupt* and include c_j to $\mathcal{F}_2$, since $m_{k,j}^{\mathbf{B}} \neq m_k^{\mathbf{A}}$ will hold. Moreover, $\mathbf{A}$ will *not* include c_i to $\mathcal{F}_2$ since $m_{k,i}^{\mathbf{B}} = m_k^{\mathbf{A}}$ will hold. Since $\mathcal{F}_2$ is communicated to $\mathbf{B}$ through Π_{RMT} and the implicit termination condition is not triggered for $\mathbf{B}$, it implies that $\mathbf{B}$ (as $\mathbf{R}$) outputs $\mathcal{F}_2$ through the instance $\Pi_{\mathsf{RMT}}^{(\mathbf{A},2)}$ at the end of Round 5. Consequently, $\mathbf{B}$ will identify that c_j is a faulty channel at the end of Round 5.

- **Case 3:** $t_a < d < t_s$ **and** $(t_s - d) > t_a$: In this case, $\mathbf{A}$ and $\mathbf{B}$ decide to use the degree of sharing $t_s - d$, where $\mathbf{A}$ computes the shares $m_1^{\mathbf{A}}, \ldots, m_{K_v}^{\mathbf{A}}$ for $m^{\mathbf{A}}$, and each share $m_k^{\mathbf{A}}$ is communicated over the set of channels W_k of size $n - d - (t_s - d) = n - t_s$. Now among these $n - t_s$ channels, Adv can delay communication over at most $t_s - d$ channels, as otherwise, it would trigger the implicit termination condition for $\mathbf{B}$, since in total, communication over more than t_s channels will be delayed, which is a contradiction. Hence each share m_k^A will be delivered over at least $n - t_s - (t_s - d) = n - 2t_s + d$ channels. Out of these, the adversary can corrupt at most t_a channels, and hence at least $n - 2t_s - t_a + d$ channels in W_k will correctly deliver the share m_k^A. Since $t_s < \frac{n}{2}$ and we are considering the case where $t_a < d$, it follows that each set W_k consists of at least one *non-faulty* channel, say c_i, over which $\mathbf{B}$ will receive the share $m_{k,i}^B = m_k^A$ during Round 3. Now, similar to the previous case, it can be shown that if any channel $c_j \in W_k$ delivers an incorrect share $m_{k,j}^B \neq m_k^A$ to $\mathbf{B}$ during Round 3, then at the end of Round 4 and Round 5, respectively, $\mathbf{A}$ and $\mathbf{B}$ will identify that c_j is a faulty channel.

The proof of Lemma 10 is similar to the proof of Lemma 4.

Lemma 10. *If the network behaves asynchronously and if $\mathbf{B}$ outputs $m^B \in \mathbb{G}$ in the protocol Π_{SSMT}, then $m^B = m^A$.*

Lemma 11, follows from the fact that the most expensive step of the protocol involves reliably sending the list of conflicting shares through Π_{RMT}.

Lemma 11. *Π_{SSMT} incurs a communication of $\mathcal{O}(K \cdot n^3 \cdot \log |\mathbb{G}|)$ bits, where $K \stackrel{def}{=} \max(K_s, K_u, K_v)$ and $K_s \stackrel{def}{=} \binom{n}{t_s}$, $K_u \stackrel{def}{=} \binom{n-d}{t_a}$, $K_v \stackrel{def}{=} \binom{n-d}{t_s-d}$ and $d = |\mathcal{D}|$.*

4 Asynchronous SMT

In this section, we outline our instantiation of Π_{ASMT}, the second component required for the generic compiler of [5] (see Sect. 2.2). The protocol achieves t_a-correctness, t_a-privacy, and t_a-termination in an *asynchronous* network. The existing instantiations of Π_{ASMT} [4,5] are based on error-correcting codes over a field $\mathbb{F}$ and will *not* over a group $\mathbb{G}$. Consequently, we use a different instantiation of Π_{ASMT} over $\mathbb{G}$. The idea is similar to that of our instantiation of Π_{SSMT}. On having the input m^A, the sender $\mathbf{A}$ generates secret-shares for m^A, as per RSS, where each share is replicated over $n - t_a$ channels. To recover m^A, $\mathbf{B}$ extracts out each share using the "majority rule". Since there will be at least one share unknown to the adversary, the privacy of m^A is guaranteed. The protocol incurs a communication of $\mathcal{O}(n \cdot K_a \cdot \log |\mathbb{G}|)$ bits, where $K_a = \binom{n}{t_a}$.

5 Conclusion and Open Problems

In this paper, we presented the *first* network-agnostic PSMT protocol with *optimal* threshold conditions and which requires a *constant* number of communication rounds in the *synchronous* network. There are several interesting research directions to pursue in this domain. Here, we list a few of them.

- The computation and communication complexity of our protocol is *exponential*. It is indeed a challenging open problem to get a network-agnostic PSMT protocol with a *constant* number of rounds in the *synchronous* network, and with *polynomial* computation and communication complexity.

- In this work, we have considered *threshold* adversaries. It will be interesting to see if we can generalize the results of this paper against *non-threshold* adversaries. Such a generalization, together with the recent results of [2], will give a complete characterization of network-agnostic MPC protocols with perfect security in *incomplete* networks against non-threshold adversaries.
- It will be interesting to derive *tight* bounds on the communication complexity and round complexity of network-agnostic PSMT protocols.

References

1. Agarwal, S., Cramer, R., de Haan, R.: Asymptotically optimal two-round perfectly secure message transmission. In: Dwork, C. (ed.) CRYPTO 2006. LNCS, vol. 4117, pp. 394–408. Springer, Heidelberg (2006). https://doi.org/10.1007/11818175_24
2. Appan, A., Chandramouli, A., Choudhury, A.: Network agnostic perfectly secure MPC against general adversaries. In: DISC. LIPIcs, vol. 281, pp. 3:1–3:19. Schloss Dagstuhl - Leibniz-Zentrum für Informatik (2023). https://doi.org/10.4230/LIPIcs.DISC.2023.3
3. Choudhury, A., Patra, A.: Secure Multi-party Computation Against Passive Adversaries. SLDCT, Springer, Cham (2023). https://doi.org/10.1007/978-3-031-12164-7
4. Choudhury, A., Patra, A., Ashwinkumar, B.V., Srinathan, K., Pandu Rangan, C.: Secure message transmission in asynchronous networks. J. Parallel Distrib. Comput. **71**(8), 1067–1074 (2011). https://doi.org/10.1016/j.jpdc.2011.03.004
5. Deligios, G., Liu-Zhang, C.: Synchronous perfectly secure message transmission with optimal asynchronous fallback guarantees. Cryptology ePrint Archive, Paper 2022/1397 (2022). Published in FC2023
6. Dolev, D., Dwork, C., Waarts, O., Yung, M.: Perfectly secure message transmission. J. ACM **40**(1), 17–47 (1993). https://doi.org/10.1145/138027.138036
7. Ito, M., Saito, A., Nishizeki, T.: Secret sharing schemes realizing general access structures. In: Globecom, pp. 99–102. IEEE (1987). https://doi.org/10.1002/ecjc.4430720906
8. Kurosawa, K., Suzuki, K.: Truly efficient 2-round perfectly secure message transmission scheme. In: Smart, N. (ed.) EUROCRYPT 2008. LNCS, vol. 4965, pp. 324–340. Springer, Heidelberg (2008). https://doi.org/10.1007/978-3-540-78967-3_19
9. Patra, A., Choudhary, A., Srinathan, K., Rangan, C.P.: Constant phase bit optimal protocols for perfectly reliable and secure message transmission. In: Barua, R., Lange, T. (eds.) INDOCRYPT 2006. LNCS, vol. 4329, pp. 221–235. Springer, Heidelberg (2006). https://doi.org/10.1007/11941378_16
10. Yao, A.C.: Protocols for secure computations (extended abstract). In: FOCS, pp. 160–164. IEEE Computer Society (1982). https://doi.org/10.1109/SFCS.1982.38

Explicit Lower Bounds
for Communication Complexity of PSM
for Concrete Functions

Kazumasa Shinagawa[1,3]($\boxtimes$) and Koji Nuida[2,3]

[1] Ibaraki University, Hitachi, Japan
kazumasa.shinagawa.np92@vc.ibaraki.ac.jp
[2] Institute of Mathematics for Industry (IMI), Kyushu University, Fukuoka, Japan
nuida@imi.kyushu-u.ac.jp
[3] National Institute of Advanced Industrial Science and Technology (AIST), Tokyo,
Japan

Abstract. Private Simultaneous Messages (PSM) is a minimal model
of secure computation, where the input players with shared randomness
send messages to the output player simultaneously and only once. In this
field, finding upper and lower bounds on communication complexity of
PSM protocols is important, and in particular, identifying the optimal
one where the upper and lower bounds coincide is the ultimate goal.
However, up until now, functions for which the optimal communication
complexity has been determined are few: An example of such a function is the two-input AND function where $(2 \log_2 3)$-bit communication
is optimal. In this paper, we provide new upper and lower bounds for several concrete functions. For lower bounds, we introduce a novel approach
using combinatorial objects called abstract simplicial complexes to represent PSM protocols. Our method is suitable for obtaining non-asymptotic
explicit lower bounds for concrete functions. By deriving lower bounds
and constructing concrete protocols, we show that the optimal communication complexity for the equality and majority functions with three
input bits are $3 \log_2 3$ bits and 6 bits, respectively. We also derive new
lower bounds for the n-input AND function, three-valued comparison
function, and multiplication over finite rings.

Keywords: secure multiparty computation · private simultaneous
messages · communication complexity · lower bounds · concrete
functions

1 Introduction

1.1 Background

Private Simultaneous Messages (PSM) is a minimal model of secure computation, initially proposed by Feige, Kilian, and Naor (hereafter, FKN) [8] and later
generalized by Ishai and Kushilevitz [9]. A PSM protocol involves the input players $P_1, \ldots, P_n$ and the output player called the referee. Each input player P_i with

© The Author(s), under exclusive license to Springer Nature Switzerland AG 2024
A. Chattopadhyay et al. (Eds.): INDOCRYPT 2023, LNCS 14460, pp. 45–61, 2024.
https://doi.org/10.1007/978-3-031-56235-8_3

input x_i sends a message m_i to the referee simultaneously and only once, and the referee computes the output value y based on the received messages $m_1, \ldots, m_n$. Here, all players share a randomness r in advance, which is independent of the inputs and inaccessible to the referee, and each message m_i is computed from the input value x_i and the randomness r only. For a function f to be computed, a PSM protocol is said to be correct if $y = f(x_1, \ldots, x_n)$ holds with probability 1, and said to be secure if, when the output value y is fixed, the distribution of the tuple of the messages is independent from the input distribution. The communication complexity of a PSM protocol is defined as $\sum_{i=1}^{n} \log_2 |M_i|$, where M_i denotes the i-th message space and $|\cdot|$ denotes the cardinality of the set.

It is important to determine the upper and lower bounds for communication complexity of PSM protocols. So far, there is an exponential gap between these bounds for general two-input functions $f : \{0,1\}^k \times \{0,1\}^k \to \{0,1\}$: The best-known upper bound is $O(2^{k/2})$ from the PSM protocol constructed by Beimel, Ishai, Kumaresan, and Kushilevitz [5], while the lower bound for a random f is $3k - O(\log k)$ by Applebaum, Holenstein, Mishra, and Shayevitz [1] (see also Vaikuntanathan's survey [11]). The situation is similar for general n-input functions $f : (\{0,1\}^k)^n \to \{0,1\}$: The best-known upper bound is $O(poly(n) \cdot 2^{\frac{nk}{2}})$ by Beimel, Kushilevitz, and Nissim [6], the upper bound for infinitely many n is $O(poly(n) \cdot 2^{\frac{(n-1)k}{2}})$ by Assouline and Liu [2], and the lower bound with $n = \omega(k)$ for a random f is $\Omega(n^2 k / \log(nk))$ by Ball, Holmgren, Ishai, Liu, and Malkin [3] for $k = 1$ and by Ball and Randolph [4] for general k.

On the other hand, for concrete functions, various PSM protocols had been proposed: the AND function [8], the three-valued comparison function [8], branching programs [9], symmetric functions [10], and so on. However, up until now, there are only few functions for which the optimal communication complexity has been determined. An example of such a function is the two-input AND function $x_1 \wedge x_2$, where the optimal communication complexity is shown to be $2 \log_2 3$ bits: The protocol is constructed by FKN [8], and the lower bound is given by Data, Prabhakaran, and Prabhakaran (hereafter, DPP) [7]. Another example is the multiplication $x_1 x_2 \cdots x_n$ over a finite group G, where the optimal communication complexity is shown to be $n \log_2 |G|$ bits: The protocol is constructed by FKN [8], and the lower bound follows from the trivial bound, i.e., the communication complexity without security. As the latter result on the lower bound did not concern the security, a new approach that fully utilizes the security condition is demanded towards non-trivial lower bounds for other functions.

1.2 Our Contribution

In this paper, we derive new upper and lower bounds on communication complexity of PSM protocols for several concrete functions, aiming to further identify the optimal communication complexity. The main technical contribution of this paper is to introduce a novel approach for proving lower bounds using combinatorial objects called abstract simplicial complexes (hereafter, simplicial complexes) to represent PSM protocols. Based on this approach, we derive lower

Table 1. Summary of our results and existing results

	upper/lower	communication complexity	condition
o AND: $x_1 \wedge \cdots \wedge x_n$			
FKN [8]	construction	$|M_i| = p$	$p > n$: prime
DPP [7]	lower bound	$|M_i| \geq 3$	$n = 2$
Section 4.2	lower bound	$|M_1| = 3 \Rightarrow |M_{i(\neq 1)}| \geq 6$	$n \geq 3$
o Equality: $(x_1 = \cdots = x_n)$?			
Section 4.3	construction	$|M_i| = p$	$p \geq n$: prime
Section 4.3	lower bound	$|M_i| \geq 3$	–
o Majority: $(x_1 + \cdots + x_n \geq \lceil n/2 \rceil)$?			
Section 4.4	construction	$|M_i| = 4$	$n = 3$
Section 4.4	lower bound	$|M_i| \geq 4$	–
o $(k+1)$-valued comparison: $(x_1 > x_2$ or $x_1 = x_2$ or $x_1 < x_2)$?			
FKN [8]	construction	$|M_i| = 7$	$k = 2$
Section 4.5	lower bound	$|M_i| \geq 2k + 1$	–
Section 4.5	lower bound	$|M_1| \geq 6$ or $|M_2| \geq 6$	$k = 2$
o Multiplication over a finite ring S: $x_1 \cdot x_2$			
Beaver Triple	construction	$|M_i| = |S|^2$	any S
Section 4.6	lower bound	$|M_i| \geq 2q - 1$	$S = \mathbb{F}_q$
Section 4.6	lower bound	$|M_i| \geq \sum_{j=1}^q \gcd(j, q)$	$S = \mathbb{Z}/q\mathbb{Z}$

bounds for the AND function, equality function, majority function, comparison function, and multiplication over finite rings. At the same time, we also provide upper bounds for the equality function, majority function, and multiplication over finite rings by constructing new protocols. As a result, we identify the optimal communication complexity for the equality function and majority function in the case of $n = 3$. Regarding the three-input AND function and three-valued comparison function, we specify all possibilities (eight for each function) for the optimal communication complexity. Our results are summarized in Table 1. In the following, we explain the details of each item.

– For the AND function $x_1 \wedge \cdots \wedge x_n$, FKN [8] proposed a PSM protocol with $|M_i| = p$ $(1 \leq i \leq n)$, where p is any prime number satisfying $p > n$, which currently provides the best-known upper bound. When $n = 2$, DPP [7] showed a lower bound $|M_i| \geq 3$ $(i \in \{1, 2\})$, which proves the optimality of the FKN protocol for $n = 2$. Based on our new approach for proving lower bounds, we provide an alternative proof for this result. When $n \geq 3$, as a new lower bound, we show that if $|M_1| = 3$, then $|M_i| \geq 6$ for all $i \neq 1$. In particular, when $n = 3$, there are only eight possibilities of the tuple $(|M_1|, |M_2|, |M_3|)$ (up to symmetry) for the protocol with the minimum value of $\sum_{i=1}^3 \log_2 |M_i|$; we specify them explicitly.
– For the equality function, we propose a PSM protocol with $|M_i| = p$, where p is any prime number satisfying $p \geq n$, and prove a lower bound of $|M_i| \geq 3$ $(1 \leq i \leq n)$. When $n = 3$, the upper and lower bounds coincide, hence, the optimal communication complexity is determined as $3 \log_2 3$ bits.

- For the majority function, we prove a lower bound of $|M_i| \geq 4$ $(1 \leq i \leq n)$. When $n = 3$, we propose a PSM protocol with $|M_i| = 4$ $(1 \leq i \leq 3)$. In this case, the upper and lower bounds coincide, hence, the optimal communication complexity for $n = 3$ is determined as 6 bits.
- For the $(k+1)$-valued comparison function $f : \{0, 1, \ldots, k\} \times \{0, 1, \ldots, k\} \to \{-1, 0, 1\}$, we prove a lower bound of $|M_i| \geq 2k+1$ $(i \in \{1, 2\})$. When $k = 2$, FKN [8] proposed a PSM protocol with $|M_i| = 7$ $(i \in \{1, 2\})$, which currently provides the best-known upper bound. In this case, as a new lower bound, we show that either $|M_1| \geq 6$ or $|M_2| \geq 6$, which implies that there are only eight possibilities of $(|M_1|, |M_2|)$ (up to symmetry) attaining the minimum value of $\sum_{i=1}^{2} \log_2 |M_i|$; we specify them explicitly.
- For the multiplication function $x_1 \cdot x_2$ over a finite ring S, we propose a PSM protocol with $|M_i| = |S|^2$ $(i \in \{1, 2\})$ using the idea of Beaver multiplication triples. As for lower bounds, we prove that $|M_i| \geq 2q - 1$ if S is the field of order q and $|M_i| \geq a(q) := \sum_{j=1}^{q} \gcd(j, q)$ if S is the integer residue ring modulo q, where $a(q)$ is known as the Pillai's arithmetic function.

1.3 Technical Overview

Our new method of deriving lower bounds, which we call the *embedding method*, is a multi-input generalization of the idea by FKN [8] using edge-colored bipartite graphs. We start with recalling their idea. For a function $f : \{0, 1\}^k \times \{0, 1\}^k \to \{0, 1\}$, the bipartite graph G_f is constructed by joining vertices x_1 and x_2 with a black edge if $f(x_1, x_2) = 0$, and with a red edge if $f(x_1, x_2) = 1$. Similarly, the decoding function $\mathsf{Dec} : M_1 \times M_2 \to \{0, 1\}$ of a PSM protocol is represented by the bipartite graph G_{Dec}. In this context, a shared randomness of the PSM protocol can be regarded as an embedding map from G_f to G_{Dec} that preserves the coloring of edges. Based on this idea, FKN constructed PSM protocols and proved lower bounds.

To extend the idea of FKN to n-input functions $f : X_1 \times \cdots \times X_n \to Y$ with $n \geq 2$, we needed, instead of an edge joining two elements x_1 and x_2, a "higher-dimensional edge" joining n elements $x_1, \ldots, x_n$. We found that a combinatorial object called *simplicial complexes* (see Sect. 2.2) is suitable for the purpose. In our setting, the simplicial complex Δ_f representing the function f consists of $(n-1)$-dimensional faces $\{x_1, \ldots, x_n\}$ (called *facets*) having n vertices $x_i \in X_i$ (and their subfaces $\{x_{i_1}, \ldots, x_{i_d}\}$, $i_1 < \cdots < i_d$). Then a facet $\{x_1, \ldots, x_n\}$ of Δ_f has color $y \in Y$ if $f(x_1, \ldots, x_n) = y$. The simplicial complex Δ_{Dec} representing the decoding function Dec is defined in the same way. Now similarly to the case of $n = 2$, a key observation here is that any shared randomness can be interpreted as an embedding map from Δ_f to Δ_{Dec} that preserves the coloring of facets.

In order to explain our idea, here we demonstrate an alternative proof for DPP's lower bound $|M_i| \geq 3$ $(i \in \{1, 2\})$ for the two-input AND function $f(x_1, x_2) = x_1 \wedge x_2$. First, as mentioned above, any shared randomness $r = (r_1, r_2)$ can be regarded as a pair of injections $r_i : X_i = \{0, 1\} \to M_i$ $(i \in \{1, 2\})$. Take a randomness $r = (r_1, r_2)$ and write $\hat{b} := r_i(b)$ $(b \in \{0, 1\}, i \in \{1, 2\})$. Here, the security of PSM protocols can be interpreted as stating that for any edge $\hat{e}$ of G_{Dec}, the probability that an edge e of G_f with the same color as $\hat{e}$ is mapped to $\hat{e}$ by some randomness is independent of e. Thus, since the black edge $e = (0,0)$ in G_f is mapped to the black edge $\hat{e} = (\hat{0},\hat{0})$ in G_{Dec}, there must exist another randomness $r' = (r_1', r_2')$ that maps another black edge $e' = (0,1)$ in G_f to $\hat{e}$. In this case, r' maps a red edge $e'' = (1,1)$ to $\hat{e'} = (r_1'(1),\hat{0})$, which should be a red edge and hence be different from $(\hat{0},\hat{0})$ and $(\hat{1},\hat{0})$. Therefore, we have $r_1'(1) \neq \hat{0},\hat{1}$ and hence M_1 must have at least three distinct elements, i.e., $|M_1| \geq 3$. By symmetry, we have $|M_2| \geq 3$.

In a general case, we establish useful lemmas for proving lower bounds (Lemmas 3 and 4) which we call *embedding lemmas*. These two versions have a trade-off that a strong version is only applicable to some restricted kind of functions f while a weak version is applicable to any f. In the previous paragraph, we derived the existence of a new embedding r' from the fact "the vertex 0 in X_1 is joined to two black edges" and the existence of the red edge $\hat{e'}$ from the fact "the vertex 1 in X_2 is joined to both black and red edges". The idea of embedding lemmas is to derive a lower bound on the number of facets of Δ_{Dec} around a lower-dimensional face based on distributions of colors for facets in Δ_f.

Basic Notations. For an integer $n \geq 1$, we write $[n] := \{1, 2, \ldots, n\}$. For any integer $q \geq 2$, we write $\mathbb{Z}_q := \mathbb{Z}/q\mathbb{Z}$ and $\mathbb{Z}_q^{\times} := (\mathbb{Z}/q\mathbb{Z})^{\times}$ identified with $\{0, 1, \ldots, q-1\}$ and $\{1, 2, \ldots, q-1\}$, respectively. For a set S, we denote by $|S|$ the cardinality of S. For a bit string $m \in \{0,1\}^*$, we denote by $|m|$ the bit length of m. For two probability distributions $\mathcal{X}, \mathcal{Y}$, we write $\mathcal{X} \equiv \mathcal{Y}$ if they are the same probability distribution.

2 PSM Protocols and Simplicial Complexes

2.1 PSM Protocols

Definition 1. *Let $n \geq 2$ be a positive integer. Let X_i, M_i, R_i $(i \in [n])$, and Y be finite sets. Write $\vec{X} = X_1 \times \cdots \times X_n$, $\vec{M} = M_1 \times \cdots \times M_n$, and $\vec{R} = R_1 \times \cdots \times R_n$. Let $\mathsf{Enc}_i \colon X_i \times R_i \to M_i$ $(i \in [n])$ be functions and $\mathsf{Dec} \colon \vec{M} \to Y$ be a partial function. Let $\mathcal{R} = (\mathcal{R}_1, \ldots, \mathcal{R}_n)$ be a random variable over $\vec{R}$. A private simultaneous messages (PSM) protocol for a function $f \colon \vec{X} \to Y$ is a tuple $\Pi = (n, (X_i)_{i \in [n]}, Y, \mathcal{R}, (M_i)_{i \in [n]}, (\mathsf{Enc}_i)_{i \in [n]}, \mathsf{Dec})$ with the following conditions:*

- *(Correctness) For any $x = (x_1, \ldots, x_n) \in \vec{X}$,*

$$\Pr[\, \mathsf{Dec}(\mathsf{Enc}_1(x_1, \mathcal{R}_1), \ldots, \mathsf{Enc}_n(x_n, \mathcal{R}_n)) = f(x) \,] = 1.$$

Note that the correctness also claims that the partial function Dec is defined over the arguments on the left-hand side.
- *(Security) For any $x = (x_i)_{i \in [n]}, x' = (x_i')_{i \in [n]} \in \vec{X}$ with $f(x) = f(x')$,*

$$(\mathsf{Enc}_1(x_1, \mathcal{R}_1), \ldots, \mathsf{Enc}_n(x_n, \mathcal{R}_n)) \equiv (\mathsf{Enc}_1(x_1', \mathcal{R}_1), \ldots, \mathsf{Enc}_n(x_n', \mathcal{R}_n)).$$

We call n the number of players, X_i the i-th input set, Y the output set, M_i the i-th message space, R_i the i-th randomness set, Enc_i the i-th encoding function, and Dec the decoding function. We also define the effectiveness as follows:

– *A random number* $r \in \vec{R}$ *is said to be* effective *if* $\Pr[r \leftarrow \mathcal{R}] > 0$. *We will denote by* R *the set of effective random numbers.*
– *A tuple of messages* $(m_{i_0}, \ldots, m_{i_d}) \in M_{i_0} \times \cdots \times M_{i_d}$ $(1 \leq i_0 < \cdots < i_d \leq n)$ *is said to be* effective *if there exist an input* $(x_1, \ldots, x_n) \in \vec{X}$ *and an effective random number* $(r_1, \ldots, r_n) \in R$ *such that* $\mathsf{Enc}_{i_j}(x_{i_j}, r_{i_j}) = m_{i_j}$ $(0 \leq j \leq d)$.

That is, a random number or a tuple of messages is said to be effective if it can appear during an execution of protocol Π. In Definition 1, an element r_i of R_i defines a function $X_i \to M_i$ that maps $x_i \in X_i$ to $\mathsf{Enc}_i(x_i, r_i) \in M_i$. This function is also denoted by r_i. We assume without loss of generality that any two elements $r_i \neq r_i'$ of R_i define different functions, since otherwise the correctness and security of the protocol are not affected by identifying r_i with r_i', hence reducing the size of R_i.

2.2 Simplicial Complexes

Let $\Delta \subseteq 2^S$ be a non-empty set of subsets of a finite set S. Δ is said to be a *simplicial complex* with *underlying set* S if $A \in \Delta$ and $B \subseteq A$ imply $B \in \Delta$ for any A, B. An element of Δ is called a *face* of Δ. The *dimension* of a face $A \in \Delta$ is defined by $\dim(A) := |A| - 1$. A maximal element A of Δ with respect to inclusion is called a *facet* of Δ. The set of all facets is denoted by $\mathsf{Facet}(\Delta)$. For a finite set C, a function $\mathsf{color} : \mathsf{Facet}(\Delta) \to C$ is said to be a *C-coloring* of Δ, and $\mathsf{color}(F)$ for a facet F is called the *color* of F.

Let $(S_1, \ldots, S_n)$ be a partition of S. A simplicial complex Δ is said to be *n-partite* with respect to $(S_1, \ldots, S_n)$ if $|A \cap S_i| \leq 1$ for any $A \in \Delta$ and $1 \leq i \leq n$. Moreover, Δ is said to be the *complete n-partite simplicial complex* with respect to $(S_1, \ldots, S_n)$ if $\{a_1, \ldots, a_n\} \in \Delta$ for any $a_i \in S_i$ $(1 \leq i \leq n)$. For an n-partite simplicial complex Δ, a face $\{a_{i_0}, a_{i_1}, \ldots, a_{i_d}\} \in \Delta$ $(1 \leq i_0 < i_1 < \cdots < i_d \leq n,$ $a_{i_j} \in S_{i_j})$ is often represented by a sequence of length n formed by placing a_{i_j} in the i_j-th position and the symbol '$\perp$' in the remaining positions. For example, when $n = 5$, a face $\{a_1, a_3, a_4\}$ is represented by $(a_1, \perp, a_3, a_4, \perp)$ or $a_1 \perp a_3 a_4 \perp$.

Let Δ and Δ' be n-partite simplicial complexes with respect to $(S_1, \ldots, S_n)$ and $(S_1', \ldots, S_n')$, respectively. Let $\phi = (\phi_1, \ldots, \phi_n)$ be a tuple of n functions $\phi_i : S_i \to S_i'$ $(1 \leq i \leq n)$. For a face $A = \{a_{i_0}, \ldots, a_{i_d}\}$ $(a_{i_j} \in S_{i_j})$ of Δ, define $\phi(A) := \{\phi_{i_0}(a_{i_0}), \ldots, \phi_{i_d}(a_{i_d})\}$. We say that ϕ is a *morphism* from Δ to Δ', denoted by $\phi : \Delta \to \Delta'$, if $\dim(\phi(A)) = \dim(A)$ for any $A \in \mathsf{Face}(\Delta)$. If ϕ is injective, ϕ is said to be an *embedding* of Δ into Δ'. If ϕ is bijective, ϕ is said to be an *isomorphism* from Δ to Δ'. When each of Δ and Δ' has a C-coloring, we consider only morphisms ϕ that are consistent with the coloring, that is, those mapping a facet A of Δ onto a facet of Δ' with the same color as A.

2.3 Simplicial Complexes for PSM Protocols

Let $f' : X_1' \times \cdots \times X_n' \to Y'$ be a partial function, and Δ a Y'-colored n-partite simplicial complex with respect to a partition $(X_1', \ldots, X_n')$. Δ is said to be *the simplicial complex defined by* f' if for any $x_i' \in X_i'$ $(1 \leq i \leq n)$, $A = \{x_1', \ldots, x_n'\}$

is a facet of Δ if and only if $f'(x'_1, \ldots, x'_n)$ is defined, and in this case, the color of A coincides with $f'(x'_1, \ldots, x'_n)$.

For a PSM protocol as in Definition 1, let Δ_f and Δ_{Dec} be the simplicial complexes defined by f and Dec, respectively. Then we can observe that the correctness of the PSM protocol is equivalent to the following condition:

For any effective randomness $r = (r_1, \ldots, r_n) \in R$ viewed as a tuple of functions $X_i \to M_i$, r is a (color-preserving) morphism from Δ_f to Δ_{Dec}.

Furthermore, the security of the PSM protocol can be rewritten as follows:

For any facets F, F' of Δ_f with the same color and any facet $\widehat{F}$ of Δ_{Dec}, the following equation holds:

$$\Pr_{r \leftarrow \mathcal{R}}[r(F) = \widehat{F}] = \Pr_{r \leftarrow \mathcal{R}}[r(F') = \widehat{F}].$$

We denote the left-hand side of the equation by $\Pr[F \mapsto \widehat{F}]$ and the right-hand side by $\Pr[F' \mapsto \widehat{F}]$. Note that we can remove any ineffective facets (and all faces included in the removed facets only) from Δ_{Dec} without affecting correctness or security. From now on, throughout this paper, we assume that all facets (and therefore all faces) of Δ_{Dec} are effective. In particular, the following is satisfied:

Lemma 1. *Let Π be a PSM protocol for a function f. Then for any facet $\widehat{F}$ of Δ_{Dec} and any facet F of Δ_f with the same color as $\widehat{F}$, there exists an effective randomness $r \in R$ satisfying that $r(F) = \widehat{F}$.*

Proof. Since $\widehat{F}$ is effective as mentioned above, there are a facet F' of Δ_f and an effective $r' \in R$ with $r'(F') = \widehat{F}$, hence $\Pr[F' \mapsto \widehat{F}] > 0$. By the correctness of Π, the color of F' is the same as that of $\widehat{F}$, hence of F as well. Now by the security of Π, we have $\Pr[F \mapsto \widehat{F}] = \Pr[F' \mapsto \widehat{F}] > 0$, implying the claim. □

3 Embedding Methods for Proving Lower Bounds

3.1 Injectivity of the Morphisms Defined by Randomness

In PSM protocols described by simplicial complexes, it is found that a morphism defined by a random number typically results in an embedding of Δ_f into Δ_{Dec}. To elaborate on this, we introduce the following definition.

Definition 2. *A function $f \colon X_1 \times \cdots \times X_n \to Y$ is said to have no redundant inputs if for any $i \in [n]$, there do not exist distinct $x_i, x'_i \in X_i$ such that*

$$f(x_1, \ldots, x_{i-1}, x_i, x_{i+1}, \ldots, x_n) = f(x_1, \ldots, x_{i-1}, x'_i, x_{i+1}, \ldots, x_n)$$

for all $x_j \in X_j$ $(j \in [n], j \neq i)$.

Note that we can remove any redundant inputs without affecting correctness or security, hence, we will focus on functions having no redundant inputs.

We also define the following notation.

Definition 3. *The type* $\mathsf{type}(Z)$ *of a d-dimensional face* Z *of* Δ_f *is defined as the tuple of indices* $(i_0, \ldots, i_d)$ *with* $1 \leq i_0 < \cdots < i_d \leq n$ *such that* $Z \cap X_{i_j} \neq \emptyset$ *for any* $0 \leq j \leq d$. *The type of a face of* Δ_{Dec} *is defined in the similar way.*

Then the following lemma for the embedding holds.

Lemma 2. *Let* Π *be a PSM protocol for a function* f *having no redundant inputs. Then for any effective* $r = (r_1, \ldots, r_n) \in R$, *each* $r_i \colon X_i \to M_i$ *is injective, and* r *is an embedding from* Δ_f *to* Δ_{Dec}.

Proof. First, we show that each r_i is injective. Assume for contradiction that $r_i(x_i) = r_i(x_i')$ for different $x_i, x_i' \in X_i$. From the correctness of Π, we have

$$f(x_1, \ldots, x_i, \ldots, x_n) = \mathsf{Dec}(r_1(x_1), \ldots, r_i(x_i), \ldots, r_n(x_n))$$
$$= \mathsf{Dec}(r_1(x_1), \ldots, r_i(x_i'), \ldots, r_n(x_n))$$
$$= f(x_1, \ldots, x_i', \ldots, x_n)$$

for any $x_j \in X_j$ ($j \in [n]$, $j \neq i$). This contradicts the assumption that f has no redundant inputs. Therefore, r_i is injective.

It remains to show that $r(Z) \neq r(Z')$ for different faces $Z, Z' \in \Delta_f$ of the same type $(i_0, \ldots, i_d)$. Since $Z \neq Z'$, there is an index i_j such that $Z \cap X_{i_j} \neq Z' \cap X_{i_j}$. Since $r_{i_j} \colon X_{i_j} \to M_{i_j}$ is injective as above, it follows that $r(Z) \cap M_{i_j} \neq r(Z') \cap M_{i_j}$. Therefore, we have $r(Z) \neq r(Z')$. $\square$

3.2   Embedding Lemmas

Let Δ be a simplicial complex with a C-coloring color. For any $j \in C$, we define a function $n_j : \Delta \to \mathbb{N}_{\geq 0}$ by $n_j(Z) := |\mathsf{Facet}(\Delta \mid j, Z)|$, where

$$\mathsf{Facet}(\Delta \mid j, Z) = \{F \in \mathsf{Facet}(\Delta) \mid Z \subseteq F, \ \mathsf{color}(F) = j\}.$$

We define a function $n : \Delta \to (\mathbb{N}_{\geq 0})^{|C|}$ as follows:

$$n(Z) = (n_j(Z))_{j \in C}.$$

We refer to this vector as the *color degree* of the face Z.

Here, we use the same notations as in Sect. 2.3. Let $Z \in \Delta_f$. We define a subset $\mathcal{F}(Z)$ of Δ_f as the set of all $Z' \in \Delta_f$ such that $\mathsf{type}(Z') = \mathsf{type}(Z)$ and there exists a color $j \in C$ such that $n_j(Z) > 0$ and $n_j(Z') > 0$.

Lemma 3 (Embedding lemma (weak form)). *Let* Π *be a PSM protocol for a function* $f \colon X_1 \times \cdots \times X_n \to Y$ *having no redundant inputs. For any face Z of Δ_f, there exists a face $\widehat{Z}$ of Δ_{Dec} such that* $\mathsf{type}(\widehat{Z}) = \mathsf{type}(Z)$ *and for any* $j \in Y$, *the following equation holds:*

$$n_j(\widehat{Z}) \geq N_j := \max\{n_j(Z') \mid Z' \in \mathcal{F}(Z)\}.$$

Proof. Let $r \in R$ be an effective embedding and set $\widehat{Z} := r(Z)$. Let $j \in Y$. Fix any face $Z' \in \mathcal{F}(Z)$. By the definition of $\mathcal{F}(Z)$, $\mathsf{type}(Z') = \mathsf{type}(Z)$ and there exists a color $j' \in Y$ such that $n_{j'}(Z) > 0$ and $n_{j'}(Z') > 0$, hence, there exist facets $F_1, F_2 \in \mathsf{Facet}(\Delta_f)$ with $Z \subseteq F_1$, $Z' \subseteq F_2$ such that $f(F_1) = j' = f(F_2)$. Writing $\widehat{F} := r(F_1)$, we have $\Pr[F_1 \mapsto \widehat{F}] > 0$. From the security of Π, we have $\Pr[F_2 \mapsto \widehat{F}] > 0$, i.e., there exists an effective embedding $r' \in R$ such that $r'(F_2) = \widehat{F}$. Since $\mathsf{type}(Z') = \mathsf{type}(Z)$ and $\widehat{Z} = r(Z) \subseteq r(F_1) = \widehat{F}$, it must hold $r'(Z') = \widehat{Z}$. Since r' is an embedding from Δ_f to Δ_{Dec} by Lemma 2, it gives an injection from $\mathsf{Facet}(\Delta_f \mid j, Z')$ to $\mathsf{Facet}(\Delta_{\mathsf{Dec}} \mid j, \widehat{Z})$, therefore $n_j(\widehat{Z}) \geq n_j(Z')$. Since $Z' \in \mathcal{F}(Z)$ is arbitrary, we have $n_j(\widehat{Z}) \geq N_j$.   □

Corollary 1. *Under the same notations as Lemma 3, if* $\mathsf{type}(Z) = (i_0, \ldots, i_d)$, *then the following equation holds:*

$$\prod_{a \in [n] \setminus \{i_0, \ldots, i_d\}} |M_a| \geq \sum_{j \in Y} N_j.$$

Proof. For the face $\widehat{Z} \in \Delta_{\mathsf{Dec}}$ as in Lemma 3, the number of facets of Δ_{Dec} containing $\widehat{Z}$ is $\prod_{a \in [n] \setminus \{i_0, \ldots, i_d\}} |M_a|$, and this value is also written as $\sum_{j \in Y} n_j(\widehat{Z})$. This relation and Lemma 3 proves the claim.   □

The following lemma is a strengthened version of Lemma 3, which holds for a certain type of function f.

Lemma 4 (Embedding lemma (strong form)). *Let Π be a PSM protocol for a function f having no redundant inputs. Fix a type t of Δ_f. Define*

$$N_j^* := \max\{n_j(Z) \mid Z \in \Delta_f, \ \mathsf{type}(Z) = t\}$$

for $j \in Y$. Furthermore, assume that there exists a face $Z \in \Delta_f$ of type t such that all components of the color degree $n(Z)$ are positive. Then, for any face $\widehat{Z}$ of Δ_{Dec} of type t, we have $n_j(\widehat{Z}) \geq N_j^*$ ($j \in Y$).

Proof. Fix a facet $\widehat{F}_1$ of Δ_{Dec} with color, say j_1, containing $\widehat{Z}$. From the assumption on Z, there exists a facet $F_1 \in \mathsf{Facet}(\Delta_f \mid j_1, Z)$. By Lemma 1, there exists an effective morphism that maps F_1 to $\widehat{F}_1$, hence maps Z to $\widehat{Z}$ since $\mathsf{type}(\widehat{Z}) = \mathsf{type}(Z)$. From the assumption that all components of $n(Z)$ are positive, all components of $n(\widehat{Z})$ are also positive. Now let Z' be any face of Δ_f of type t, and fix a facet F_2 of Δ_f with color, say j_2, containing Z'. Since $n_{j_2}(\widehat{Z}) > 0$ as above, there is a facet $\widehat{F}_2 \in \mathsf{Facet}(\Delta_{\mathsf{Dec}} \mid j_2, \widehat{Z})$. By Lemmas 1 and 2, there exists an effective embedding that maps F_2 to $\widehat{F}_2$ (and hence Z' to $\widehat{Z}$) and gives an injection $\mathsf{Facet}(\Delta_f \mid j, Z') \to \mathsf{Facet}(\Delta_{\mathsf{Dec}} \mid j, \widehat{Z})$ for any $j \in Y$, implying $n_j(\widehat{Z}) \geq n_j(Z')$. As Z' is arbitrary, $n_j(\widehat{Z}) \geq N_j^*$, which proves the claim.   □

4   Communication Complexity for Concrete Functions

In this section, using the embedding method from Sect. 3, we provide lower bounds on the communication complexity for concrete functions. For some functions, we also provide upper bounds on it by constructing PSM protocols.

Since the functions f below have no redundant inputs, Lemma 2 implies that any PSM protocol for f must satisfy $|M_i| \geq |X_i|$ for each i (since there is an injection $X_i \to M_i$). Hereafter, we refer to this as the *trivial lower bound*.

4.1   Multiplication in Groups

For a finite group G, let $f \colon G^n \to G$ be the multiplication function $f(x_1, \ldots, x_n) = x_1 \cdots x_n$. As already noted in Sect. 1.1, the optimal communication complexity for f has been determined to $n \log_2 |G|$. Here, we give another proof of this fact using the embedding method.

Since f has no redundant inputs (by setting the remaining input components to be the identity element 1_G), we have the trivial lower bound $|M_i| \geq |X_i| = |G|$. Since FKN [8] designed a PSM protocol for the function f with $|M_i| = |G|$, it is optimal in terms of the communication complexity. For a special case, by letting $G = \mathbb{Z}_2$ with the group operation $\oplus$ (XOR operation), we can construct a PSM protocol for the n-input XOR function $x_1 \oplus \cdots \oplus x_n$ with $|M_i| = 2$ $(i \in [n])$, which is optimal in terms of the communication complexity.

4.2   AND Function

Let $f \colon \{0,1\}^n \to \{0,1\}$ be the AND function $f(x_1, \ldots, x_n) = x_1 \wedge \cdots \wedge x_n$. f has no redundant inputs (by setting the remaining input components to be 1). For $n = 2$, we gave in Sect. 1.3 another proof of the lower bound given by DPP [7]. For $n \geq 3$, by using the strong form of the embedding lemma (Lemma 4), we can derive a stronger lower bound in the following.

Theorem 1. *Let $n \geq 3$. For any PSM protocol computing the n-input AND function f, we have $|M_i| \geq 3$ for any i, and if $|M_i| = 3$ for some $i \in [n]$, then $|M_{i'}| \geq 6$ for any $i' \neq i$.*

Proof. As mentioned above, f has no redundant inputs. For any $1 \leq i \leq n$ and any $(n-2)$-dimensional face $Z \in \Delta_f$ with $Z \cap X_i = \emptyset$, the coloring degree of Z is $(n_0(Z), n_1(Z)) = (1,1)$ if all components of Z are 1 and $(2,0)$ otherwise. In particular, f satisfies the assumptions of Lemma 4 with $N_0^* = 2$ and $N_1^* = 1$. Hereafter, we will use Lemma 4 without explicit mention. Then the number $|M_i|$ of facets in Δ_{Dec} including any given face of type $(1, \ldots, i-1, i+1, \ldots, n)$ is at least $N_0^* + N_1^* = 3$, therefore the former claim holds. For the remaining claim, since f is symmetric, it suffices to assume $|M_1| = 3$ and $|M_2| \leq 5$ and derive a contradiction. We denote the set of all facets of a simplicial complex Δ of color j by $\mathsf{Facet}(\Delta \mid j)$.

By Lemma 2, we take an effective embedding (called "standard embedding") and denote its image of a vertex $a \in X_i = \{0,1\}$ $(i \in [n])$ by $\widehat{a} \in M_i$. Then, for any facet $x_1 \cdots x_n$ of Δ_f, $\widehat{x_1} \cdots \widehat{x_n}$ is a facet of Δ_{Dec} of the same color. We specify the colors of the facets of Δ_{Dec}. Here, $a^{n-2} := aa \cdots a$ $(n-2$ a's).

(1) Take a face $\widehat{0} \bot \widehat{0}^{n-2}$ of Δ_{Dec}. By the standard embedding, $\widetilde{000}^{n-2}, \widetilde{010}^{n-2} \in \mathsf{Facet}(\Delta_{\mathsf{Dec}} \mid 0)$, hence there must exist another facet in $\mathsf{Facet}(\Delta_{\mathsf{Dec}} \mid 1)$ including the face $\widehat{0} \bot \widehat{0}^{n-2}$ (since $n_1(\widehat{0} \bot \widehat{0}^{n-2}) \geq N_1^* = 1$). Thus, there exists $\widehat{2} \in M_2 \setminus \{\widehat{0}, \widehat{1}\}$ such that $\widetilde{020}^{n-2} \in \mathsf{Facet}(\Delta_{\mathsf{Dec}} \mid 1)$.

(2) Take a face $\bot \widehat{20}^{n-2}$ of Δ_{Dec}. From (1), $\widetilde{020}^{n-2} \in \mathsf{Facet}(\Delta_{\mathsf{Dec}} \mid 1)$. Since $|M_1| = 3$ and $n_0(\bot \widehat{20}^{n-2}) \geq N_0^* = 2$, $\widetilde{120}^{n-2}, \widetilde{220}^{n-2} \in \mathsf{Facet}(\Delta_{\mathsf{Dec}} \mid 0)$ where we set $M_1 = \{\widehat{0}, \widehat{1}, \widehat{2}\}$.

(3) Take two faces $\bot \widehat{00}^{n-2}$ and $\bot \widehat{10}^{n-2}$ of Δ_{Dec}. By the standard embedding, $\widetilde{000}^{n-2}, \widetilde{100}^{n-2}, \widetilde{010}^{n-2}, \widetilde{110}^{n-2} \in \mathsf{Facet}(\Delta_{\mathsf{Dec}} \mid 0)$. (Here, we used the condition $n \geq 3$ for $\widetilde{110}^{n-2}$.) Thus, $\widetilde{200}^{n-2}, \widetilde{210}^{n-2} \in \mathsf{Facet}(\Delta_{\mathsf{Dec}} \mid 1)$.

(4) From (2), $\widetilde{120}^{n-2} \in \mathsf{Facet}(\Delta_{\mathsf{Dec}} \mid 0)$, hence by Lemmas 1 and 2, there exists an effective $r = (r_1, \ldots, r_n) \in R$ such that r maps $000^{n-2} \in \mathsf{Facet}(\Delta_f \mid 0)$ to $\widetilde{120}^{n-2}$. This r maps $100^{n-2} \in \mathsf{Facet}(\Delta_f \mid 0)$ to $r_1(1)\widetilde{20}^{n-2} \in \mathsf{Facet}(\Delta_{\mathsf{Dec}} \mid 0)$ different from $r(000^{n-2}) = \widetilde{120}^{n-2}$. From (1), $\widetilde{020}^{n-2} \in \mathsf{Facet}(\Delta_{\mathsf{Dec}} \mid 1)$ and $r_1(1) \neq \widehat{0}, \widehat{1}$, hence, $r_1(1) = \widehat{2}$ and $r(100^{n-2}) = \widetilde{220}^{n-2}$.

(5) r maps $110^{n-2} \in \mathsf{Facet}(\Delta_f \mid 0)$ to $\widehat{2}r_2(1)\widehat{0}^{n-2} \in \mathsf{Facet}(\Delta_{\mathsf{Dec}} \mid 0)$ different from $r(100^{n-2}) = \widetilde{220}^{n-2}$. From (3), $\widetilde{200}^{n-2}, \widetilde{210}^{n-2} \in \mathsf{Facet}(\Delta_{\mathsf{Dec}} \mid 1)$, hence, $r_2(1) \neq \widehat{0}, \widehat{1}, \widehat{2}$. Thus, there exists $\widehat{3} \in M_2 \setminus \{\widehat{0}, \widehat{1}, \widehat{2}\}$ such that $r_2(1) = \widehat{3}$. Therefore, we have $\widetilde{230}^{n-2} \in \mathsf{Facet}(\Delta_{\mathsf{Dec}} \mid 0)$, and since $r(000^{n-2}) = \widetilde{120}^{n-2}$ as in (4), we have $r(010^{n-2}) = \widetilde{130}^{n-2} \in \mathsf{Facet}(\Delta_{\mathsf{Dec}} \mid 0)$.

(6) Take a face $\bot \widehat{30}^{n-2}$ of Δ_{Dec}. From (5), $\widetilde{130}^{n-2}, \widetilde{230}^{n-2} \in \mathsf{Facet}(\Delta_{\mathsf{Dec}} \mid 0)$, hence, $\widetilde{030}^{n-2} \in \mathsf{Facet}(\Delta_{\mathsf{Dec}} \mid 1)$.

(7) Take a face $\widehat{1} \bot \widehat{0}^{n-2}$ of Δ_{Dec}. From the standard embedding and (2) and (5), we have $\widetilde{100}^{n-2}, \widetilde{110}^{n-2}, \widetilde{120}^{n-2}, \widetilde{130}^{n-2} \in \mathsf{Facet}(\Delta_{\mathsf{Dec}} \mid 0)$. Thus, there exists $\widehat{4} \in M_2 \setminus \{\widehat{0}, \widehat{1}, \widehat{2}, \widehat{3}\}$ such that $\widetilde{140}^{n-2} \in \mathsf{Facet}(\Delta_{\mathsf{Dec}} \mid 1)$. Since $|M_2| \leq 5$, we have $M_2 = \{\widehat{0}, \widehat{1}, \widehat{2}, \widehat{3}, \widehat{4}\}$.

(8) Take a face $\bot \widehat{40}^{n-2}$ of Δ_{Dec}. From (7), $\widetilde{140}^{n-2} \in \mathsf{Facet}(\Delta_{\mathsf{Dec}} \mid 1)$, hence, $\widetilde{040}^{n-2}, \widetilde{240}^{n-2} \in \mathsf{Facet}(\Delta_{\mathsf{Dec}} \mid 0)$.

(9) From (8), $\widetilde{040}^{n-2} \in \mathsf{Facet}(\Delta_{\mathsf{Dec}} \mid 0)$, hence by Lemmas 1 and 2, there exists an effective $r' = (r'_1, \ldots, r'_n) \in R$ such that r' maps $000^{n-2} \in \mathsf{Facet}(\Delta_f \mid 0)$ to $\widetilde{040}^{n-2}$. This r' maps $100^{n-2} \in \mathsf{Facet}(\Delta_f \mid 0)$ to $r'_1(1)\widetilde{40}^{n-2} \in \mathsf{Facet}(\Delta_{\mathsf{Dec}} \mid 0)$ different from $r'(000^{n-2}) = \widetilde{040}^{n-2}$. From (7), $\widetilde{140}^{n-2} \in \mathsf{Facet}(\Delta_{\mathsf{Dec}} \mid 1)$ and $r'_1(1) \neq \widehat{0}, \widehat{1}$, hence we have $r'_1(1) = \widehat{2}$.

(10) r' maps $010^{n-2} \in \mathsf{Facet}(\Delta_f \mid 0)$ into $\widehat{0}r'_2(1)\widehat{0}^{n-2} \in \mathsf{Facet}(\Delta_{\mathsf{Dec}} \mid 0)$ different from $r'(000^{n-2}) = \widetilde{040}^{n-2}$. From (1) and (6), $\widetilde{020}^{n-2}, \widetilde{030}^{n-2} \in \mathsf{Facet}(\Delta_{\mathsf{Dec}} \mid 1)$, hence, $r'_2(1) \neq \widehat{2}, \widehat{3}, \widehat{4}$ and $r'_2(1) \in \{\widehat{0}, \widehat{1}\}$. Thus, r' maps $110^{n-2} \in \mathsf{Facet}(\Delta_f \mid 0)$ to $\widehat{2}r'_2(1)\widehat{0}^{n-2} \in \mathsf{Facet}(\Delta_{\mathsf{Dec}} \mid 0)$, which must be either $\widetilde{200}^{n-2}$ or $\widetilde{210}^{n-2}$. However, from (3), we have $\widetilde{200}^{n-2}, \widetilde{210}^{n-2} \in \mathsf{Facet}(\Delta_{\mathsf{Dec}} \mid 1)$, yielding a contradiction.

This completes the proof. $\qquad\qquad\qquad\qquad\qquad\qquad\qquad\qquad\qquad\qquad\square$

Corollary 2. *For the PSM protocol for the three-input AND with $|M_1| \leq |M_2| \leq |M_3|$ attaining the minimum value of $\sum_{i=1}^{3} \log_2 |M_i|$, $(|M_1|, |M_2|, |M_3|)$ is one of $(3, 6, 6)$, $(4, 4, 4)$, $(4, 4, 5)$, $(4, 4, 6)$, $(4, 4, 7)$, $(4, 5, 5)$, $(4, 5, 6)$, and $(5, 5, 5)$.*

Proof. Since the protocol in [8] satisfies $|M_i| = 5$ for any i, the optimal case satisfies $\prod_{i=1}^{3} |M_i| \leq 5^3 = 125$. Now the claim follows from Theorem 1. □

4.3 Equality Function

Let $f: \{0,1\}^n \to \{0,1\}$ be the n-input equality function that outputs 1 if and only if all bits are the same. From the embedding method, we obtain the following lower bound for f.

Theorem 2. *Any PSM protocol for the n-input equality function f satisfies $|M_i| \geq 3$ for any $i \in [n]$.*

Proof. By symmetry, it suffices to show that $|M_n| \geq 3$. Similarly to Sect. 4.2, f has no redundant inputs. Now the color degrees of faces $Z := 0^{n-1}\perp$ and $Z' := 0^{n-2}1\perp$ of Δ_f are $n(Z) = (1,1)$ and $n(Z') = (2,0)$, respectively. Therefore $Z, Z' \in \mathcal{F}(Z)$ and we have $N_0 \geq \max\{1, 2\} = 2$ and $N_1 \geq \max\{1, 0\} = 1$ in Corollary 1. Hence $|M_n| \geq N_0 + N_1 \geq 3$ by Corollary 1. □

Let $p \geq n$ be any prime number. We design a PSM protocol for the n-input equality function as follows.

Shared randomness:
  - $r_i = (b, c_i)$ $(i \in [n])$, where $b \in \mathbb{Z}_p^{\times}$ and $c_1, \ldots, c_{n-1} \in \mathbb{Z}_p$ are chosen uniformly at random and $c_n = -\sum_{i=1}^{n-1} c_i \in \mathbb{Z}_p$.
The protocol:
  1. P_i, holding $x_i \in \{0,1\}$, computes $m_i = bx_i + c_i \pmod{p}$ for $i \in [n-1]$ and $m_n = b(p - n + 1)x_n + c_n \pmod{p}$, and sends it to the referee.
  2. The referee outputs 1 if $\sum_{i=1}^{n} m_i = 0 \pmod{p}$ and 0 otherwise.
Communication complexity: $|M_i| = p$.

Proposition 1. *The above protocol is a correct and secure PSM protocol for the n-input equality function with $|M_i| = p$ for any prime $p \geq n$.*

Proof. Let $\bar{x} := \sum_{i=1}^{n-1} x_i + (p - n + 1)x_n \in \mathbb{Z}$. Then $0 \leq \bar{x} \leq p$, and we have $\bar{x} = 0$ (resp., p) if and only if all x_i are 0 (resp., 1). Hence, since $b \in \mathbb{Z}_p^{\times}$ is uniformly random, $\sum_{i=1}^{n} m_i = b\bar{x} \pmod{p}$ is 0 if all x_i are equal, and is uniformly random over $\mathbb{Z}_p^{\times}$ otherwise, implying the correctness. Moreover, due to the uniform choices for the c_i's, $(m_1, \ldots, m_n)$ is uniformly random over all those tuples with $\sum_{i=1}^{n} m_i$ being 0 (resp., uniformly random over $\mathbb{Z}_p^{\times}$) if $f(x_1, \ldots, x_n) = 1$ (resp., 0). This implies the security. □

Corollary 3. *When $n = 3$, the protocol above with $|M_i| = p := 3$ is optimal in terms of the communication complexity for the three-input equality function.*

Proof. This follows from Theorem 2 and Proposition 1.

4.4 Majority Function

Let $f: \{0,1\}^n \to \{0,1\}$ be the n-input majority function that outputs 1 if and only if $\sum_{i=1}^n x_i \geq \lceil n/2 \rceil$. We obtain the following lower bound.

Theorem 3. *Any PSM protocol for the n-input majority function f satisfies $|M_i| \geq 4$ for any $i \in [n]$.*

Proof. By symmetry, we focus on the case $i = n$. Set $m := \lceil n/2 \rceil$. By considering the inputs where except for x_n, $m - 1$ bits are 1 and the others are 0, we see that f has no redundant inputs. Then, the color degrees of $Z := 0^{n-m}1^{m-1}\bot$, $Z' := 0^{n-1}\bot$, and $Z'' := 1^{n-1}\bot$ are $(1,1)$, $(2,0)$, and $(0,2)$, respectively. Therefore, by applying Corollary 1 to this Z, since $Z, Z', Z'' \in \mathcal{F}(Z)$, we have $N_0 \geq \max\{1,2,0\} = 2$ and $N_1 \geq \max\{1,0,2\} = 2$, hence $|M_n| \geq N_0 + N_1 \geq 4$. $\square$

We design a PSM protocol for the three-input majority function as follows.

Shared randomness:
- $r_i = (b, c_i)$ ($i \in [3]$), where $b \in \{0,1\}$ and $c_1, c_2 \in \mathbb{Z}_4$ are chosen uniformly at random and $c_3 = -c_1 - c_2 \in \mathbb{Z}_4$.

The protocol:
1. If $b = 0$, P_i, holding $x_i \in \{0,1\}$, computes $m_i = x_i + c_i$ (mod 4). If $b = 1$, P_1 computes $m_1 = 1 - x_1 + c_1$ (mod 4) and P_i ($i \in \{2,3\}$) computes $m_i = -x_i + c_i$. Each party P_i sends m_i to the referee.
2. The referee computes $m = m_1 + m_2 + m_3$ (mod 4) and outputs 0 if $m \in \{0,1\}$ and 1 if $m \in \{2,3\}$.

Proposition 2. *The above protocol is a correct and secure PSM protocol for the three-input majority function with $|M_i| = 4$ for $i \in [3]$.*

Proof. A direct calculation shows that when $\bar{x} := \sum_{i=1}^3 x_i$ is 0, 1, 2, or 3, we have $m = 0, 1, 2,$ or 3 if $b = 0$, and $m = 1, 0, 3,$ or 2 if $b = 1$, respectively. Then by the uniformly random choices of b and (c_1, c_2, c_3) with $\sum_{i=1}^3 c_i = 0$ (mod 4), if $\bar{x} \leq 1$ (resp., $\bar{x} \geq 2$), the tuple (m_1, m_2, m_3) is uniformly random over all those satisfying $m \in \{0,1\}$ (resp., $\{2,3\}$). This implies the claim. $\square$

Corollary 4. *The protocol above with $|M_1| = |M_2| = |M_3| = 4$ is optimal in terms of the communication complexity for the three-input majority function.*

Proof. This follows from Theorem 3 and Proposition 2.

4.5 Comparison Function

Let $k \geq 2$ be an integer. Let $f: \{0, 1, \ldots, k\}^2 \to \{0,1,2\}$ be the $(k+1)$-valued comparison function $f(x_1, x_2)$ that outputs 0 if $x_1 < x_2$, 1 if $x_1 = x_2$, and 2 if $x_1 > x_2$. When $k = 2$, FKN [8] constructed a PSM protocol satisfying $|M_1| = |M_2| = 7$. From the weak form of the embedding lemma (Lemma 3), we obtain the following lower bound for any $k \geq 2$.

Theorem 4. *Any PSM protocol for the $(k+1)$-valued comparison function f satisfies $|M_i| \geq 2k+1$ for any $i \in [2]$.*

Proof. By comparing inputs of the form (x, x) with (x', x) where $x \neq x'$, we see that f has no redundant inputs. The color degrees of $Z := 0\bot$, $Z' := 1\bot$, and $Z'' := k\bot$ are $n(Z) = (k, 1, 0)$, $n(Z') = (1, 1, k-1)$, and $n(Z'') = (0, 1, k)$, respectively. Therefore, by applying Corollary 1 to Z, since $Z, Z', Z'' \in \mathcal{F}(Z)$, we have $N_0 \geq \max\{k, 1, 0\} = k$, $N_1 \geq \max\{1, 1, 1\} = 1$, and $N_2 \geq \max\{0, k-1, k\} = k$, hence $|M_2| \geq N_0 + N_1 + N_2 = 2k+1$. The case of M_1 is similar. $\square$

When $k = 2$, by using the strong form of the embedding lemma (Lemma 4), we can derive a stronger lower bound in the following.

Theorem 5. *When $k = 2$, any PSM protocol for the three-valued comparison function f satisfies either $|M_1| \geq 6$ or $|M_2| \geq 6$.*

Proof. Assume for contradiction that $|M_1| = |M_2| = 5$. As in the proof of Theorem 4, f has no redundant inputs. For Δ_f, since $n(0\bot) = (2, 1, 0)$, $n(1\bot) = (1, 1, 1)$, and $n(2\bot) = (0, 1, 2)$, it satisfies the assumptions of Lemma 4 with $(N_0^*, N_1^*, N_2^*) = (2, 1, 2)$. Thus, from $|M_2| = 5$, it must hold that $n(Z) = (2, 1, 2)$ for any face Z of type (1). Similarly, we have $n(Z) = (2, 1, 2)$ for any face Z of type (2). We use the same notation $\mathsf{Facet}(\Delta \mid j)$ as the proof of Theorem 1.

Similarly to the proof of Theorem 1, for each $i \in [2]$, we write the image of $a \in X_i = \{0, 1, 2\}$ by a fixed ("standard") effective embedding as $\widehat{a} \in M_i$. By Lemmas 1 and 2, there exists an effective $r = (r_1, r_2) \in R$ such that r maps $01 \in \mathsf{Facet}(\Delta_f \mid 0)$ to $\widehat{02} \in \mathsf{Facet}(\Delta_{\mathsf{Dec}} \mid 0)$. This maps $11 \in \mathsf{Facet}(\Delta_f \mid 1)$ to $r_1(1)\widehat{2} \in \mathsf{Facet}(\Delta_{\mathsf{Dec}} \mid 1)$. From $n_1(\bot 2) = 1$ as in the previous paragraph and $\widehat{22} \in \mathsf{Facet}(\Delta_{\mathsf{Dec}} \mid 1)$, we have $r_1(1) = \widehat{2}$. Also, r maps $02 \in \mathsf{Facet}(\Delta_f \mid 0)$ to $\widehat{0}r_2(2) \in \mathsf{Facet}(\Delta_{\mathsf{Dec}} \mid 0)$, which is different from $r(01) = \widehat{02}$. From $n_0(\widehat{0}\bot) = 2$, $\widehat{0}r_2(2)$ must be $\widehat{01}$, i.e., $r_2(2) = \widehat{1}$. Then r maps $12 \in \mathsf{Facet}(\Delta_f \mid 0)$ to $r_1(1)r_2(2) = \widehat{21} \in \mathsf{Facet}(\Delta_{\mathsf{Dec}} \mid 2)$, a contradiction. This implies the claim. $\square$

Corollary 5. *For the PSM protocol for the three-valued comparison function with $|M_1| \leq |M_2|$ attaining the minimum value of $\sum_{i=1}^{2} \log_2 |M_i|$, we have*

$$(|M_1|, |M_2|) \in \{(5, 6), (5, 7), (5, 8), (5, 9), (6, 6), (6, 7), (6, 8), (7, 7)\}.$$

Proof. Since the protocol in [8] mentioned above satisfies $|M_1| = |M_2| = 7$, the optimal case satisfies $|M_1| \cdot |M_2| \leq 7^2 = 49$. Now the claim follows from Theorem 4 (with $k = 2$) and Theorem 5. $\square$

4.6   Multiplication over Finite Rings

For any (not necessarily commutative) finite ring S, let $f: S^2 \to S$ be the multiplication function $f(x_1, x_2) = x_1 x_2$. In the following, we design a PSM protocol for f by using the idea of Beaver multiplication triples.

Shared randomness:
 - $r_1 = (a, b, c_1)$, $r_2 = (a, b, c_2)$, where a, b, c_1 are uniformly random elements of S and $c_2 = -c_1$.

The protocol:
1. P_1, holding $x_1 \in S$, computes $m_1 = (m_{1,1}, m_{1,2}) = (x_1 - a, x_1 b - ab + c_1)$. P_2, holding $x_2 \in S$, computes $m_2 = (m_{2,1}, m_{2,2}) = (x_2 - b, ax_2 + c_2)$. Each party P_i sends m_i to the referee.
2. The referee outputs $m_{1,1} m_{2,1} + m_{1,2} + m_{2,2} \in S$.

Proposition 3. *The above protocol is a correct and secure PSM protocol for the multiplication function f over a finite ring S with $|M_i| = |S|^2$ for $i \in [2]$.*

Proof. The correctness follows from the following computation:

$$
\begin{aligned}
m_{1,1} m_{2,1} + m_{1,2} + m_{2,2} &= (x_1 - a)(x_2 - b) + (x_1 b - ab + c_1) + (ax_2 + c_2) \\
&= x_1 x_2 - x_1 b - ax_2 + ab + x_1 b - ab + c_1 + ax_2 + c_2 \\
&= x_1 x_2 + c_1 + c_2 = x_1 x_2.
\end{aligned}
$$

To prove the security, we compute the probability that given $m'_{1,1}, m'_{1,2}, m'_{2,1}, m'_{2,2} \in S$ with $m'_{1,1} m'_{2,1} + m'_{1,2} + m'_{2,2} = x_1 x_2$, both $m_1 = (m'_{1,1}, m'_{1,2})$ and $m_2 = (m'_{2,1}, m'_{2,2})$ hold. First, from $m_{1,1} = m'_{1,1}$ and $m_{2,1} = m'_{2,1}$, we must have $a = x_1 - m'_{1,1}$ and $b = x_2 - m'_{2,1}$, which hold with probability $|S|^{-2}$. Then, under the condition that these a, b are chosen, we have

$$
\begin{aligned}
m_{1,2} &= x_1 b - ab + c_1 \\
&= x_1(x_2 - m'_{2,1}) - (x_1 - m'_{1,1})(x_2 - m'_{2,1}) + c_1 \\
&= x_1 x_2 - x_1 m'_{2,1} - x_1 x_2 + x_1 m'_{2,1} + m'_{1,1} x_2 - m'_{1,1} m'_{2,1} + c_1 \\
&= m'_{1,1} x_2 - m'_{1,1} m'_{2,1} + c_1, \\
m_{2,2} &= ax_2 + c_2 = (x_1 - m'_{1,1})x_2 + c_2 = x_1 x_2 - m'_{1,1} x_2 + c_2.
\end{aligned}
$$

Therefore, from $m_{1,2} = m'_{1,2}$ and $m_{2,2} = m'_{2,2}$, we must have

$$
c_1 = m'_{1,2} - m'_{1,1} x_2 + m'_{1,1} m'_{2,1}, \quad c_2 = m'_{2,2} - x_1 x_2 + m'_{1,1} x_2.
$$

Since they satisfy

$$
c_1 + c_2 = m'_{1,1} m'_{2,1} + m'_{1,2} + m'_{2,2} - x_1 x_2 = 0,
$$

the probability that these c_1, c_2 are chosen is $|S|^{-1}$. In summary, the probability that $m_1 = (m'_{1,1}, m'_{1,2})$ and $m_2 = (m'_{2,1}, m'_{2,2})$ is $|S|^{-3}$, which does not depend on the inputs (x_1, x_2). This proves the security. $\square$

Let $q \geq 2$ be a prime power. When $S = \mathbb{F}_q$, the field of order q, we obtain the following lower bound.

Theorem 6. *Any PSM protocol for the multiplication function f over the finite field $\mathbb{F}_q$ satisfies $|M_1|, |M_2| \geq 2q - 1$.*

Proof. The color degrees of faces $Z := 1\perp$ and $Z' := 0\perp$ of Δ_f are $n(Z) = (1, 1, \ldots, 1)$ and $n(Z') = (q, 0, \ldots, 0)$, respectively. From the strong version of the embedding lemma, since $N_0^* \geq \max\{q, 1\} = q$ and $N_j^* \geq \max\{1, 0\} = 1$ $(j \neq 0)$, we have $|M_2| \geq q + (q - 1) \cdot 1 = 2q - 1$. By symmetry, $|M_1| \geq 2q - 1$. $\square$

Let $q \geq 2$ be any integer. When $S = \mathbb{Z}_q$, the integer residue ring modulo q, we obtain the following lower bound.

Theorem 7. *Any PSM protocol for the multiplication function f over $\mathbb{Z}_q$ satisfies $|M_1|, |M_2| \geq \sum_{i=1}^{q} \gcd(i, q)$.*

Proof. By symmetry, we focus on M_2. For $j \in \mathbb{Z}_q$ and a face $x\perp$ of Δ_f with $x \in \mathbb{Z}_q$, we have $n_j(x\perp) = |A_{j,x}|$ where $A_{j,x} := \{c \in \mathbb{Z}_q \mid c \cdot x = j \pmod{q}\}$. In particular, $n_j(1\perp) = 1$ for any j. Hence by the strong version of the embedding lemma, we have $|M_2| \geq \sum_{j \in \mathbb{Z}_q} N_j^* \geq \sum_{j \in \mathbb{Z}_q} \max_{x \in \mathbb{Z}_q} |A_{j,x}|$. Therefore, it suffices to show that $\gcd(j, q) = \max_{x \in \mathbb{Z}_q} |A_{j,x}|$ for any $j \in \mathbb{Z}_q$.

We write $d := \gcd(j, q)$ and $\delta_x := \gcd(x, d)$. Then any $c \in A_{j,x}$ satisfies that $c \cdot x = 0 \pmod{d}$ and hence $c \cdot (x/\delta_x) = 0 \pmod{d/\delta_x}$, therefore $c = 0 \pmod{d/\delta_x}$ since $\gcd(d/\delta_x, x/\delta_x) = 1$. Hence $c \mapsto c/(d/\delta_x)$ gives an injection $A_{j,x} \to A_{j,x \cdot d/\delta_x}$, and d divides $x \cdot d/\delta_x = d \cdot x/\delta_x$. Therefore, to show the claim, it suffices to consider $x \in \mathbb{Z}_q$ that is a multiple of d. Write $q = dq_0$, $j = dj_0$, and $x = dx_0$. Now $c \in \mathbb{Z}_q$ belongs to $A_{j,x}$ if and only if $c \cdot x_0 = j_0 \pmod{q_0}$. Since $\gcd(j_0, q_0) = 1$ by the definition of d, the condition for c is equivalent to $\gcd(x_0, q_0) = 1$ and $c = j_0 \cdot (x_0)^{-1} \pmod{q_0}$ where $(x_0)^{-1}$ is the inverse of x_0 modulo q_0. Hence we have $\max_{x \in \mathbb{Z}_q} |A_{j,x}| = q/q_0 = d = \gcd(j, q)$, as desired. $\square$

Acknowledgments. This work was supported by JSPS KAKENHI Grant Numbers JP19H01109, JP21K17702, JP22K11906, and JP23H00479, and JST CREST Grant Number JPMJCR22M1, Japan. This work was supported by Institute of Mathematics for Industry, Joint Usage/Research Center in Kyushu University. (FY2022 Short-term Visiting Researcher "On Minimal Construction of Private Simultaneous Messages Protocols" (2022a006) and FY2023 Short-term Visiting Researcher "On the Relationship between Physical and Non-physical Secure Computation Protocols" (2023a009)).

References

1. Applebaum, B., Holenstein, T., Mishra, M., Shayevitz, O.: The communication complexity of private simultaneous messages, revisited. J. Cryptol. **33**(3), 917–953 (2020)
2. Assouline, L., Liu, T.: Multi-party PSM, revisited. In: Nissim, K., Waters, B. (eds.) TCC 2021. LNCS, vol. 13043, pp. 194–223. Springer, Cham (2021). https://doi.org/10.1007/978-3-030-90453-1_7
3. Ball, M., Holmgren, J., Ishai, Y., Liu, T., Malkin, T.; On the complexity of decomposable randomized encodings, or: how friendly can a garbling-friendly PRF be? In: ITCS 2020. Schloss Dagstuhl-Leibniz-Zentrum für Informatik (2020)

4. Ball, M., Randolph, T.: A note on the complexity of private simultaneous messages with many parties. In: ITC 2022. Schloss Dagstuhl-Leibniz-Zentrum für Informatik (2022)
5. Beimel, A., Ishai, Y., Kumaresan, R., Kushilevitz, E.: On the cryptographic complexity of the worst functions. In: Lindell, Y. (ed.) TCC 2014. LNCS, vol. 8349, pp. 317–342. Springer, Heidelberg (2014). https://doi.org/10.1007/978-3-642-54242-8_14
6. Beimel, A., Kushilevitz, E., Nissim, P.: The complexity of multiparty PSM protocols and related models. In: Nielsen, J.B., Rijmen, V. (eds.) EUROCRYPT 2018. LNCS, vol. 10821, pp. 287–318. Springer, Cham (2018). https://doi.org/10.1007/978-3-319-78375-8_10
7. Data, D., Prabhakaran, M.M., Prabhakaran, V.M.: On the communication complexity of secure computation. In: Garay, J.A., Gennaro, R. (eds.) CRYPTO 2014. LNCS, vol. 8617, pp. 199–216. Springer, Heidelberg (2014). https://doi.org/10.1007/978-3-662-44381-1_12
8. Feige, U., Killian, J., Naor, M.: A minimal model for secure computation. In: Proceedings of the 26th ACM STOC, pp. 554–563 (1994)
9. Ishai, Y., Kushilevitz, E.: Private simultaneous messages protocols with applications. In: Proceedings of the 5th Israeli Symposium on Theory of Computing and Systems (ISTCS 1997), pp. 174–183. IEEE (1997)
10. Shinagawa, K., Eriguchi, R., Satake, S., Nuida, K.: Private simultaneous messages based on quadratic residues. Designs Codes Cryptogr. (to appear)
11. Vaikuntanathan, V.: Some open problems in information-theoretic cryptography. In: 37th IARCS Annual Conference on Foundations of Software Technology and Theoretical Computer Science (FSTTCS 2017). Schloss Dagstuhl-Leibniz-Zentrum fuer Informatik (2018)

Distributed Protocols for Oblivious Transfer and Polynomial Evaluation

Aviad Ben Arie[ID] and Tamir Tassa[(✉)][ID]

The Open University of Israel, Ra'anana, Israel
`tamirta@openu.ac.il`

Abstract. A secure multiparty computation (MPC) allows several parties to compute a function over their inputs while keeping their inputs private. In its basic setting, the protocol involves only parties that hold inputs. In *distributed* MPC, there are also external servers who perform a distributed protocol that executes the needed computation, without learning information on the inputs and outputs. Here we propose distributed protocols for several fundamental MPC functionalities. We begin with a Distributed Scalar Product (DSP) protocol for computing scalar products of private vectors. We build upon DSP in designing various protocols for Oblivious Transfer (OT): k-out-of-N OT, Priced OT, and Generalized OT. We also use DSP for Oblivious Polynomial Evaluation (OPE) and Oblivious Multivariate Polynomial Evaluation (OMPE). All those problems involve a sender and a receiver that hold private vectors and they wish to compute their scalar product. However, in each of these problems the receiver must submit a vector of a specified form. Hence, a crucial ingredient in our protocols is a sub-protocol for validating that the receiver's vector complies with the relevant restrictions, without learning anything else on that vector. Therefore, while previous studies presented distributed protocols for 1-out-of-N OT and OPE, our protocols are the first ones that are secure against malicious receivers. Our distributed protocols for the other OT variants and for OMPE are the first ones that handle such problems. Our protocols offer information-theoretic security, under the assumption that the servers are semi-honest and have an honest majority, and they are very efficient.

Keywords: Multiparty Computation · Distributed Protocols · Oblivious Transfer · Oblivious Polynomial Evaluation

1 Introduction

Secure multiparty computation (MPC) [35] is a central field of study in cryptography that aims at designing methods for several parties to jointly compute some function over their inputs while keeping those inputs private. In the basic setting of MPC, there are n mutually distrustful parties, $P_1, \ldots, P_n$, that hold private inputs, $x_1, \ldots, x_n$, and they wish to compute some joint function on their inputs, $f(x_1, \ldots, x_n)$. (The function can be sometimes multi-valued and issue different outputs to different designated parties.) No party should gain any information

© The Author(s), under exclusive license to Springer Nature Switzerland AG 2024
A. Chattopadhyay et al. (Eds.): INDOCRYPT 2023, LNCS 14460, pp. 62–83, 2024.
https://doi.org/10.1007/978-3-031-56235-8_4

on other parties' inputs, beyond what can be inferred from their own input and the output.

Typically, the only parties that participate in the protocol are those that hold the inputs or those who need to receive the outputs. However, some studies considered a model of computation that is called *the mediated model* [2,3,11,12,16,29,32], *the client-server model*, [6,10,18,27], or *the distributed model* [4,8,9,22,24,25]. Protocols in that model involve also external *servers* (or *mediators*), $M_1, \ldots, M_D$, $D \geq 1$, to whom the parties outsource some of the needed computations. The servers perform the computations while remaining oblivious to the private inputs and outputs. It turns out that such a distributed model of computation offers significant advantages: it may facilitate achieving the needed privacy goals; it does not require the parties to communicate with each other (a critical advantage in cases where the parties cannot efficiently communicate among themselves, or do not even known each other); in some settings it reduces communication costs; and it allows the parties, that may run on computationally-bounded devices, to outsource costly computations to dedicated servers [29].

In this work we focus on basic MPC problems that involve two ($n = 2$) parties, Alice (the sender) and Bob (the receiver), and propose distributed MPC protocols for their solution. In each of the studied problems, Alice's and Bob's private inputs may be encoded as vectors in a vector space over a finite field $\mathbb{Z}_p$; specifically, $\mathbf{a} = (a_1, \ldots, a_N) \in \mathbb{Z}_p^N$ is Alice's private vector and $\mathbf{b} = (b_1, \ldots, b_N) \in \mathbb{Z}_p^N$ is Bob's, for some integer N. Alice and Bob delegate to a set of $D > 2$ servers, $M_1, \ldots, M_D$, secret shares in their private vectors. Subsequently, the servers perform a multiparty computation on the received secret shares in order to validate the legality of the inputs, if the problem at hand dictates rules by which the input vectors must abide. If the inputs were validated, the servers proceed to compute secret shares in the required output and then they send those shares to Alice and/or Bob who use those shares in order to reconstruct the required output. The computational burden on Alice and Bob is thus reduced to secret sharing computations in the initial and final stages.

Our Contribution. We begin by discussing the generic problem of scalar product, in which the required output is the scalar product, $\mathbf{a} \cdot \mathbf{b}$, of the two private input vectors [13,14,34]. We propose a simple protocol in which Alice and Bob only perform secret sharing computations while the servers perform only local computations, without needing to communicate among themselves. Our distributed scalar product protocol is then used in the subsequent problems that we tackle.

Next, we consider the problem of oblivious transfer (OT) [15,26], which is a fundamental building block in MPC [19] and in many application scenarios such as Private Information Retrieval (PIR) [7]. We consider several variants of OT: 1-out-of-N OT [1,20,21,23], k-out-of-N OT [5], Priced OT [1], and Generalized OT [17,30]. While several previous studies proposed distributed protocols for 1-out-of-N OT, $N \geq 2$, ours is the first one that does not rely on Bob's honesty. Specifically, while previous distributed 1-out-of-N OT protocols enabled Bob to learn any single linear combination of Alice's N secret messages, our protocol restricts Bob to learning just a single message, as mandated in OT (see our discussion in Sect. 7). As for the other OT variants that we consider, we are the first to propose distributed protocols for their solution.

Then we deal with the problem of Oblivious Polynomial Evaluation (OPE) [23,33]. Here, Alice holds a private uni- or multi-variate polynomial $f(\cdot)$ and Bob holds a private value α. The goal is to let Bob have $f(\alpha)$ so that Alice learns nothing on α while Bob learns nothing on f beyond what is implied by α and $f(\alpha)$. Here too, while existing distributed OPE protocols allow Bob to learn any single linear combination of f's coefficients (and thus amount to protocols of distributed scalar product) ours is the first one that restricts Bob to learning only point values of f, at a point of his choice. We are also the first to propose a distributed protocol for OMPE — Oblivibious *Multivariate* Polynomial Evaluation.

Our OT and OPE protocols demonstrate the advantages that the distributed model offers. The delegation of computation to dedicated servers significantly simplifies computations that are typically more involved when Alice and Bob are on their own. The bulk of the computation is carried out by the servers, while Alice and Bob are active only in the initial and final stages, that are computationally lean. Another prominent advantage of the distributed model is that it enables carrying out all of the MPC problems that we consider even when Alice and Bob do not know each other and thus cannot communicate among themselves. In fact, Alice can complete her part in the protocol well before Bob starts his. For example, if Alice is a data custodian that holds some database, her private vector could hold decryption keys for the items in her database. The other party, Bob, can be any client that wishes to retrieve one of the items in that database, while keeping Alice oblivious of his choice, which is encoded in his private vector. Alice and Bob can use our various OT protocols for that purpose. But as they need to communicate only with the servers, Bob may perform his retrieval long time after Alice had already uploaded all information relating to her database. Moreover, in such an application scenario there is a single Alice but many "Bobs". While other protocols (non-distributed or even distributed) require Alice to be responsive to each Bob, our protocols allow Alice to act just once, at the initialization stage, while from that point onward only the servers deal with each of the future requests of potential clients (Bob). Our distributed OMPE protocol also offers such advantages.

We assume that the servers are semi-honest and have an honest majority. Namely, the servers follow the prescribed protocol, but a minority of the servers may collude among themselves or with Alice or Bob and share their views in the protocol. Under these assumptions our protocols are information-theoretic secure and provide unconditional security to both Alice and Bob, even when some of the parties collude.

Outline of the Paper. Section 2 provides the relevant cryptographic preliminaries and assumptions. In Sect. 3 we describe our distributed scalar product protocol. Section 4 is devoted to the various distributed OT protocols. In Sect. 5 we present the OMPE protocol. We report experimental results in Sect. 6. In

Sect. 7 we review the prior art on distributed OT and OPE protocols and compare those protocols to ours. We conclude in Sect. 8.

2   Preliminaries

Secret Sharing. The main idea in our protocols for solving the various MPC problems discussed herein is to use secret sharing. Alice and Bob distribute among the D servers, $M_1, \ldots, M_D$, shares in each entry of their private vectors, using t-out-of-D Shamir's secret sharing scheme [28], with

$$t = \lfloor (D+1)/2 \rfloor. \tag{1}$$

(Hereinafter we shall refer to such sharing as (t, D)-sharing.) Namely, Alice generates for each entry a_n, $n \in [N] := \{1, \ldots, N\}$, a polynomial $f_n^A(x) = a_n + \sum_{i=1}^{t-1} \alpha_i x^i$, where α_i are secret random field elements, and then she sends to M_d the value $[a_n]_d := f_n^A(d)$, $d \in [D] := \{1, \ldots, D\}$. Bob acts similarly. The servers then execute some distributed computation on the received shares in order to arrive at secret shares in the needed output. At the end, they distribute to Alice and/or Bob shares in the desired output from which Alice and/or Bob may reconstruct that output. The underlying field $\mathbb{Z}_p$ is selected so that p is larger than all values in the underlying computation.

Computing Arithmetic Expressions in Shared Secrets. In our protocols we will need to securely compute arithmetic expressions of shared secrets, where the expressions are degree two polynomials in the secrets (namely, they are sums of addends, each involving at most one multiplication of two secrets). We proceed to describe how we execute such computations.

First, we recall that secret sharing is affine in the following sense: if s_1 and s_2 are two secrets that are independently (t, D)-shared among $M_1, \ldots, M_D$, and a, b, c are three public field elements, then the servers can compute shares in $as_1 + bs_2 + c$. Specifically, if $[s_i]_d$ is M_d's share in s_i, $i = 1, 2$, $d \in [D]$, then $\{a[s_1]_d + b[s_2]_d + c : d \in [D]\}$ is a proper (t, D)-sharing of $as_1 + bs_2 + c$.

We turn to discuss the multiplication of shared secrets. Assume that the servers hold (t, D)-shares in s_i, $i = 1, 2$, where M_d's share in s_i is $[s_i]_d$. Assume that each server M_d, $d \in [D]$, multiplies the two shares that he holds and gets $c_d = [s_1]_d[s_2]_d$. It is easy to see that the set $\{c_d : d \in [D]\}$ is a $(2t-1, D)$-sharing of $s_1 s_2$. Therefore, the servers can recover $s_1 s_2$ by computing $c_d = [s_1]_d[s_2]_d$, then interpolate a polynomial F of degree $2t - 2$ based on $\{c_1, \ldots, c_D\}$, and consequently infer that $s_1 s_2 = F(0)$. For simplicity, we will assume hereinafter that D is odd, in which case $2t - 1 = D$. Hence, $\{c_d = [s_1]_d[s_2]_d : d \in [D]\}$ constitute a (D, D)-sharing in $s_1 s_2$.

Scrambling Shares. In some cases we shall perform the above described multiplication procedure when s_1 and s_2 are related (specifically, when $s_2 = s_1 - 1$). In such cases, the above described practice is problematic since each server M_d would need to expose to his peers the product of his secret shares $[s_1]_d[s_2]_d$, and due to the known relation between s_1 and s_2, that product of shares may reveal information on $[s_1]_d$ and $[s_2]_d$, and consequently also information on the value of s_1 and s_2.

To avoid such potential information leakage, the servers perform a *scrambling* of the shares $\{c_1, \ldots, c_D\}$, in the sense that they generate a new random set of shares $\{c'_1, \ldots, c'_D\}$ that are also (D, D)-shares in $s_1 s_2$. They do that in the following manner. Each server M_d, $d \in [D]$, generates a random (D, D)-sharing of 0 and distributes the resulting shares to all servers. Subsequently, each server adds up the zero shares that he had received from all D servers. As a result, the mediators will hold (D, D)-shares of 0, denoted $\{[0]_1, \ldots, [0]_D\}$, where each share distributes uniformly in $\mathbb{Z}_p$. Finally, each server M_d sets $c'_d = c_d + [0]_d$, $d \in [D]$. Clearly, $\{c'_1, \ldots, c'_D\}$ are also (D, D)-shares in $s_1 s_2$, and their values do not leak any information on the original shares in s_1 and s_2.

We note that it is essential to generate a new set of zero shares, $[0]_d$, $d \in [D]$, for each operation of scrambling. However, it is possible to prepare such shares offline, before running the protocol in which scrambling is needed.

Security Assumptions. The servers are assumed to be semi-honest, i.e., they follow the prescribed protocol, but try to extract from their view in the protocol information on the private inputs. We also assume them to have an *honest majority*, in the sense that if some of them are corrupted by a malicious adversary, their number is smaller than $t = \lfloor (D+1)/2 \rfloor$ (Eq. (1)). Hence, our protocols are immune against a coalition of up to $t-1$ servers who collude among themselves or with Alice or Bob.

3   Distributed Scalar Product

Here we deal with the following MPC problem.

Definition 1. *(DSP) Assume that Alice has a private vector $\mathbf{a} = (a_1, \ldots, a_N) \in \mathbb{Z}_p^N$, and Bob has a private vector $\mathbf{b} = (b_1, \ldots, b_N) \in \mathbb{Z}_p^N$. They wish to compute their scalar product $\mathbf{a} \cdot \mathbf{b}$ without revealing any other information on their private vectors.*

Protocol 1 solves that problem. In the first loop (Lines 1–3), Alice and Bob distribute to the servers (t, D)-shares in each entry of their vectors. Then, each server M_d computes a (D, D)-share in $a_n \cdot b_n$ for each $n \in [N]$, and subsequently he computes a (D, D)-share in the scalar product into s_d (Line 5). He then sends that share to Alice and Bob (Line 6). So now Alice and Bob have a full set of (D, D)-shares in $\mathbf{a} \cdot \mathbf{b}$ so they can recover the needed scalar product by means of interpolation (Line 7).

The protocol is correct and secure as we state next.

Theorem 1. *Protocol 1 is correct and provides information-theoretic security to both Alice and Bob when all servers are semi-honest and have an honest majority. Moreover, a coalition of one of the parties (Alice or Bob) with any subset of $t-1$ servers does not yield any information beyond what is implied by that party's input and the output.*

Protocol 1: Distributed Scalar Product

Parameters: p - field size, N - the dimension of the vectors, D - number of servers, $t = \lfloor (D+1)/2 \rfloor$.

Inputs: Alice has a private vector $\mathbf{a} = (a_1, \ldots, a_N) \in \mathbb{Z}_p^N$, Bob has a private vector $\mathbf{b} = (b_1, \ldots, b_N) \in \mathbb{Z}_p^N$.

1 **forall** $n \in [N]$ **do**
2   Alice sends to M_d, $d \in [D]$, a (t, D)-share in a_n, denoted $[a_n]_d$.
3   Bob sends to M_d, $d \in [D]$, a (t, D)-share in b_n, denoted $[b_n]_d$.
4 **forall** $d \in [D]$ **do**
5   M_d computes $s_d \leftarrow \sum_{n \in [N]} ([a_n]_d \cdot [b_n]_d)$.
6   M_d sends s_d to Alice and Bob.
7 Alice and Bob use $\{s_1, \ldots, s_D\}$ to reconstruct $\mathbf{a} \cdot \mathbf{b}$.

Output: Alice and Bob get $\mathbf{a} \cdot \mathbf{b}$.

Due to the page limitation we omit all proofs and will provide them in the full version of this paper.

4 Distributed Oblivious Transfer

In this section we consider several variants of the Oblivious Transfer (OT) protocol. We begin with the basic variants of 1-out-of-N and k-out-of-N OT in Sect. 4.1. We then discuss Priced OT (Sect. 4.2). The case of Generalized OT is introduced in Sect. 4.3; the detailed discussion is deferred to the full version of this paper.

4.1 k-out-of-N Oblivious Transfer

The problem that we consider here is the following:

Definition 2. (OT_k^N) Assume that Alice has a set of N messages, $m_1, \ldots, m_N \in \mathbb{Z}_p$. Bob wishes to learn k of those messages, say $m_{j_1}, \ldots, m_{j_k}$, for some $j_1, \ldots, j_k \in [N]$. A k-out-of-N Oblivious Transfer (OT_k^N) protocol allows Bob to learn $m_{j_1}, \ldots, m_{j_k}$, and nothing beyond those messages, while preventing Alice from learning anything about Bob's selection.

We begin by considering the case $k = 1$ and then we address the general case. The OT_1^N problem can be reduced to DSP (Sect. 3) if Alice sets $\mathbf{a} := (m_1, \ldots, m_N)$ and Bob sets $\mathbf{b} := \mathbf{e}_j$ (the unit vector that consists of $N-1$ zeros and a single 1 in the jth entry, where j is the index of the message that Bob wishes to retrieve). However, the DSP protocol cannot be executed naïvely, since Bob may cheat and send to the servers shares in a vector that is not a unit vector and, consequently, he may obtain some linear combination of the messages, and not just a single message as dictated by the OT definition. Such an abuse of the protocol may sometimes enable a malicious Bob to learn more than just one message. For example, if Bob happens to know that m_1 belongs to some one-dimensional subspace of $\mathbb{Z}_p^N$ while m_2 belongs to another one-dimensional subspace of $\mathbb{Z}_p^N$, then by choosing to learn the linear combination $m_1 + m_2$ he will be able to infer both m_1 and m_2. To that end, the DSP protocol can be executed only after the servers apply some preliminary validation protocol:

Definition 3. *(DVV) Assume that the servers $M_1, \ldots, M_D$ hold (t, D)-shares in a vector $\mathbf{v} \in \mathbb{Z}_p^N$. Let W be a subset of $\mathbb{Z}_p^N$. A Distributed Vector Validation (DVV) protocol is a protocol that the servers may execute on their shares that outputs 1 if $\mathbf{v} \in W$ and 0 otherwise, and reveals no further information on $\mathbf{v}$ in the case where $\mathbf{v} \in W$.*

In our case $W = \{\mathbf{e}_j : j \in [N]\}$. The servers can validate that $\mathbf{b} \in W$ by verifying the following two conditions: (1) $b_n \cdot (b_n - 1) = 0$ for all $n \in [N]$; and (2) $\sum_{n \in [N]} b_n = 1$. Indeed, the first condition implies that all entries in $\mathbf{b}$ are either 0 or 1, while the second condition ascertains that exactly one of the entries equals 1. Note that if the two conditions are verified, then the servers may infer that Bob's vector is legal, but nothing more than that, as desired. Namely, if Bob is honest then his privacy is fully protected. However, if Bob is dishonest and distributed shares in a vector $\mathbf{b} \notin W$, then the above described DVV protocol will reveal some additional information on $\mathbf{b}$; however, that is acceptable since by acting dishonestly Bob looses his right for privacy.

Protocol 2 implements those ideas. After Alice and Bob set their vectors and distribute shares in them to the servers (Lines 1–5), the servers validate Bob's vector for compliance with conditions 1 (Lines 6–12) and 2 (Lines 13–17). (The scrambling operation in Line 8 is as discussed in Sect. 2.) If Bob's vector was validated, they compute (D, D)-shares in the scalar product and send them to Bob so that he can recover the scalar product that equals his message of choice (Lines 18–21).

For a general $k > 1$, it is possible to solve OT_k^N by running Protocol 2 k times, with one exception: Alice needs to distribute shares in her vector only once (Lines 1 and 4 in Protocol 2). We proceed to describe another solution that is more efficient in terms of communication complexity.

Protocol 3 multiplies Alice's vector $\mathbf{a} := (m_1, \ldots, m_N)$ with the vector $\mathbf{b} = \sum_{i=1}^k \mathbf{e}_{j_i}$ where $1 \leq j_1 < \ldots < j_k \leq m$ are the indices of the k messages that Bob wishes to retrieve. But instead of computing their scalar product, $\sum_{n=1}^N a_n b_n$, the protocol computes shares in the products $a_n b_n$ for all $n \in [N]$ and sends them to Bob. Bob then uses the shares of $a_n b_n$ only for $n \in \{j_1, \ldots, j_k\}$ in order to recover the requested messages.

Here, the DVV sub-protocol consists of verifying two conditions: that $b_n \cdot (b_n - 1) = 0$ for all $n \in [N]$, and that $\sum_{n \in [N]} b_n = k$. The first condition implies that all entries in $\mathbf{b}$ are either 0 or 1, while the second condition ascertains that exactly k of the entries equal 1.

After Alice and Bob set their vectors and distribute shares in them to the servers (Lines 1–5), the servers validate Bob's vector for compliance with conditions 1 (Lines 6–12) and 2 (Lines 13–17). If Bob's vector was validated, they compute (D, D)-shares in each of the N products between the components of the two vectors and send them to Bob (Lines 18–21) for him to recover the requested k messages (Lines 22–23).

Protocol 2: 1-out-of-N Oblivious Transfer

Parameters: p - field size, N - number of messages, D - number of servers, $t = \lfloor (D+1)/2 \rfloor$.

Inputs: Alice has $U = \{m_1, \ldots, m_N\}$; Bob has a selection index $j \in [N]$.

1 Alice sets $\mathbf{a} = (m_1, \ldots, m_N)$.

2 Bob sets $\mathbf{b} = \mathbf{e}_j$.

3 **forall** $n \in [N]$ **do**

4   Alice sends to M_d, $d \in [D]$, a (t, D)-share in a_n, denoted $[a_n]_d$.

5   Bob sends to M_d, $d \in [D]$, a (t, D)-share in b_n, denoted $[b_n]_d$.

6 **forall** $1 \leq n \leq N$ **do**

7   Each M_d, $d \in [D]$, sets $c_d = [b_n]_d \cdot ([b_n]_d - 1)$.

8   The servers perform scrambling of $(c_1, \ldots, c_D)$ and compute a new set of (D, D)-shares in $b_n \cdot (b_n - 1)$, denoted $(c'_1, \ldots, c'_D)$.

9   Each M_d, $d \in [D]$, broadcasts c'_d.

10   The servers use $(c'_1, \ldots, c'_D)$ in order to compute $\omega := b_n \cdot (b_n - 1)$.

11   **if** $\omega \neq 0$ **then**

12 | Abort

13 **forall** $d \in [D]$ **do**

14   M_d computes $c_d \leftarrow \sum_{n \in [N]} [b_n]_d$.

15 The servers use any t shares out of $\{c_1, \ldots, c_D\}$ to compute $\omega := \sum_{n \in [N]} b_n$.

16 **if** $\omega > 1$ **then**

17 | Abort

18 **forall** $d \in [D]$ **do**

19   M_d computes $s_d \leftarrow \sum_{n \in [N]} ([a_n]_d \cdot [b_n]_d)$.

20   M_d sends s_d to Bob.

21 Bob uses $\{s_1, \ldots, s_D\}$ to reconstruct $\mathbf{a} \cdot \mathbf{b} = m_j$.

Output: Bob gets m_j.

Theorem 2. *Protocols 2 and 3 are correct and provide information-theoretic security to both Alice and an honest Bob when all servers are semi-honest and have an honest majority. Moreover, a coalition of one of the parties (Alice or Bob) with any subset of $t-1$ servers does not yield any information beyond what is implied by that party's input and the output.*

In the full version of this paper we describe an alternative 1-out-of-N Oblivious Transfer protocol that is also based on DSP. In that protocol, the DVV process is replaced by another mechanism that is based on an idea that was presented by Naor and Pinkas in [24] for their 1-out-of-2 OT protocol. The advantage in that protocol is that it does not require the servers to communicate with each other. However, on the down side, it enforces Alice to be responsive to any OT request of any client (Bob), as opposed to Protocol 2 in which Alice finishes her part in the initial phase.

Protocol 3: k-out-of-N Oblivious Transfer

Parameters: p - field size, N - number of messages, D - number of servers,
$t = \lfloor (D+1)/2 \rfloor$.
Inputs: Alice has $U = \{m_1, \ldots, m_N\}$; Bob has selection indices
$1 \leq j_1 < \ldots < j_k \leq N$.

1 Alice sets $\mathbf{a} = (m_1, \ldots, m_N)$.
2 Bob sets $\mathbf{b} = (b_1, \ldots, b_N)$, where $b_n = 1$ for $n \in \{j_1, \ldots, j_k\}$ and $b_n = 0$
   otherwise.
3 **forall** $n \in [N]$ **do**
4   Alice sends to M_d, $d \in [D]$, a (t, D)-share in a_n, denoted $[a_n]_d$.
5   Bob sends to M_d, $d \in [D]$, a (t, D)-share in b_n, denoted $[b_n]_d$.
6 **forall** $1 \leq n \leq N$ **do**
7   Each M_d, $d \in [D]$, sets $c_d = [b_n]_d \cdot ([b_n]_d - 1)$.
8   The servers perform scrambling of $(c_1, \ldots, c_D)$ and compute a new set of
       (D, D)-shares in $b_n \cdot (b_n - 1)$, denoted $(c'_1, \ldots, c'_D)$.
9   Each M_d, $d \in [D]$, broadcasts c'_d.
10   The servers use $(c'_1, \ldots, c'_D)$ in order to compute $\omega := b_n \cdot (b_n - 1)$.
11   **if** $\omega \neq 0$ **then**
12   **Abort**
13 **forall** $d \in [D]$ **do**
14   M_d computes $c_d \leftarrow \sum_{n \in [N]} [b_n]_d$.
15 The servers use any t shares out of $\{c_1, \ldots, c_D\}$ to compute $\omega := \sum_{n \in [N]} b_n$.
16 **if** $\omega \neq k$ **then**
17   **Abort**
18 **forall** $d \in [D]$ **do**
19   **forall** $n \in [N]$ **do**
20   M_d computes $[c_n]_d \leftarrow [a_n]_d \cdot [b_n]_d$.
21   M_d sends $[c_n]_d$ to Bob.
22 **forall** $n \in \{j_1, \ldots, j_k\}$ **do**
23   Bob uses $\{[c_n]_1, \ldots, [c_n]_D\}$ to reconstruct $c_n = a_n \cdot b_n = m_n$.
   Output: Bob gets $m_{j_1}, \ldots, m_{j_k}$.

4.2   Priced Oblivious Transfer

Consider a setting of OT in which each of Alice's messages has a weight and
the retrieval policy allows Bob to learn any subset of messages in which the sum
of weights does not exceed some given threshold. For example, if Alice holds a
database of movies and each movie has a price tag, then if Bob had prepaid
some amount, Alice wishes to guarantee that he retrieves movies of aggregated
cost that does not exceed what he had paid, while Bob wishes to prevent Alice
from knowing what movies he chose to watch.

Definition 4. *Let $U = \{m_1, \ldots, m_N\}$ be the set of messages that Alice has.
Assume that each massage m_n has a weight $w_n \geq 0$, $n \in [N]$, and let $T > 0$
be some given threshold. Then a Priced OT protocol allows Bob to retrieve any
subset $B \subseteq U$ for which $\sum_{m_n \in B} w_n \leq T$. Bob cannot learn any information on
the messages in $U \backslash B$, while Alice has to remain oblivious of Bob's choice.*

We assume that the weights $w_1, \ldots, w_N$ are publicly known, since they represent information that is supposed to be known to all. The threshold T, on the other hand, represents the amount that Bob had paid and, therefore, it is private and should remain so.

Protocol 4 executes Priced OT. It coincides with Protocol 3 except for the second part of the DVV sub-protocol (Lines 13–17). If in Protocol 3 the servers obliviously verified in that part that $\sum_{n \in [N]} b_n \leq k$, then here it is necessary to obliviously verify that $\sum_{m_n \in B} w_n = \sum_{n \in [N]} w_n b_n \leq T$. (Recall that in Lines 6–12 in Protocol 3 we have already verified that $b_n \in \{0, 1\}$, for all $n \in [N]$.) To enable that verification, the protocol starts by publishing the vector of weights (Line 1). Then, both Alice and Bob distribute to the servers (t, D)-shares in T (Lines 2–3) and then the servers verify that the two underlying thresholds equal, without recovering that threshold (Lines 4–7). Those steps are necessary in order to ascertain that Alice and Bob agree on the same value of the threshold, before using that value in the DVV sub-protocol. (Namely, Bob is ascertained that Alice did not provide a too low value of T while Alice is ascertained that Bob did not provide a too high value of T).

The core of the protocol is the execution of the OT_k^N protocol - Protocol 3 (Line 8). That protocol is executed as is except for the replacement of Lines 13–17 there with Sub-protocol 5. The sub-protocol begins with the servers computing (t, D)-shares in the difference $e := T - \sum_{n \in [N]} w_n b_n$ (Lines 1–2). Then, any subset of t servers can recover e (Line 3). Finally, if $e \neq 0$ the protocol aborts (Line 4), while otherwise it proceeds towards completing the transfer.

Note that Bob is allowed to retrieve any subset of messages of aggregated weight at most T. Sub-protocol 5, however, assumes that Bob had requested a subset of messages of aggregated weight that equals exactly T. Such an equality can be guaranteed as we proceed to describe. First, Bob can add to his list of requested messages additional redundant messages that he will ignore later on. By adopting such a practice, the difference $e = T - \sum_{n \in [N]} w_n b_n$ can be made a nonnegative number smaller than $\overline{w} := \max_{n \in [N]} w_n$. Assume that $\overline{w} < 2^\ell$, for some $\ell > 0$. Then Alice may add ℓ phantom messages $\hat{m}_i$, $0 \leq i < \ell$, with the weights 2^i, to her list of messages. Consequently, Bob will add to his list of requested messages also the subset of phantom messages of which the sum of weights equals exactly e. That way, the servers will always recover in Line 3 in Sub-protocol 5 the value 0.

The Case of Secret Weights. Even though the weights of messages are typically public, it is possible to modify the protocol so that also the weights remain hidden from the servers. To do that, instead of publishing the vector of weights $\mathbf{w}$ (as done in Line 1 of Protocol 4), Alice would distribute to the servers (t, D)-shares in them. Let $[w_n]_d$ denote M_d's share in w_n, $d \in [D]$, $n \in [N]$. Then, in Sub-protocol 5, Line 2 will be replaced with $[e]_d \leftarrow [T]_d - \sum_{n \in [N]} [w_n]_d [b_n]_d$. As discussed in Sect. 2, the set $\{[e]_1, \ldots, [e]_D\}$ is a set of (D, D)-shares in e. The servers may use those shares in order to reconstruct $e = T - \sum_{n \in [N]} w_n b_n$. No further changes are required.

Protocol 4: Priced Oblivious Transfer

Parameters: p - field size, N - number of messages, D - number of servers,
$t = \lfloor (D+1)/2 \rfloor$.

Inputs: Alice has $U = \{m_1, \ldots, m_N\}$, and corresponding weights $w_n \geq 0$,
$n \in [N]$; Bob has a set of selection indices $j_1, \ldots, j_k \in [N]$; Alice and
Bob have $T \geq 0$.

1 Alice publishes the vector of weights $\mathbf{w} = (w_1, \ldots, w_N)$.
2 Alice sends to M_d, $d \in [D]$, a (t, D)-share in T, denoted $[T]_d$.
3 Bob sends to M_d, $d \in [D]$, a (t, D)-share in T, denoted $[T']_d$.
4 **forall** $d \in [D]$ **do**
5 $\quad \mid \quad M_d$ computes $[e]_d \leftarrow [T]_d - [T']_d$.
6 The servers use any t shares out of $\{[e]_1, \ldots, [e]_D\}$ to compute $e = T - T'$.
7 **if** $e \neq 0$ **then Abort**.
8 Alice, Bob and the servers execute Protocol 3 in which Lines 13-17 are replaced
with Sub-protocol 5.

Output: Bob gets $\{m_{j_1}, \ldots, m_{j_k}\}$ iff $\sum_{i=1}^{k} w_{j_i} \leq T$.

Sub-protocol 5: Priced OT: verifying that $\sum_{i=1}^{k} w_{j_i} \leq T$.

1 **forall** $d \in [D]$ **do**
2 $\quad \mid \quad M_d$ computes $[e]_d \leftarrow [T]_d - \sum_{n \in [N]} w_n [b_n]_d$.
3 The servers use any t shares out of $\{[e]_1, \ldots, [e]_D\}$ to compute
$e = T - \sum_{n \in [N]} w_n b_n$.
4 **if** $e \neq 0$ **then Abort**.

As Protocol 4 coincides with Protocol 3 where only the DVV part is slightly modified, Theorem 2 applies also to that protocol, in both cases (public or secret weights).

4.3   Generalized Oblivious Transfer

Ishai and Kushilevitz [17] presented an extension of OT called *Generalized Oblivious Transfer* (GOT). While in OT_k^N Bob was restricted to learn any subset of at most k out of the N messages that Alice has, in GOT the policy is extended as described below.

Definition 5. *Let $U = \{m_1, \ldots, m_N\}$ be the set of messages that Alice has. An access structure is a collection of subsets of U, $\mathcal{A} \subseteq 2^U$, which is monotone decreasing in the sense that if $B \in \mathcal{A}$ and $B' \subset B$ then also $B' \in \mathcal{A}$.*

Bob is allowed to retrieve any subset of messages $B \subset U$ provided that $B \in \mathcal{A}$. As before, Bob cannot learn any information on messages in $U \backslash B$, while Alice must remain oblivious of Bob's selection. The retrieval policy can be determined by any access structure on the set of messages, e.g. a multipartite access structure [31], in which the set of messages is partitioned into distinct compartments and the question of whether Bob may retrieve $B \subset U$ is determined only by the number of messages that B has in each of the compartments.

In the full version of this paper we present a protocol for that purpose that is based on the GOT protocol that was presented in [30], and it invokes the OT_k^N protocol, Protocol 3.

5   Oblivious Polynomial Evaluation

The oblivious polynomial evaluation problem was presented in [23], and was extended to the case of multivariate polynomials in [33]. We devise herein a distributed protocol for the multivariate problem.

We begin by defining multivariate polynomials (Definitions 6 and 7) and then define the corresponding MPC problem (Definition 8).

Definition 6. *(Monomial) Let $\mathbb{Z}_p$ be a finite field , $\mathbf{x} = (x_1, \ldots, x_k)$ be a k-dimensional vector over $\mathbb{Z}_p$ and $\mathbf{j} = (j_1, \ldots, j_k)$ be a k-dimensional vector of nonnegative integers. Then the monomial $\mathbf{x}^{\mathbf{j}}$ is defined as $\mathbf{x}^{\mathbf{j}} := \prod_{i=1}^{k} x_i^{j_i}$.*

Definition 7. *(Multivariate Polynomial) let $\mathbb{Z}_+^k := \{\mathbf{j} = (j_1, \ldots, j_k) : j_i \in \mathbb{Z}_+ = \{0, 1, 2, \ldots\} : 1 \leq i \leq k\}$ be the set of all k-tuples of nonnegative integers, and $\mathbb{Z}_+^{k,N}$ be the subset of $\mathbb{Z}_+^k$ consisting of all tuples of which the sum of components is at most N, i.e.: $\mathbb{Z}_+^{k,N} := \{\mathbf{j} \in \mathbb{Z}_+^k : |\mathbf{j}| := \sum_{i=1}^{k} j_i \leq N\}$. An N-degree k-variate polynomial $f(\mathbf{x})$ over the field $\mathbb{Z}_p$, where $\mathbf{x} = (x_1, \ldots, x_k) \in \mathbb{Z}_p^k$, is defined as:*

$$f(\mathbf{x}) = \sum_{\mathbf{j} \in \mathbb{Z}_+^{k,N}} a_{\mathbf{j}} \cdot \mathbf{x}^{\mathbf{j}}, \quad a_{\mathbf{j}} \in \mathbb{Z}_p. \tag{2}$$

Definition 8. *(OMPE) Assume that Alice has an N-degree multivariate polynomial $f(\mathbf{x}) = f(x_1, \ldots, x_k)$, while Bob has a point $\boldsymbol{\alpha} = (\alpha_1, \ldots, \alpha_k) \in \mathbb{Z}_p^k$. They wish to enable Bob to learn $f(\boldsymbol{\alpha})$, and nothing else on f, while keeping Alice oblivious to $\boldsymbol{\alpha}$.*

OMPE can be solved by reducing it to DSP, with the needed prior validations. The vector that Alice will submit to the protocol consists of the coefficients of her polynomial, $\mathbf{a} = (a_{\mathbf{j}} : \mathbf{j} \in \mathbb{Z}_+^{k,N})$. The vector that Bob will submit to the protocol is the following:

$$\mathbf{b} = (b_{\mathbf{j}} : \mathbf{j} \in \mathbb{Z}_+^{k,N}), \quad \text{where} \quad b_{\mathbf{j}} := \boldsymbol{\alpha}^{\mathbf{j}}. \tag{3}$$

It is easy to see that the dimension of these vectors is $\binom{N+k}{k}$.

First, it is necessary to agree upfront on an ordering of $\mathbb{Z}_+^{k,N}$ so that in the scalar product between the two vectors, each power of $\boldsymbol{\alpha}$ will be multiplied by the corresponding polynomial coefficient. We suggest ordering the set $\mathbb{Z}_+^{k,N}$ by arranging its monomials into $N+1$ tiers, as follows. The 0th tier would be $T_0 := \mathbb{Z}_+^{k,0}$, and then the nth tier, $n = 1, \ldots, N$, would be $T_n := \mathbb{Z}_+^{k,n} \backslash \mathbb{Z}_+^{k,n-1}$; namely,

---

Protocol 6: Oblivious Multivariate Polynomial Evaluation

Parameters: p - field size, k-number of variables, N - the degree of the secret
polynomial f, D - number of servers, $t = \lfloor (D+1)/2 \rfloor$.

Inputs: Alice has a secret N-degree k-variate polynomial $f(\mathbf{x})$, Eq. (2); Bob has
a secret point $\boldsymbol{\alpha} = (\alpha_1, \ldots, \alpha_k) \in \mathbb{Z}_p^k$.

1 Alice sets $\mathbf{a} = (a_{\mathbf{j}} : \mathbf{j} \in \mathbb{Z}_+^{k,N})$, according to the ordering convention.

2 Bob sets $\mathbf{b} = (b_{\mathbf{j}} = \boldsymbol{\alpha}^{\mathbf{j}} : \mathbf{j} \in \mathbb{Z}_+^{k,N})$, according to the ordering convention.

3 **forall** $\mathbf{j} \in \mathbb{Z}_+^{k,N}$ **do**

4 $\quad$ Alice sends to M_d, $d \in [D]$, a (t, D)-share in $a_{\mathbf{j}}$, denoted $[a_{\mathbf{j}}]_d$.

5 $\quad$ Bob sends to M_d, $d \in [D]$, a (t, D)-share in $b_{\mathbf{j}}$, denoted $[b_{\mathbf{j}}]_d$.

6 **forall** $2 \leq n \leq N$ **do**

7 $\quad$ **forall** $\mathbf{j} \in T_n$ **do**

8 $\quad\quad$ Select a monomial $\mathbf{h} \in T_{n-1}$ such that $\mathbf{j} = \mathbf{h} + \mathbf{e}_i$ for some $1 \leq i \leq k$,
$\quad\quad$ where $\mathbf{e}_i$ is the i-th unit vector.

9 $\quad\quad$ The servers compute $\omega := b_{\mathbf{h}} \cdot b_{\mathbf{e}_i} - b_{\mathbf{j}}$.

10 $\quad\quad$ **if** $\omega \neq 0$ **then**

11 $\quad\quad\quad$ **Abort**

12 **forall** $d \in [D]$ **do**

13 $\quad$ M_d computes $s_d \leftarrow \sum_{\mathbf{j} \in \mathbb{Z}_+^{k,N}} ([a_{\mathbf{j}}]_d \cdot [b_{\mathbf{j}}]_d)$.

14 $\quad$ M_d sends s_d to Bob.

15 Bob uses $\{s_1, \ldots, s_D\}$ to reconstruct $\mathbf{a} \cdot \mathbf{b} = f(\boldsymbol{\alpha})$.

Output: Bob gets $f(\boldsymbol{\alpha})$.

---

the nth tier T_n consists of all monomials of degree exactly $n \in \{0, 1, \ldots, N\}$.
The order within each tier would be lexicographical.

Protocol 6 starts with Alice and Bob setting their input vectors $\mathbf{a}$ and $\mathbf{b}$ in
accord with the ordering convention (Lines 1–2). Then they distribute to the
servers (t, D)-shares in them (Lines 3–5). Observe that the first entry in $\mathbf{b}$, i.e.
$b_{\mathbf{j}}$ for $\mathbf{j} = (0, \ldots, 0)$, equals 1 (see Eq. (3)). Hence, in Line 5 for $\mathbf{j} = (0, \ldots, 0)$
Bob does not generate and distribute shares; instead, each server M_d, $d \in [D]$,
sets $[b_{\mathbf{j}}]_d = 1$.

After completing the distribution of shares, the servers perform the relevant
DVV sub-protocol in order to validate that the secret input vector $\mathbf{b}$ is of the
form as in Eq. (3) (Lines 6–11). To that end we state the following lemma.

Lemma 1. *The vector* $\mathbf{b} = (b_{\mathbf{j}} : \mathbf{j} \in \mathbb{Z}_+^{k,N})$, *where* $b_{\mathbf{j}} = 1$ *for* $\mathbf{j} = (0, \ldots, 0)$, *is
of the form as in Eq. (3) if and only if* $\omega = 0$ *in all stages of the validation loop
in Lines 6–11 of Protocol 6.*

In the final stage of Protocol 6, the servers compute (D, D)-shares in the
scalar product and send them to Bob (Lines 12–14) who uses them in order to
recover the scalar product (Line 15).

Example. We illustrate the validation process when $k = 2$ and $N = 2$. Bob is expected to submit here vectors of the form

$$\mathbf{b} = (b_{(0,0)}, b_{(1,0)}, b_{(0,1)}, b_{(2,0)}, b_{(1,1)}, b_{(0,2)}) = (1, \alpha_1, \alpha_2, \alpha_1^2, \alpha_1\alpha_2, \alpha_2^2).$$

Since the first entry is always 1, and the next two entries can be anything, validation is applied only on the last three entries — $b_{(2,0)}$, $b_{(1,1)}$, and $b_{(0,2)}$:

- To validate $b_{(2,0)}$, we observe that there is only one way to represent the multi-index $\mathbf{j} = (2,0)$ as a sum $\mathbf{h} + \mathbf{e}_i$, namely, $(2,0) = (1,0) + (1,0)$. Hence, the DVV sub-protocol checks whether $b_{(2,0)} = b_{(1,0)} \cdot b_{(1,0)}$. Therefore, validation of this entry succeeds if and only if $b_{(2,0)} = \alpha_1^2$.
- Similarly, $b_{(0,2)}$ is validated if and only if $b_{(0,2)} = \alpha_2^2$.
- To validate $b_{(1,1)}$, we observe that $\mathbf{j} = (1,1) = \mathbf{h} + \mathbf{e}_i$ with $\mathbf{h} = (1,0)$ and $\mathbf{e}_i = (0,1)$ or with $\mathbf{h} = (0,1)$ and $\mathbf{e}_i = (1,0)$. In either case, the DVV sub-protocol checks whether $b_{(1,1)} = b_{(1,0)} \cdot b_{(0,1)} = \alpha_1 \cdot \alpha_2$.

We conclude by noting that the security guarantees of Protocol 6 are as stated in Theorem 2.

6 Experiments

Implementation Setails. We implemented our protocols in Java on a Lenovo Ideapad Gaming 3 laptop, powered by an AMD Ryzen 7 5800H processor and 16 GB of RAM. The operating system was Windows 11 64-bit, and the environment was Eclipse-Workspace. A 64-bit prime number p was chosen at random for the size of the underlying field $\mathbb{Z}_p$. To enable computations modulo such prime, we used the BigInteger Java class. The code is available at https://github.com/b1086960/Distributed_OT_OPE.

All experiments were conducted on randomly generated vectors (or sets of messages or polynomials). Each experiment was repeated ten times and the average runtimes for Alice, Bob and the servers are reported (where the runtimes for the servers are averaged over the ten runs as well as over the D servers). The standard deviation is omitted from the graphical display of our results since it is barely noticeable.

Results. In the first experiment we tested our basic protocol that solves DSP, Protocol 1. Figure 1 shows the runtimes for Alice and Bob and the average runtimes for the servers as a function of N (the dimension of the two vectors). The runtimes in Fig. 1 grow linearly in N. Figure 2 displays those runtimes as a function of D. The runtimes for Alice and Bob grow quadratically in D since they need to perform D polynomial evaluations where the polynomial is of degree $t - 1 = O(D)$. The servers' runtime, on the other hand, is not affected by D and only slightly fluctuates randomly between 125 and 150 milliseconds for all tested values of D.

In the next experiment we tested Protocol 3 that solves the OT_k^N problem. Here we focus only on the servers, since Bob's computations in that protocol

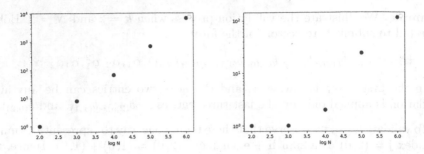

Fig. 1. Runtimes (milliseconds) for Protocol 1 (DSP), as a function of $\log_{10}(N)$, for $D = 7$. The left plot shows the runtimes for Alice and Bob; the right plot shows the average runtimes for the servers. The runtimes are presented on a logarithmic scale.

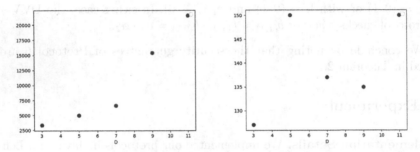

Fig. 2. Runtimes (milliseconds) for Protocol 1 (DSP), as a function of D, for $N = 10^6$. The left plot shows the runtimes for Alice and Bob; the right plot shows the average runtimes for the servers. The runtimes are presented on a linear scale.

are the same as in Protocol 1, while Alice's computations are the same as in the beginning of Protocol 1. The servers' runtimes are shown in Fig. 3. The dependence on N is linear. As for D, while in Protocol 1 the servers' runtimes do not depend on D, here they do depend on D, linearly, due to the DVV part of the protocol. Their runtimes are not affected by k.

We turn our attention to Protocol 4 (Priced OT). Like in Protocol 3, we ignore the runtimes of Alice and Bob and focus on the servers' average runtime and demonstrate its linear dependence on N and on D, see Fig. 4.

Finally, we consider Protocol 6 (OMPE). We ran that protocol with random polynomials of degrees $N \in \{5, 10, 20, 30, 40, 50\}$, where the number of variables was set to $k = 3$ — see Fig. 5. The shown runtimes grow linearly with $\binom{N+k}{k}$, since that is the size of the two vectors in the scalar product.

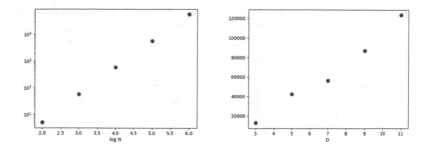

Fig. 3. Average runtimes (milliseconds) for the servers in Protocol 3 (OT_k^N). Left: runtimes, on a logarithmic scale, as a function of $\log_{10}(N)$, for $D = 7$ and $k = 10$. Right: runtimes as a function of D, for $N = 1000000$ and $k = 10$.

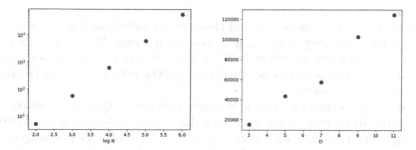

Fig. 4. Average runtimes (milliseconds) for the servers in Protocol 4 (Priced OT)). Left: runtimes as a function of N, for $T = 100$ and $D = 7$; the runtimes are presented on a logarithmic scale. Right: runtimes as a function of D, for $T = 100$ and $N = 1000000$.

7   Related Work

Naor and Pinkas [24] introduced the first version of a distributed OT. Their setting is similar to the one that we consider here: (a) apart from the sender (Alice) and the receiver (Bob) there are external servers that participate in the computation; (b) Alice sends information only to the servers and her role ends after doing so; (c) Bob can perform his part in a later time by communicating solely with the servers.

They considered OT_1^2: namely, Alice has m_1 and m_2, Bob has a selection index $j \in \{1, 2\}$, and the goal is to let Bob have m_j and nothing else, while Alice should remain oblivious of j. Their protocols are referred to as ℓ-out-of-D DOT_1^2, meaning that Bob has to communicate with ℓ out of the D servers in order to receive his message of choice.[1].

The two protocols that are proposed in [24] are based on secret sharing of some univariate polynomial. Specifically, Alice chooses a random bivariate polynomial $Q(x, y)$ that encodes m_1 and m_2, Bob chooses some random univariate

---

[1] In our discussion of related work we replace the original parameter notations with the ones that we used in the present work, for consistency and clarity.

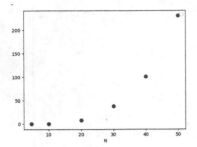 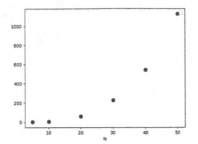

Fig. 5. Runtimes (milliseconds) for Protocol 6 (OMPE), as a function of N, the polynomial degree, for $k = 3$. Left: runtimes for Bob; right: average runtimes for the servers.

polynomial $S(x)$ that encodes j, and then, by carefully selecting the degrees of those polynomials, they induce a univariate polynomial $R(x) = Q(x, S(x))$ of degree $\ell - 1$. The free coefficient in $R(x)$ is m_j and, consequently, Bob can get that value by obtaining the value of $R(x)$ in ℓ points. Bob does that by receiveing information from ℓ servers.

The first protocol uses a simple bivariate polynomial $Q(x, y)$. It suffers from two shortcomings: each server learns the difference $m_2 - m_1$ and, in addition, if a single server colludes with Bob, they obtain both of Alice's messages. The second protocol uses a more involved bivariate polynomial, that prevents the above described breach in Alice's privacy. However, that protocol still allows Bob to learn any linear combination of the two messages, rather than just m_1 or m_2. Later on they outline a manner which enforces Bob to learn just m_1 or m_2 but not any other linear combination of the two messages. The idea is to perform the protocol twice: in one execution Alice submits her two messages masked by random multipliers, $c_1 m_1$, and $c_2 m_2$; in the second execution Alice submits the two multipliers, c_1 and c_2. They then argue that if $m_1 \neq m_2$, such a course of action disables Bob from inferring any linear combination of m_1 and m_2 which is not one of the two messages.

Blundo et al. [4] generalized the protocols of [24] to distributed OT_1^N. In their generalization, Alice uses an N-variate polynomial. $Q(x, y_1, \ldots, y_{N-1})$ that encodes her N messages, $m_1, \ldots, m_N$. Bob, on the other hand, encodes his index j by $N - 1$ univariate polynomials, $Z_1, \ldots, Z_{N-1}$. Those polynomials implicitly induce a univariate polynomial of degree $\ell - 1$, $R(x) = Q(x, Z_1(x), \ldots, Z_{N-1}(x))$, such that $R(0) = m_j$. As in [24], Bob contacts ℓ servers in order to get ℓ point values of R that enable him to recover $R(0) = m_j$. They showed that any coalition of up to $\ell - 1$ servers cannot obtain any information on j, and that any coalition of up to $\ell - 1$ servers with Bob cannot obtain any information on Alice's messages. However, their protocol has the same vulnerability as that of [24]: each server learns the differences $m_n - m_1$ for all $1 \leq n \leq N$; and a coalition of Bob with a single server enables the recovery of *all* N messages.

Hence, the protocols of [24] and [4] are vulnerable to a collusion of Bob with just a single server. Blundo et al. defined the following privacy goal: a coalition of Bob with any subset of $\ell - 1$ servers should not be able to infer any information on Alice's messages, beyond the message that Bob had selected. They proved that such a goal cannot be achieved in a one-round DOT protocol.

Nikov et al. [25] presented an analysis of the ℓ-out-of-D DOT_1^N framework used in the above described studies. Namely, they considered protocols that involve a sender (Alice), a receiver (Bob) and D servers, through which Bob can retrieve a single message out of Alice's N messages by contacting ℓ of the D servers. They considered such a scheme to be (t, k)-secure if (a) any coalition of $t - 1$ servers cannot infer anything on Bob's selection index, and (b) a coalition of Bob with k corrupt servers does not yeild to Bob any further information. They then showed [25, Corollary 1] that such a scheme can exist iff $\ell \geq t + k$. They continued to demonstrate a construction of such a scheme with a minimal threshold of $\ell = t + k$. Later on, they considered settings in which not all servers enjoy the same level of trust and presented a DOT_1^N protocol in which Bob can recover his message of choice by contacting an authorized subset of servers, where the authorized subsets are defined by a general access structure.

We note that the protocols of [4, 25] enable Bob to learn any single linear combination of Alice's messages, and not just a single message; hence, they implement only a weaker version of OT.

Corniaux and Ghodosi [9] took a different approach in their solution of the distributed OT_1^N problem. As opposed to the above described works, they allow the servers to communicate with each other, thus breaching out of the framework of one-round DOT. Their protocol is similar to our DOT_1^N protocol (Protocol 2): Alice distributes to the servers secret shares in her vector of messages, while Bob distributes secret shares in the binary vector that encodes his selection index. The requested message is the scalar product between those two private vectors. However, the protocol in [9] lacks the DVV part, which is at the heart of our Protocol 2 (Lines 6–15). Consequently, Bob can create any selection vector and hence can recover any linear combination of the messages $m_1, \ldots, m_N$. Hence, the protocol in [9] too does not implement OT but a weaker form of that problem. (We note that there are other technical differences between our Protocol 2 and the one in [9], e.g., the fact that we do not need to perform a transformation from one threshold scheme to another, as they do; we omit further details.)

The problem of OPE (Oblivious Polynomial Evaluation) was introduced by Naor and Pinkas in [23]. It is closely related to OT: here, too, Alice has a set of secrets and Bob is allowed to get a single linear combination of those secret while Alice should remain oblivious of his choice. While in OT the secrets are messages and the allowed linear combinations are the ones that consist of a single message, in OPE the secrets are the coefficients of a private polynomial, $f(x)$, and the allowed linear combinations are those that relate to a point value of that polynomial, $f(\alpha)$. In the OPE protocol of [23] Alice hides her secret polynomial $f(x)$ in some bivariate polynomial while Bob hides his secret point α in some univariate polynomial. Those two polynomials induce a univariate polynomial $R(x)$ such that $R(0) = f(\alpha)$. Bob then learns $d_R + 1$ point values of R, where d_R is the degree of R, and then proceeds to recover $R(0)$. He does that by invoking $d_R + 1$ instances of 1-out-of-m OT, where m is a small security parameter.

We are interested here with distributed protocols for OPE. The first such protocol was introduced by Li et al. [22]. They suggested three protocols for that matter, which are based on secret sharing and polynomial interpolation. In the first and simplest method, Alice secret shares each of her polynomial coefficients among the servers, while Bob distributes secret shares in the corresponding powers of his selected point. The desired value is then obtained by computing the scalar product between the two shared vectors. The two subsequent versions of this basic protocol are designed in order to increase the immunity of the protocol to collusion between the servers and Bob. The protocols assume that all parties are semi-honest. Since Bob is also assumed to be semi-honest, Bob can submit to the protocol secret shares in any vector, not necessarily one of the form $(0, \alpha, \alpha^2, \ldots, \alpha^N)$ (where N is the degree of Alice's polynomial f). Hence, their protocols amount to protocols of distributed scalar product.

Cianciullo and Ghodosi [8] described another DOPE protocol that offers better security and complexity than the protocols of Li et al. [22]. Specifically, their protocol offers security for both Alice and Bob against collusion of up to $t-1$ out of the D servers, for some threshold t that can be tuned by the degrees of the secret sharing polynomials that the protocol uses. Despite the advantages that their protocol offers with respect to that of Li et al. [22], it too does not restrict Bob to learning only point values of $f(x)$, as it allows Bob to learn any linear combination of f's coefficients. In addition, it requires Alice to communicate with Bob and generate a new set of secret shares per each request. Protocol 6 that we presented herein allows Alice to act just once and by thus serve an unlimited number of future queries of "Bobs"; it allows the computation only of point values of f; and it is the first protocol that is designed for multi-variate polynomials.

8   Conclusion

We presented here distributed MPC protocols for three fundamental MPC functionalities: scalar product, oblivious transfer (k-out-of-N, Priced, and Generalized OT), and oblivious (multivariate) polynomial evaluation (OMPE). While previous studies offered distributed MPC protocols for 1-out-of-N OT and for (univariate) OPE, ours are the first ones that consider malicious receivers and restrict them to receive only the outputs that the MPC problem dictates. To the best of our knowledge, our study is also the first one that suggests distributed MPC protocols for k-out-of-N OT, Priced OT, Generalized OT, and OMPE.

Our OT and OMPE protocols demonstrate the advantages that the distributed model offers: the existence of external servers enables much simpler and more efficient MPC protocols; it allows the MPC parties (the sender Alice and the receiver Bob) to delegate the bulk of the computation to the dedicated servers; and it completely disconnects Alice from Bob so that they do not need to communicate with each other, or even to know each other or to be active at the same time. Moreover, in cases where the sender wishes to serve a multitude of receivers, she can perform her part just once, and from that point onward only the servers attend to any request of any future receiver.

When the servers are semi-honest and have an honest majority, our protocols are information-theoretic secure and provide unconditional security to both Alice and Bob, even when some of the parties collude.

An interesting future research direction would be to strengthen our protocols in order to render them secure against malicious servers.

While OT and OPE can serve as building blocks for general MPC problems [19,22], it would be interesting to use the ideas presented here in order to develop distributed protocols for the following fundamental two-party MPC problems:

- Oblivious Function Evaluation (OFE): Alice has a function that is represented by a Boolean circuit and Bob has a suitable input binary vector. The goal is to let Bob learn the output of Alice's circuit over his input and nothing else, while Alice remains oblivious of Bob's input.
- Oblivious Automaton Evaluation (OAE): Alice has a deterministic finite or pushdown automaton $\mathcal{A}$ with an input alphabet Σ; Bob has a word $w \in \Sigma^*$. The goal is to let Bob learn whether w is a word that $\mathcal{A}$ accepts without learning any other information on $\mathcal{A}$, while Alice remains oblivious of w.
- Oblivious Turing Machine Evaluation (OTME): Alice has a Turing Machine M with an input alphabet Σ and Bob has a word $w \in \Sigma^*$. The goal is to let Bob know the output $M(w)$ without learning any other information on M, while Alice remains oblivious of w.

We believe that the distributed model can be most effective in designing solutions to such fundamental problems of multiparty computation as well as in practical problems that arise in privacy-preserving distributed computation.

References

1. Aiello, B., Ishai, Y., Reingold, O.: Priced oblivious transfer: how to sell digital goods. In: Pfitzmann, B. (ed.) EUROCRYPT 2001. LNCS, vol. 2045, pp. 119–135. Springer, Heidelberg (2001). https://doi.org/10.1007/3-540-44987-6_8
2. Alwen, J., Katz, J., Lindell, Y., Persiano, G., shelat, A., Visconti, I.: Collusion-free multiparty computation in the mediated model. In: Halevi, S. (ed.) CRYPTO 2009. LNCS, vol. 5677, pp. 524–540. Springer, Heidelberg (2009). https://doi.org/10.1007/978-3-642-03356-8_31
3. Alwen, J., Shelat, A., Visconti, I.: Collusion-free protocols in the mediated model. In: Wagner, D. (ed.) CRYPTO 2008. LNCS, vol. 5157, pp. 497–514. Springer, Heidelberg (2008). https://doi.org/10.1007/978-3-540-85174-5_28
4. Blundo, C., D'Arco, P., De Santis, A., Stinson, D.R.: On unconditionally secure distributed oblivious transfer. J. Cryptol. 20(3), 323–373 (2007)
5. Brassard, G., Crepeau, C., Robert, J.-M.: All-or-nothing disclosure of secrets. In: Odlyzko, A.M. (ed.) CRYPTO 1986. LNCS, vol. 263, pp. 234–238. Springer, Heidelberg (1987). https://doi.org/10.1007/3-540-47721-7_17

6. Catrina, O., Kerschbaum, F.: Fostering the uptake of secure multiparty computation in e-commerce. In: ARES, pp. 693–700 (2008)
7. Chor, B., Kushilevitz, E., Goldreich, O., Sudan, M.: Private information retrieval. J. ACM **45**(6), 965–981 (1998)
8. Cianciullo, L., Ghodosi, H.: Unconditionally secure oblivious polynomial evaluation: a survey and new results. J. Comput. Sci. Technol. **37**(2), 443–458 (2022)
9. Corniaux, C.L.F., Ghodosi, H.: Scalar product-based distributed oblivious transfer. In: Rhee, K.-H., Nyang, D.H. (eds.) ICISC 2010. LNCS, vol. 6829, pp. 338–354. Springer, Heidelberg (2011). https://doi.org/10.1007/978-3-642-24209-0_23
10. Damgård, I., Ishai, Y.: Constant-round multiparty computation using a black-box pseudorandom generator. In: Shoup, V. (ed.) CRYPTO 2005. LNCS, vol. 3621, pp. 378–394. Springer, Heidelberg (2005). https://doi.org/10.1007/11535218_23
11. Dery, L., Tassa, T., Yanai, A.: Fear not, vote truthfully: secure multiparty computation of score based rules. Expert Syst. Appl. **168**, 114434 (2021)
12. Dery, L., Tassa, T., Yanai, A., Zamarin, A.: Demo: a secure voting system for score based elections. In: CCS, pp. 2399–2401 (2021)
13. Du, W., Atallah, M.J.: Privacy-preserving cooperative statistical analysis. In: ACSAC, pp. 102–110 (2001)
14. Du, W., Zhan, J.Z.: A practical approach to solve secure multi-party computation problems. In: NSPW, pp. 127–135 (2002)
15. Even, S., Goldreich, O., Lempel, A.: A randomized protocol for signing contracts. Commun. ACM **28**(6), 637–647 (1985)
16. Ben Horin, A., Tassa, T.: Privacy preserving collaborative filtering by distributed mediation. In: RecSys, pp. 332–341 (2021)
17. Ishai, Y., Kushilevitz, E.: Private simultaneous messages protocols with applications. In: ISTCS, pp. 174–184 (1997)
18. Kamara, S., Mohassel, P., Raykova, M.: Outsourcing multi-party computation. IACR Cryptology ePrint Archive (2011). 272
19. Kilian, J.: Founding cryptography on oblivious transfer. In: STOC, pp. 20–31 (1988)
20. Kolesnikov, V., Kumaresan, R., Rosulek, M., Trieu, N.: Efficient batched oblivious PRF with applications to private set intersection. In: CCS, pp. 818–829 (2016)
21. Kushilevitz, E., Ostrovsky, R.: Replication is NOT needed: SINGLE database, computationally-private information retrieval. In: FOCS, pp. 364–373 (1997)
22. Li, H.-D., Yang, X., Feng, D., Li, B.: Distributed oblivious function evaluation and its applications. J. Comput. Sci. Technol. **19**(6), 942–947 (2004)
23. Naor, M., Pinkas, B.: Oblivious transfer and polynomial evaluation. In: STOC, pp. 245–254 (1999)
24. Naor, M., Pinkas, B.: Distributed oblivious transfer. In: Okamoto, T. (ed.) ASIACRYPT 2000. LNCS, vol. 1976, pp. 205–219. Springer, Heidelberg (2000). https://doi.org/10.1007/3-540-44448-3_16
25. Nikov, V., Nikova, S., Preneel, B., Vandewalle, J.: On unconditionally secure distributed oblivious transfer. In: Menezes, A., Sarkar, P. (eds.) INDOCRYPT 2002. LNCS, vol. 2551, pp. 395–408. Springer, Heidelberg (2002). https://doi.org/10.1007/3-540-36231-2_31
26. Rabin, M.O.: How to exchange secrets by oblivious transfer. Technical Report TR-81, Aiken Computation Laboratory, Harvard University (1981)
27. Schneider, J.: Lean and fast secure multi-party computation: minimizing communication and local computation using a helper. In: SECRYPT, pp. 223–230 (2016)
28. Shamir, A.: How to share a secret. Commun. ACM **22**(11), 612–613 (1979)

29. Shmueli, E., Tassa, T.: Mediated secure multi-party protocols for collaborative filtering. ACM Trans. Intell. Syst. Technol. **11**, 1–25 (2020)
30. Tassa, T.: Generalized oblivious transfer by secret sharing. Des. Codes Cryptogr. **58**(1), 11–21 (2011)
31. Tassa, T., Dyn, N.: Multipartite secret sharing by bivariate interpolation. In: Bugliesi, M., Preneel, B., Sassone, V., Wegener, I. (eds.) ICALP 2006. LNCS, vol. 4052, pp. 288–299. Springer, Heidelberg (2006). https://doi.org/10.1007/11787006_25
32. Tassa, T., Grinshpoun, T., Yanai, A.: PC-SyncBB: a privacy preserving collusion secure DCOP algorithm. Artif. Intell. **297**, 103501 (2021)
33. Tassa, T., Jarrous, A., Ben-Ya'akov, Y.: Oblivious evaluation of multivariate polynomials. J. Math. Cryptol. **7**(1), 1–29 (2013)
34. Vaidya, J., Clifton, C.: Privacy preserving association rule mining in vertically partitioned data. In: SIGKDD, pp. 639–644 (2002)
35. Yao, A.C.: Protocols for secure computation. In: FOCS, pp. 160–164 (1982)

Obfuscating Evasive Decision Trees

Shalini Banerjee$^{(\boxtimes)}$ ⓘ, Steven D. Galbraith ⓘ, and Giovanni Russello ⓘ

University of Auckland, Auckland, New Zealand
{shalini.banerjee,s.galbraith,g.russello}@auckland.ac.nz

Abstract. We present a new encoder for hiding parameters in an interval membership function. As an application, we design a simple and efficient *virtual black-box obfuscator* for *evasive* decision trees. The security of our construction is proved in the random oracle model. Our goal is to increase the class of programs that have practical and cryptographically secure obfuscators.

Keywords: program obfuscation · cryptographic hash functions · decision trees

1 Introduction

Program obfuscation has received considerable attention by the cryptographic community in recent years. An obfuscator $\mathcal{O}$ is a probabilistic polynomial-time algorithm that transforms a program C to a semantically equivalent counterpart $\tilde{C}$, such that a secret that is efficiently computable from C, is hard to extract given $\tilde{C}$.

The definitional framework of program obfuscation was given by Barak *et al.* in their seminal work [4] using a simulation-based security paradigm. They established the notion of *virtual black-box* (VBB) obfuscation, where a polynomial-time adversary $\mathcal{A}$ that takes input $\tilde{C}$ has a negligible advantage over a polynomial-time simulator $\mathcal{S}$ who only has oracle access to C; in short, anything that is efficiently computable from $\tilde{C}$, can also be computed efficiently from the input-output access to the program. Their main results rule out the possibility of designing efficient obfuscators for a generic class of programs. However, obfuscators for *specific* families of programs may be achievable. Canetti [14] shows the construction of an efficient obfuscator for point functions, that achieves a relaxed notion of virtual black-box security using a probabilistic hashing algorithm $\mathcal{R}$, which imitates the 'useful' properties of a random oracle. The obfuscated program stores $\mathcal{R}(x)$, where x is sampled uniformly from a superlogarithmic min-entropy distribution, such that on input y, outputs 1 if $\mathcal{R}(x) = \mathcal{R}(y)$. Such favoring assertions were followed by designing efficient VBB obfuscators for a special family of functions (*evasive functions* [3]) which achieve the goal that a PPT adversary cannot distinguish between the obfuscation of C drawn randomly from the function family and obfuscation of a function that always outputs zero. Notable works in this direction include obfuscation of point-functions [30],

A. Chattopadhyay et al. (Eds.): INDOCRYPT 2023, LNCS 14460, pp. 84–104, 2024.
https://doi.org/10.1007/978-3-031-56235-8_5

pattern-matching with wildcards [7,8], compute-and-compare programs [21,31], fuzzy-matching for Hamming distance [19], hyperplane membership [15], etc.

In this paper, we focus on a new technique for *encoding interval membership functions*. This is motivated by designing an efficient *virtual black-box obfuscator* for evasive decision trees (see Definition 3).

1.1 Privacy-Preserving Classification Using Decision Trees

In the interest of establishing the usefulness and significance of obfuscating decision trees, we provide a brief overview on privacy-preserving classification using decision tree classifiers.

Decision tree classifiers are extensively used for prediction and analysis in sensitive applications such as spam detection, medical or genomics, stock investment, etc. [9,17,28].

Consider an example of a medical facility (model-provider) who designs a model from sensitive profiles of patients to diagnose certain disease. The model is then outsourced to a cloud server to provide classification to a user who wants to make a prediction about her health. If the model is leaked, the sensitive training data will be disclosed [1,18], breaching the HIPAA[1] compliance. What's more, the user does not want to reveal her queries and classification results to the cloud server. This calls for *privacy-preserving classification techniques*, where the model should be hidden from anyone but the model-provider, and prediction queries/classification must remain private to the user, such that no leakage of useful information happen during the classification phase.

The state-of-art privacy-preserving classification solutions employ an *interactive* approach: encrypt and outsource the model to cloud server, where it processes encrypted queries and forwards encrypted classification to the users. The solutions involve multiple rounds of communication and rely upon expensive cryptographic computations using fully-homomorphic encryption (FHE) [10,11], garbled circuits [5,6], etc. Brickell *et al.* [13] suggest an interactive two-party protocol employing additive homomorphic encryption and ml oblivious transfers (where l is the bit-length of each input feature and m is the number of decision nodes), restricting the user from performing multiple queries on the encrypted tree. In their well-known work, Bost *et al.* [11] present a comparison protocol between model-provider and the user for each node in the decision tree using FHE methods. Tai *et al.* [25] make use of multiple communication rounds to transfer the path costs and encrypted labels to the client. The authors of [16] design an FHE-based solution (SortingHat) to secure prediction queries and classification results with reduced communication costs, but do not guarantee the privacy of the model.

Our motivation for obfuscating decision trees is to eliminate interaction between user and the model-provider/cloud server. In particular, we aim to construct an efficient non-interactive solution to privacy-preserving classification with evasive decision trees. We now explain why we do not consider obfuscating

---

[1] Health Insurance Portability and Accountability Act of 1996.

arbitrary decision trees. If a decision tree can be learned from the input-output behaviour of the model, then protecting the privacy of the model is impossible. Note that, learning a decision tree means identifying the decision nodes and input attributes associated with them, and identifying the accepting nodes. Tramer *et al.* [29] show that a decision tree can be learned through $m \cdot \log_2(b/\epsilon)$ oracle queries, where m is the number of internal nodes and ϵ is the minimum width of an interval in a node; they call it *model extraction attack*. To prevent such attacks, the existing literature observes API calls to issue warnings [23,27] or adds perturbations [24,32]. However, since there are no theoretical restrictions on the number of prediction queries made by a user [26], limiting them is not reasonable approach towards thwarting such attacks. We define a special class of decision trees, for which it is hard to find an accepting input, such that an efficient algorithm cannot extract the model except with negligible probability; we call such decision trees *evasive*, and claim that if a decision tree is not evasive, then it is impossible to protect the privacy of the model, and hence there is no choice but to restrict to evasive decision trees.

Lockable obfuscation (also called compute-and-compare obfuscation) [21,31] is a very general tool which encodes a class of branching programs under learning-with-errors (LWE) assumption. It could be employed to build a decision tree obfuscator, by writing the decision tree as a circuit. Nevertheless, we focus on solutions that are simpler and potentially more practical. In [12], the authors initiate a theoretical investigation on decision tree obfuscation based on indistinguishability obfuscation [20] which is the 'best-possible' obfuscation from the point of view of VBB, but does not guarantee the privacy of the decision tree model. We aim to achieve stronger notions of security that allow us to protect the privacy of the model.

Our work has been used in a recent paper by Banerjee *et al.* [2]. The authors aim to provide security against reverse-engineering of PLC programs concerning critical industrial facilities (nuclear-enrichment etc.) in order to prevent infamous attacks such as Stuxnet [22]. They extend our scheme to design a platform ObfCP, with empirical results that reflect its efficiency in real-time applications.

1.2 Our Contributions

We present a new technique for encoding interval membership functions. As an application, we construct an efficient VBB obfuscator for *evasive decision trees* (see Definition 6). We focus on trees of bounded number of inputs and depth.

Note that, we do not consider privacy-preserving methods to construct the model. How the decision tree is constructed is out of the scope of this study. A technical briefing of our construction follows.

Technical Overview. We consider decision trees that perform binary classification based on the values of n attributes. Attributes are represented as ℓ-bit strings x_i, and are interpreted as integers in $[0, 2^\ell)$. A decision tree is a full binary tree of depth d. Internal node v_j (also called *decision* node) associates threshold t_j, where t_j is an integer between 0 and $2^\ell - 1$. Each decision node tests $x_i \leq t_j$ for some i. The leaf nodes $(s_1, \ldots, s_{2^d})$ are labelled 0 (reject) or 1 (accept). Hence the decision tree is represented by the pairs $[\![t_j, i]\!]$, and the labels on the leaf nodes.

Without loss of generality we may assume that, for any specific path from the root to an accepting leaf, x_i is compared at most twice. Hence each accepting path corresponds to a sequence of interval membership predicates $x_i \in (c_i, c_i + w_i]$. The key observation is that membership $x_i \in (c_i, c_i + w_i]$ can be expressed as a union of distinct predicates $x_i \in [a, a + 2^p)$ for certain pairs (a, p). Each such predicate can be turned into a point function predicate and hence be obfuscated using hashing. We explain the details in the next paragraph.

Let $f_i : \{0,1\}^\ell \to \{0,1\}^{\ell-i}$ such that $f_i(y) = \lfloor \frac{y}{2^i} \rfloor$ for $i \in \{0, 1, \ldots, \ell - 1\}$. Calculate intersection of sub-intervals $\mathcal{I}^i$ corresponding to $(c_i, c_i + w_i)$ (of the form $[a, a + 2^p)$). Encode each entry in $\mathcal{I}^i$ using $H(f_p(a))$, where $H : \{0,1\}^* \to \{0,1\}^\omega$ is a cryptographic hash function; call the set $\mathcal{B}^i$. Note that, this method converts interval membership predicate $x_i \in [a, a + 2^p)$ into a point function predicate that determines whether $H(f_p(a))$ is equal to $H(f_q(x_i))$ for some $q \in \{0, \ldots, \ell - 1\}$. Finally, for each encoding in $\mathcal{B}^i$, concatenate n entries sorted in the order of i, apply cryptographic hash function $H_c : \{0,1\}^* \to \{0,1\}^q$, and publish the set of hashes. Reordering the nodes in the order of i along each accepting path hides the structure, though the size of the obfuscated program may reveal the number of different accepting paths. To classify input $(x_i)_{i \in [n]}$, compute the set of encodings $\mathcal{E}^i$ by calculating $H(f_q(x_i))$, where $q \in \{0, \ldots, \ell - 1\}$ and for each encoding, concatenate n entries sorted in order of i, and apply H_c. For an accepting input, one of the hashes computed by the evaluation procedure will be contained in the set of hashes published by the obfuscator.

2   Preliminaries

We denote by $|S|$, size of a set S. We use the standard notations to denote intervals as (a, b), $(a, b]$, $[a, b)$ and $[a, b]$, for $a, b \in \mathbb{N}$. We denote by $\lfloor x \rfloor$ the integral part of x, where $x \in \mathbb{R}$. We use $\log_2(n)$ to denote the power to which 2 should be raised to obtain the value $n \in \mathbb{N}$. We denote the binary encoding of n as $r_{\ell-1} \cdot 2^{\ell-1} + \cdots + r_0 \cdot 2^0$ for $n \in \mathbb{N}^+$, where $r_i \in \{0, 1\}$. We denote Hamming weight of n as $wt(n) = \sum_{i=0}^{\ell-1} r_i$. For a program C, we denote its size by $|C|$.

We provide the honest parties and the adversaries with a security parameter $\lambda \in \mathbb{N}$. We model the adversaries as a family of probabilistic polynomial time (PPT) programs, running in time $a \cdot \lambda^c$, for some constants a, c. A function $\mu : \mathbb{N} \to \mathbb{R}^+$ is called negligible in n, if it grows slower than n^{-c}, for every constant c. We measure negligibility with respect to the security parameter λ. We use $x \leftarrow\!\!\$\, X$ to denote x is drawn uniformly at random from the space X. Finally, we let $\|_{i=1}^n (a_i)$ denote concatenation $a_1 \| a_2 \| \cdots \| a_n$ of the sequence $(a_i)_{i \in [n]}$.

3 Obfuscation Definitions

In this section, we present the standard definition of obfuscation and *distributional virtual black box* (DVBB) security that our obfuscator satisfies.

Definition 1 (Distributional Virtual Black-Box Obfuscation [3,4]). *Let $\lambda \in \mathbb{N}$ be the security parameter. Let $\mathcal{C} = \{\mathcal{C}_\lambda\}$ be a family of polynomial-size programs parameterized by inputs of length $n(\lambda)$, and let $\mathcal{D} = \{\mathcal{D}_\lambda\}$ be a class of distribution ensembles, where $\mathcal{D}_\lambda$ is a distribution over $\mathcal{C}_\lambda$. A PPT algorithm $\mathcal{O}$ is an obfuscator for the distribution $\mathcal{D}$, if it satisfies the following conditions:*

- *Correctness: For every $\lambda \in \mathbb{N}$ and for every $x \in \{0,1\}^{n(\lambda)}$, there exists a negligible function $\mu(\lambda)$, such that:*

$$\Pr_{\mathcal{O},C \leftarrow \mathcal{D}_\lambda}[\mathcal{O}(C)(x) = C(x)] > 1 - \mu(\lambda)$$

  where the probability is over the sampling of the program and coin tosses of $\mathcal{O}$.
- *Polynomial Slowdown: There exists a polynomial q such that for every $\lambda \in \mathbb{N}$ and for every $C \in \mathcal{D}_\lambda$, the running time of $\mathcal{O}(C)$ is bounded by $q(|C|)$, where $|C|$ denotes the size of the program.*
- *Virtual Black-Box: For every (non-uniform) polynomial size adversary $\mathcal{A}$, there exists a (non-uniform) polynomial size simulator $\mathcal{S}$ with oracle access to C, such that for every λ:*

$$\left| \Pr_{C \leftarrow \mathcal{D}_\lambda, \mathcal{O}, \mathcal{A}}[\mathcal{A}(\mathcal{O}(C)) = 1] - \Pr_{C \leftarrow \mathcal{D}_\lambda, \mathcal{S}}[\mathcal{S}^C(1^\lambda) = 1] \right| \leq \mu(\lambda)$$

  where $\mu(\lambda)$ is a negligible function.

Definition 2 (Evasive Program Collection [3]). *A distribution of programs $\mathcal{D} = \{\mathcal{D}_\lambda\}_{\lambda \in \mathbb{N}}$ parameterized by inputs of length $n(\lambda)$ is called evasive, if there exists a negligible function $\mu(\lambda)$, such that for every $\lambda \in \mathbb{N}$, and for every input $x \in \{0,1\}^{n(\lambda)}$*

$$\Pr_{C \leftarrow \mathcal{D}_\lambda}[C(x) = 1] \leq \mu(\lambda)$$

where the probability is taken over the random sampling of C from $\mathcal{D}_\lambda$.

4 Decision Trees

In this section, we formalize binary decision trees. Without loss of generality, we consider decision trees to be full binary trees and restrict to binary classification. Following this, we introduce *evasive decision trees* and what it means to *learn* a decision tree.

Definition 3 (Decision Trees). *Let $n, d, \ell \in \mathbb{N}$ and $(x_i)_{i=1}^n = (x_1, \ldots x_n) \in \mathbb{N}^n$ be a finite sequence of input elements, where x_i is an integer between 0 and $2^\ell - 1$ that represents the value of some attribute.*

A decision tree is a representation of a function $C : [0, 2^\ell)^n \to \{0, 1\}$. It is a full binary tree of depth d with internal nodes $(v_1, \ldots, v_{2^d-1})$ (where v_1 is the root, v_2, v_3 are nodes at the second level, and so on) and leaf nodes $(s_1, \ldots, s_{2^d})$. Leaf node $s_k \in \{0, 1\}$ gives the value of the function C. Internal nodes v_j are labelled by a pair $[\![t_j, i]\!]$ that defines a predicate g_j as $g_j(x_i) = 1$ if and only if $x_i \leq t_j$. To evaluate the tree on an input $(x_1, \ldots, x_n)$, one follows a path from the root to a leaf, by taking the left child if $g_j = 0$ and the right child if $g_j = 1$. The output is the value of the leaf s_k at the end point of this walk in the tree (Fig. 1).

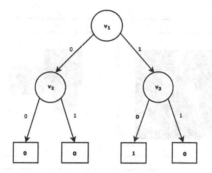

Fig. 1. Binary classification with a decision tree: the circular nodes represent decision nodes, and the square nodes represent leaf nodes. Decision nodes are numbered in level-order sequence. The path in brown represents the accepting path with leaf node labeled 1.

To define model extraction resistance we define the *assets* of a decision tree. Note that a decision tree does not necessarily have a unique representation.

Definition 4 (Asset of Decision Tree). *Let C be the function represented by a decision tree. We define* asset(C) *to be a sequence of pairs $[\![t_j, i]\!]$ and a sequence of leaf nodes $(s_1, \ldots, s_{2^d})$ such that the corresponding decision tree from Definition 3 defines the same function C.*

Without loss of generality we may assume x_i to be compared at most twice along an accepting path. (We stress that different accepting paths may arise from different comparisons of x_i) It follows that each accepting path is checking $x_i \in (c_i, c_i + w_i]$ for some c_i and w_i. As we now explain, for evasiveness we need the w_i to be not too large, so we introduce an upper bound $w_{max} \in \mathbb{N}$.

Definition 5 (Decision Region). *Let $n, \ell, w_{max} \in \mathbb{N}$. Let $c_i, c_i + w_i$ be integers between 0 and $2^\ell - 1$ and $w_i \in (0, w_{max}]$. Define a decision region as the hyper-rectangle formed by n intervals $(c_i, c_i + w_i]$.*

Evasive Function Family. As already explained, it is impossible to hide the assets in a learnable decision tree. Hence we study *evasive decision trees*. Throughout the paper, we assume that an adversary knows the domain of inputs, but not the accepting inputs.

Definition 6 (Evasive Decision Tree Distribution). *Let* $\mathcal{D} = \{D_\lambda\}_{\lambda \in \mathbb{N}}$ *be a distribution of polynomial-size classification functions represented as decision trees of depth* $d(\lambda)$ *on* $n(\lambda)$ *variables. We say* $\mathcal{D}$ *is evasive, if there exists a negligible function* μ *such that for every* $\lambda \in \mathbb{N}$, *for every input* $(x_i)_{i \in [n(\lambda)]}$

$$\Pr_{C \leftarrow \mathcal{D}_\lambda} [\, P((x_i)_{i \in [n(\lambda)]}) = 1\,] \leq \mu(\lambda)$$

In short, Definition 6 requires that for every $\{x_i\}_{i \in [n]}$, a program C chosen randomly from the distribution evaluates to 1 with negligible probability.

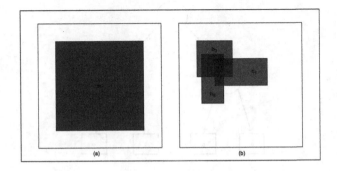

Fig. 2. Two example cases of distributions which lead towards non-evasiveness: (a) Decision region is very big. (b) Overlapping decision regions h_1, h_2 and h_3 always accept a point x in their common intersection.

A distribution $X_n \in [0, 2^\ell)^n$ defines a distribution $\mathcal{D}_\lambda$, such that $C \leftarrow \mathcal{D}_\lambda$ computes whether an input $(x_i)_{i \in [n]}$ is accepted or not as follows: sample $(c_1, \ldots, c_n) \leftarrow X_n$ and $(w_1, \ldots, w_n) \leftarrow (0, w_{max}]^n$. The accepted inputs satisfy $x_i \in (c_i, c_i + w_i]$ for every $i \in [n]$. For the program collection to be evasive, it is necessary that, for fixed $(x_1, \ldots, x_n)$, the probability is negligible that $x_i \in (c_i, c_i + w_i]$ for every i. Thus we require X_n to have large entropy. As uniform distributions provide the highest entropy, this is the best case. However, real-world applications may have accepting regions that are less uniform. The scenarios that lead to non-evasiveness are: (1) the decision regions are too big (so a random x is likely to be in the set); (2) the number of points in the space $[0, 2^l)^n$ representing $(c_1, \ldots, c_n)$ is too few (not enough entropy); (3) the decision regions overlap each other (so one can choose x from the intersection of the regions). Figure 2 shows two example distributions that give non-evasive decision trees. Hence for an evasive collection, the distribution X_n needs to have high entropy and it needs to be "well-spread" (meaning that a set of randomly chosen accepting regions should have empty intersection).

We now calculate some parameters that suffice for evasiveness. We start with uniform distribution on $[0, 2^l)^n$ where n, ℓ are polynomials in λ.

Lemma 1. *Let $n, \ell \in \mathbb{N}$. Let c_i be an integer between 0 and $2^\ell - 1$ and w_i be an integer between 1 and w_{max}. The number of elements in the decision region $(c_1, c_1 + w_1] \times \cdots \times (c_n, c_n + w_n]$ is at most $(w_{max})^n$.*

Proof. Each interval has length at most w_{max}. □

Lemma 2. *Let $\lambda \in \mathbb{N}$ be the security parameter and $n, \ell, w_{max} \in \mathbb{N}$, where $w_{max} \leq 2^{(\ell - \frac{\lambda}{n})}$. Fix an input $(x_i)_{i \in [n]}$. Choose uniformly $c_i \in [0, 2^\ell)$ and $w_i \in (0, w_{max}]$. Then the probability that $(x_i)_{i \in [n]}$ belongs to the decision region defined by $(c_i, c_i + w_i]$, for $i \in [n]$ is not more than $2^{-\lambda}$.*

Proof. The total number of points $(x_i)_{i \in [n]}$ in the space $[0, 2^\ell)^n$ is $2^{\ell n}$. By Lemma 1, the decision region defined by $(c_i, c_i + w_i]$, $i \in [n]$ has size $(w_{max})^n$. Hence uniformly sampled input $(x_i)_{i \in [n]}$ is contained in the decision region with probability $\frac{(w_{max})^n}{2^{\ell n}}$. For $w_{max} \leq 2^{(\ell - \frac{\lambda}{n})}$, the above probability is at most $2^{-\lambda}$. □

The result shows that if the intervals $(c_i, c_i + w_i]$'s are uniformly chosen with $w_i \leq 2^{(\ell - \frac{\lambda}{n})}$, then the probability that an input (fixed a priori) belongs to the decision region is negligible in λ. We now prove that the class of decision tree functions defined by uniform distributions that follow the above mentioned parameter restrictions, forms an evasive program collection.

Lemma 3. *Let $\lambda \in \mathbb{N}$ be the security parameter and ℓ, n be polynomials in λ. Let $w_{max} = w_{max}(\lambda)$ be a function such that $w_{max}(\lambda) \leq 2^{\ell(\lambda) - \frac{\lambda}{n(\lambda)}}$. Let $2^{-\lambda}$ be a negligible function. Let X_n be the uniform distribution in $[0, 2^\ell)^n$, and let $\mathcal{D}_\lambda$ be the corresponding distribution on decision trees that determines if $(x_i)_{i \in [n]}$ belongs to the decision region defined by the c_i's and w_i's. Then $\mathcal{D}_\lambda$ is an evasive program distribution.*

Proof. The uniform distribution on $[0, 2^\ell)^n$ defines $\mathcal{D}_\lambda$. We need to show that for every $\lambda \in \mathbb{N}$ and every $(x_i)_{i \in [n]}$, $\Pr_{C \leftarrow \mathcal{D}_\lambda} [C((x_i)_{i \in [n]}) = 1] \leq \mu(\lambda)$. For $C \leftarrow \mathcal{D}_\lambda$, the probability that $(x_i)_{i \in [n]}$ is accepted by C is equal to the probability that (x_i) lies in the product of uniformly chosen intervals $(c_i, c_i + w_i]$ as above. Lemma 2 shows the probability is at most $2^{-\lambda}$, which is a negligible function in λ. □

We next discuss what it means to *learn* an evasive decision tree. If a decision tree is unlearnable, then there is no model extraction attack.

Definition 7. (Unlearnable Decision Trees). *A collection of classification functions $\mathcal{C}$ is unlearnable, if for every polynomial time algorithm $\mathcal{A}$ with oracle access to C, there exists a negligible function μ, such that for every $\lambda \in \mathbb{N}$:*

$$\Pr_{C \leftarrow \mathcal{D}_\lambda} [\mathcal{A}^{C(1^\lambda)} = \mathsf{asset}(C)] \leq \mu(\lambda)$$

Remark 1. Note that evasiveness implies unlearnability, because evaluating an evasive function always returns 0 with overwhelming probability and so no information about the function is provided by these queries.

5 Obfuscating Evasive Decision Trees

In this section, we introduce a *new technique for encoding interval membership functions*. We follow this with a description of our decision tree obfuscator.

5.1 Setup

Without loss of generality, we assume decision trees to be full binary trees that perform binary classification on an input $(x_i)_{i \in [n]}$, where $x_i \in \{0,1\}^\ell$. We consider a decision tree function $C \in \mathcal{D}_\lambda$ with a depth d. We denote decision nodes by $(v_1, \ldots, v_{2^d-1})$ and leaf nodes by $\mathcal{S} = (s_1, \ldots, s_{2^d})$. An accepting path path_{s_τ} is defined to be the sequence of tuples $[\![t_j, i, b]\!]$, such that v_j is an ancestor node of $s_\tau \in \mathcal{S}$, where $s_\tau = 1$ and $b \in \{0,1\}$ denotes the output of the predicate $g_j(x_i)$. We assume each element in the input sequence to be compared at most twice along an accepting path. This assumption in reasonable, since any collection of inequalities in the form $x_i \leq t_j$ and $x_i > t_j$ defines an interval, and so is defined by a pair of comparisons. We assume that the evaluation procedure (Algorithm 6) is oblivious to the tuples in path_{s_τ}. This implies the depth is at most two times the number of input elements.

5.2 Encoding Intervals

We now describe our technique to encode integer intervals. We then extend this to construct our decision tree obfuscator.

Converting an Inequality into Intervals. A decision node tests $x \leq t$ for some fixed $t \in [0, 2^\ell)$. In other words, the node partitions $[0, 2^\ell)$ into $\mathcal{X} = [0, t+1)$ and $\mathcal{X}' = [t+1, 2^\ell)$. We further divide $\mathcal{X}$ and $\mathcal{X}'$ into disjoint sub-intervals of the form $[a, a + 2^p)$. This is the primary building block of our construction. The formal procedure is given in Algorithms 1 and 2.

Algorithm 1. $\mathsf{GenInt}_\mathcal{X}(t)$

Input: $\ell \in \mathbb{N}, t \in [0, 2^\ell)$
Output: $\mathcal{I}_\mathcal{X} = \{[a_j, a_j + 2^{p_j})\}_{j \in [k]}$
1: $\mathcal{I}_\mathcal{X} = \emptyset$; $\mathsf{temp} = 0$
2: Compute $k = wt(t+1)$
3: Compute $p_1, \ldots, p_k$, such that $t + 1 = \sum_{j=1}^k 2^{p_j}$ and $p_j < p_{j-1}$
4: $\mathcal{I}_\mathcal{X} = \{[0, 2^{p_1})\}$
5: **for** $j = 2$ to k **do**
6:    $a_j = \mathsf{temp} + 2^{p_{j-1}}$
7:    $\mathcal{I}_\mathcal{X} = \mathcal{I}_\mathcal{X} \cup \{[a_j, a_j + 2^{p_j})\}$
8:    $\mathsf{temp} = a_j$
9: **end for**
10: **return** $\mathcal{I}_\mathcal{X}$

Algorithm 2. $\mathsf{GenInt}_{\mathcal{X}'}(t)$

Input: $\ell \in \mathbb{N}, t \in [0, 2^\ell)$
Output: $\mathcal{I}_{\mathcal{X}'} = \{[b_j, b_j + 2^{p'_j})\}_{j \in [k']}$
 1: $\mathcal{I}_{\mathcal{X}'} = \emptyset$; $\mathrm{temp} = t + 1$
 2: Compute $k' = wt(2^\ell - t - 1)$
 3: Compute $p'_1, \ldots, p'_{k'}$, such that $2^\ell - t - 1 = \sum_{j=1}^{k'} 2^{p'_j}$ and $p'_j > p'_{j-1}$
 4: $\mathcal{I}_{\mathcal{X}'} = [\mathrm{temp}, 2^{p'_1})$
 5: **for** $j = 2$ to k' **do**
 6: $\quad b_j = \mathrm{temp} + 2^{p'_{j-1}}$
 7: $\quad \mathcal{I}_{\mathcal{X}'} = \mathcal{I}_{\mathcal{X}'} \cup \{[b_j, b_j + 2^{p'_j})\}$
 8: $\quad \mathrm{temp} = b_j$
 9: **end for**
10: **return** $\mathcal{I}_{\mathcal{X}'}$

Lemma 4. *Let* $\ell \in \mathbb{N}$ *and* $t \in [0, 2^\ell)$. *Consider algorithms* $\mathsf{GenInt}_{\mathcal{X}}$ *(Algorithm 1) and* $\mathsf{GenInt}_{\mathcal{X}'}$ *(Algorithm 2). Let* $\mathcal{I}_{\mathcal{X}} \leftarrow \mathsf{GenInt}_{\mathcal{X}}(t)$ *and* $\mathcal{I}_{\mathcal{X}'} \leftarrow \mathsf{GenInt}_{\mathcal{X}'}$. *Then* $\mathcal{I}_{\mathcal{X}}$ *defines sub-intervals of the form* $[a, a + 2^p)$ *for some a and p, whose union is* $[0, t+1)$, *and* $\mathcal{I}_{\mathcal{X}'}$ *defines sub-intervals of the form* $[b, b + 2^{p'})$ *for some b and p', whose union is* $[t+1, 2^\ell)$.

Proof. $\mathsf{GenInt}_{\mathcal{X}}(t)$ divides $[0, t+1)$ into k disjoint sub-intervals $\{[a_j, a_j + 2^{p_j})\}_{j \in [k]}$ such that $\sum_{j=1}^{k} 2^{p_j} = t + 1$ and $p_j < p_{j-1}$. Since $a_1 = 0$, $a_j = a_{j-1} + 2^{p_{j-1}}$, the output is $\mathcal{I}_{\mathcal{X}} = \{[0, 2^{p_1}), [2^{p_1}, 2^{p_1} + 2^{p_2}), \ldots, [\sum_{j=1}^{k-1} 2^{p_j}, \sum_{j=1}^{k} 2^{p_j})\}$ and the union of these intervals is $[0, t+1)$.

$\mathsf{GenInt}_{\mathcal{X}'}(t)$ divides $[t+1, 2^\ell)$ into k' disjoint sub-intervals $\{[b_j, b_j + 2^{p'_j})\}_{j \in [k']}$ such that $\sum_{j=1}^{k'} 2^{p'_j} = 2^\ell - t - 1$ and $p'_j > p'_{j-1}$. Since $b_1 = t+1$, $b_j = b_{j-1} + 2^{p'_{j-1}}$, we can write $\mathcal{I}_{\mathcal{X}'} = \{[t+1, t+1+2^{p'_1}), [t+1+2^{p'_1}, t+1+2^{p'_1}+2^{p'_2}), \ldots, [t+1+\sum_{j=1}^{k'-1} 2^{p'_j}, t+1+\sum_{j=1}^{k'} 2^{p'_j})\}$ and the union of these intervals is $[t+1, 2^\ell)$. $\square$

Intersection of Intervals. Let $\mathcal{I}_{\mathcal{X}}$ be a set of sub-intervals, all of the form $[a, a + 2^p)$ for various a and p. Let $\mathcal{I}_{\mathcal{X}'}$ be a set of sub-intervals, all of the form $[b, b + 2^r)$ for some b and r. Define the intersection of $\mathcal{I}_{\mathcal{X}}$ and $\mathcal{I}_{\mathcal{X}'}$ as $\mathcal{I}_{\mathcal{X}} \cap \mathcal{I}_{\mathcal{X}'} := \{I \cap J : I \in \mathcal{I}_{\mathcal{X}}, J \in \mathcal{I}_{\mathcal{X}'}\} \setminus \emptyset$.

Lemma 5. *Let* $\ell \in \mathbb{N}$. *Consider algorithms* $\mathsf{GenInt}_{\mathcal{X}}$ *(Algorithm 1) and* $\mathsf{GenInt}_{\mathcal{X}'}$ *(Algorithm 2). Let* $c, c + w \in \mathbb{Z}$ *be such that* $0 \leq c < c + w < 2^\ell$. *Let* $\mathcal{I}_{\mathcal{X}} \leftarrow \mathsf{GenInt}_{\mathcal{X}}(c + w)$, $\mathcal{I}_{\mathcal{X}'} \leftarrow \mathsf{GenInt}_{\mathcal{X}'}(c)$. *Let* $\mathcal{I} = \mathcal{I}_{\mathcal{X}} \cap \mathcal{I}_{\mathcal{X}'} = \{I \cap J : I \in \mathcal{I}_{\mathcal{X}}, J \in \mathcal{I}_{\mathcal{X}'}\} \setminus \emptyset$. *Then every interval in* $\mathcal{I}$ *is of the form* $[a, a + 2^i)$, *for some i, and* $|\mathcal{I}| \leq 2\ell - 2$.

Proof. We will prove that if $I \in \mathcal{I}_{\mathcal{X}}$ and $J \in \mathcal{I}_{\mathcal{X}'}$ are such that $I \cap J \neq \emptyset$, then $I \subseteq J$ or $J \subseteq I$. $\mathsf{GenInt}_{\mathcal{X}}(c + w)$ divides $[0, c + w + 1)$ into k disjoint sub-intervals $[a_j, a_j + 2^{p_j})$, where k is the Hamming weight of ℓ-bit binary encoding of $c + w + 1$. Since $p_j < p_{j-1}$, we can conclude that $2^{p_1} \leq c + w + 1 < 2^{p_1+1}$ and $\mathcal{I}_{\mathcal{X}} = \{[0, 2^{p_1}), \ldots, [2^{p_1} + \cdots + 2^{p_{k-1}}, 2^{p_1} + \cdots + 2^{p_k})\}$, where $\ell > p_1$. Since $c \in [0, c + w + 1)$, we consider the following:

- If $c + 1 = 2^q$, where $q \leq p_1$, then $\mathcal{I}_{\mathcal{X}'} = \{[2^i, 2^{i+1})\}_{i \in \{q, q+1, \ldots, \ell-1\}}$, and it can be clearly seen that for every non-empty intersection $I \cap J$, either $I \subseteq J$ or $J \subseteq I$.

- If $2^{q-1} < c + 1 < 2^q$, where $q \leq p_1$. Let k' be the Hamming weight of ℓ-bit binary encoding of $2^q - (c+1)$. Then, for $J \in \{[c+1, c+1+2^{p'_1}), \ldots, [c+1+\cdots+2^{p'_{k'-1}}, c+1+\cdots+2^{p'_{k'}})\}$, where $2^q = c+1+\cdots+2^{p'_{k'}}$, $J \subseteq I$, where $I = [0, 2^{p_1})$. Also note, for $J = [2^{p_1}, 2^{p_1+1})$ and $I \in \mathcal{I}_{\mathcal{X}} \setminus \{[0, 2^{p_1})\}$, $I \subseteq J$.

- If $2^{p_1} + \cdots + 2^{p_{m-1}} \leq c+1 < 2^{p_1} + \cdots + 2^{p_m}$, where $m \leq k$. Since $\mathsf{GenInt}_{\mathcal{X}'}(c)$ divides $[c+1, 2^{\ell})$ into k' sub-intervals $\{[b_r, b_r + 2^{p'_r})\}_{r \in [k']}$, then let $p'_1 < \cdots < p'_s \leq p_m < p'_{s+1} < \cdots < p'_{k'}$ for some $s < k'$. Let $I = [2^{p_1} + \cdots + 2^{p_{m-1}}, 2^{p_1} + \ldots, +2^{p_m}) \in \mathcal{I}_{\mathcal{X}}$. Then, for a non-empty intersection J, let $J \in \mathcal{I}_{\mathcal{X}'}$ be such that $I \cap J \neq \emptyset$. Then J is an element of the set $\{[c+1, c+1+2^{p'_1}), \ldots, [c+1+\cdots+2^{p'_{s-1}}, c+1+\cdots+2^{p'_s})\}$. Note that, $2^{p_1} + \cdots + 2^{p_m} = c+1+\cdots+2^{p'_s}$ as 0 to $m-1$ bits of $2^{\ell} - (c+1)$ and $2^{p_1} + \cdots + 2^{p_m} - (c+1)$ are equal, and thus $J \subseteq I$. For all other non-empty intersections, $I \subseteq J$.

It is clear from Algorithm 1 and 2 that $|\mathcal{I}_{\mathcal{X}}|, |\mathcal{I}_{\mathcal{X}'}| \leq \ell$ (when the Hamming weight of $|\mathcal{X}|$ and $|\mathcal{X}'|$ are ℓ). Let $\mathcal{I} = \mathcal{I}_{\mathcal{X}} \cap \mathcal{I}_{\mathcal{X}'}$ and consider any $H \in \mathcal{I}$. By the above, either $H = I$ for some $I \in \mathcal{I}_{\mathcal{X}}$ or $H = J$ for some $J \in \mathcal{I}_{\mathcal{X}'}$. It follows that $|\mathcal{I}| \leq |\mathcal{I}_{\mathcal{X}}| + |\mathcal{I}_{\mathcal{X}'}|$. In the case $H = I$, there is some $J \in \mathcal{I}_{\mathcal{X}'}$ such that $I \subseteq J$, and so J itself is not counted in $|\mathcal{I}|$. Similarly in the case $H = J$ there is some $I \in \mathcal{I}_{\mathcal{X}}$ such that $J \subseteq I$, and so I is not counted in $|\mathcal{I}|$. Hence $|\mathcal{I}| \leq |\mathcal{I}_{\mathcal{X}}| + |\mathcal{I}_{\mathcal{X}'}| - 2 \leq 2\ell - 2$.   □

Calculating the Encodings. Let $\mathcal{I}$ be a set of sub-intervals, all of the form $[a, a + 2^j)$. The encoder (Algorithm 3) receives $\mathcal{I}$ as input, and outputs the set of encodings $E = \{h_1, \ldots, h_{|\mathcal{I}|}\}$. For $i \in \{0, \ldots, \ell - 1\}$ define $f_i(y) = \lfloor \frac{y}{2^i} \rfloor$, which defines a function $f_i : \{0,1\}^{\ell} \to \{0,1\}^{\ell-i}$. Let $H : \{0,1\}^* \to \{0,1\}^{\omega}$ be a cryptographic hash function with $\omega > 2\ell$ so that H is injective when restricted to inputs of length at most ℓ. A random oracle is not generally injective, but when the output length is large enough compared to the input then it will be. Our argument behind imposing this additional constraint on H is that every $\lfloor \lfloor \frac{y}{2^i} \rfloor \rfloor \leq \ell$ maps to a unique encoding in E. Algorithm 4 receives as input $E \leftarrow \mathsf{IntEnc}(\mathcal{I})$ and $x \in \{0,1\}^{\ell}$ and outputs 1, if x belongs to any of the sub-intervals in $\mathcal{I}$.

Lemma 6 (Correctness). *Let $c, c+w \in [0, 2^{\ell})$. Consider algorithms $\mathsf{GenInt}_{\mathcal{X}}$ (Algorithm 1), $\mathsf{GenInt}_{\mathcal{X}'}$ (Algorithm 2), IntEnc (Algorithm 3), Dec (Algorithm 4) and input $x \in \{0,1\}^{\ell}$. Let $\mathcal{I}_{\mathcal{X}} \leftarrow \mathsf{GenInt}_{\mathcal{X}}(c+w)$ and $\mathcal{I}_{\mathcal{X}'} \leftarrow \mathsf{GenInt}_{\mathcal{X}}(c)$ and $\mathcal{I} \leftarrow \mathcal{I}_{\mathcal{X}} \cap \mathcal{I}_{\mathcal{X}'}$. Let $H : \{0,1\}^* \to \{0,1\}^{\omega}$ be injective when restricted to inputs of length at most ℓ (e.g., $\omega > 2\ell$). Then $x \in (c, c+w]$ if and only if Dec outputs 1.*

Algorithm 3. IntEnc($\{[a_j, a_j + 2^{p_j})\}_{j\in[k]}$)

1: $E = \emptyset$
2: **for** $j = 1$ to k **do**
3:     Compute $\mu_j = f_{p_j}(a_j)$
4:     Compute $h_j = H(\mu_j)$.
5:     $E = E \cup \{h_j\}$
6: **end for**
7: **return** E

Algorithm 4. Dec (with embedded data E)

Input: $\ell \in \mathbb{N}$, $x \in \{0,1\}^\ell$
Output: 0 or 1.
1: **for** $i = 0$ to $\ell - 1$ **do**
2:     Compute $H(f_i(x))$
3:     **if** $H(f_i(x)) \in E$ **then**
4:         **return** 1
5:     **end if**
6: **end for**
7: **return** 0

Proof. Let $\mathcal{I} = \{[a_j, a_j + 2^{p_j})\}_{j\in[k]}$, then from Lemma 5 we have $(c, c + w] = \bigcup_{j=1}^k [a_j, a_j + 2^{p_j})$. Let x be an integer in $(c, c + w]$, then it must belong to one of the sub-intervals in $\mathcal{I}$. Algorithm 3 computes $f_{p_j}(y) = \mu_j$, for every $[a_j, a_j + 2^{p_j}) \in \mathcal{I}$. If $x \in [a_j, a_j + 2^{p_j})$, then there exists an $i \in \{0, \dots, \ell-1\}$ such that $f_i(x) = f_{p_j}(a_j) = \mu_j$. Hence $H(f_i(x)) \in E \leftarrow$ IntEnc$(\mathcal{I})$, and Dec outputs 1. If $x \notin (c, c + w]$ then x does not lie in any of the intervals $[a_j, a_j + 2^{p_j})$. It follows that that $f_i(x) \notin \{\mu_j : 1 \leq j \leq |\mathcal{I}|\}$ and therefore, since H is injective, E will not contain $H(f_i(x))$. Hence, Dec will correctly reject the input. □

We illustrate the interval encoding technique with the following concrete setting: consider the interval membership predicate $x \in (10, 14]$. To encode the interval, calculate $\mathcal{I}_\mathcal{X} \leftarrow$ GenInt$_\mathcal{X}(14)$ and $\mathcal{I}_{\mathcal{X}'} \leftarrow$ GenInt$_{\mathcal{X}'}(10)$ which gives the set of sub-intervals in $[0, 15)$ and $[15, 256)$ for $\ell = 8$. Finally, calculate IntEnc$(\mathcal{I})$, where $\mathcal{I} \leftarrow \mathcal{I}_\mathcal{X} \cap \mathcal{I}_{\mathcal{X}'}$. The sets are indicated as follows:

$\mathcal{I}_\mathcal{X} = \{[0, 8), [8, 12), [12, 14), [14, 15)\}$
$\mathcal{I}_{\mathcal{X}'} = \{[11, 12), [12, 16), [16, 32), [32, 64), [64, 128), [128, 256)\}$
$\mathcal{I} = \{[11, 12), [12, 14), [14, 15)\}$
IntEnc$(\mathcal{I})$ = $\{H(f_0(00001011), H(f_1(00001100), H(f_0(00001110))\}$ = $\{H(0000\ 1011), H(0000110), H(00001110)\}$

5.3 Obfuscator $\mathcal{O}$

We now present our decision tree obfuscator $\mathcal{O}$. Let $\mathcal{S} = (s_1, \dots, s_{2^d})$ be the list of leaf nodes and path$_{s_\tau}$ denotes an accepting path through the tree to leaf $s_\tau = 1$.

Each accepting path is a conjunction of inequalities, and our objective is to obfuscate the conjunctions using our encoding technique for interval membership functions (see Sect. 5.2). Our construction relies on the fact that the terms in a conjunction can be reordered. Recall that x_i is compared at most twice along an accepting path, and hence the accepting path corresponds to $x_i \in (c_i, c_i + w_i]$ for every $i \in [n]$.

Precisely, the obfuscator works as follows: to encode $(c_i, c_i + w_i]$, calculate $\mathcal{I}_\mathcal{X}^i \leftarrow \mathsf{GenInt}_\mathcal{X}(c_i + w_i)$ and $\mathcal{I}_{\mathcal{X}'}^i \leftarrow \mathsf{GenInt}_{\mathcal{X}'}(c_i)$, which gives the set of sub-intervals in $[0, c_i + w_i + 1)$ and $[c_i + 1, 2^\ell)$. Next, determine $\mathcal{I}^i \leftarrow \mathcal{I}_\mathcal{X}^i \cap \mathcal{I}_{\mathcal{X}'}^i$ which is a set of sub-intervals whose union is $(c_i, c_i + w_i]$. To encode each sub-interval in $\mathcal{I}^i$, calculate encodings $\mathcal{B}^i \leftarrow \mathsf{IntEnc}(\mathcal{I}^i)$. Let $H_c : \{0,1\}^* \rightarrow \{0,1\}^q$ be a cryptographic hash function with $q \geq 2\omega n$ (and hence injective on restricted inputs). The main idea is to concatenate each combination of n hashes sorted in ascending order of i for each entry in $\mathcal{B}^i$, and apply H_c; call the set of hashes $\mathcal{B}$. If $|\mathcal{B}| < 2^d(2\ell - 2)^n$, add dummy entries drawn uniformly at random from $\{0,1\}^q$. Finally, output $\mathcal{B}$. We give the formal details in Algorithm 5.

On input $(x_i)_{i \in [n]}$, the evaluation procedure calculates all possible encodings by evaluating $H(f_p(x_i))$ for every $p \in \{0, \ldots, \ell - 1\}$; call it $\mathcal{E}^i = \{h_1^i, \ldots, h_\sigma^i\}$. Finally compute all possible $H_c(\|_{i \in [n]} h_{q_i}^i)$, where encodings are listed in ascending order of i. For an accepting input, one of the computed hash values belongs to the set $\mathcal{B}$ published by $\mathcal{O}$. Formally, the evaluation procedure is specified in Algorithm 6.

5.4 Correctness and Efficiency

We now analyze the correctness and efficiency of the obfuscator.

Lemma 7 (Correctness.). *Let* $\lambda \in \mathbb{N}$ *be the security parameter, and let* n, ℓ, ω *and* q *be polynomials in* λ. *Consider algorithms* $\mathcal{O}$ *(Algorithm 5) and* Eval *(Algorithm 6), and an input* $(x_i)_{i \in [n]}$ *with* $x_i \in \{0,1\}^\ell$. *Let* $H : \{0,1\}^* \rightarrow \{0,1\}^\omega$ *with* $\omega > 2\ell$, *and let* $H_c : \{0,1\}^* \rightarrow \{0,1\}^q$ *with* $q > 2\omega n$. *Then for every* $[\![t_j, i, b]\!] \in \mathsf{path}_{s_\tau}$, *where* $t_j \in [0, 2^\ell)$, $i \in \{1, \ldots, n\}$, $b \in \{0,1\}$, *for every* $\mathcal{S} = (s_1, \ldots, s_{2^d})$ *with* $s_\tau \in \{0,1\}$, *for every* $\mathcal{B} \leftarrow \mathcal{O}$ *and for every input* $(x_i)_{i \in [n]}$, *if* $C((x_i)_{i \in [n]}) = 1$, *then* Eval *outputs* 1, *else it outputs* 0 *with overwhelming probability in* λ.

Proof. Algorithm 5 calculates set of encodings $\mathcal{B}^i \leftarrow \mathsf{IntEnc}(\mathcal{I}^i)$ for every $i \in [n]$. If there is some i such that $[\![t_j, i, b]\!] \notin \mathsf{path}_{s_\tau}$ (which means x_i is not compared in the path) then $\mathcal{B}^i = \{1^\ell\}$. Let $(x_i)_{i \in [n]}$ be an accepting input. From Definition 5, $x_i \in (c_i, c_i + w_i]$, for every $i \in [n]$. From Lemma 6, if $x_i \in (c_i, c_i + w_i]$, then there exists a unique $h_k^i \in \mathcal{E}^i$, such that $h_k^i \in \mathcal{B}^i$. If $\exists i$, such that $[\![t_j, i, b]\!] \notin \mathsf{path}_{s_\tau}$, then $h_k^i = 1^\ell \in \mathcal{B}^i$. Thus, for an accepting input $(x_i)_{i \in [n]}$, there exists a unique $(h_{k_1}^1, \ldots, h_{k_n}^n)$, where $h_{k_i}^i \in \mathcal{E}^i$ and $H_c(h_{k_1}^1 \| \ldots \| _{k_n}^n)$ will be contained in $\mathcal{B}$, and Algorithm 6 will correctly output 1.

If $C((x_i)_{i \in [n]}) \neq 1$, then by Lemma 6, hash values computed from (x_i) will not match the hash values input to H_c. As we choose parameters such that H_c

---

Algorithm 5. $\mathcal{O}\left(d, n, \ell \in \mathbb{N}, \mathsf{asset}(C)\right)$

1: $\mathcal{B} = \emptyset$
2: $\alpha = 2^d(2\ell - 2)^n$
3: **for all** τ such that $s_\tau = 1$ **do**
4:     Compute $\mathsf{path}_{s_\tau} = (\llbracket t_j, i, b \rrbracket : v_j$ is an ancestor of s_τ, and $b = g_j(t_j))$
5:     **for** $i = 1$ to n **do**
6:        $\mathcal{I}_{\mathcal{X}}^i = \mathcal{I}_{\mathcal{X}'}^i = [0, 2^\ell)$
7:        **if** $\llbracket t_{j_1}, i, 1 \rrbracket \in \mathsf{path}_{s_\tau}$ **then**
8:           $\mathcal{I}_{\mathcal{X}}^i \leftarrow \mathsf{GenInt}_{\mathcal{X}}(t_{j_1})$
9:        **end if**
10:       **if** $\llbracket t_{j_2}, i, 0 \rrbracket \in \mathsf{path}_{s_\tau}$ **then**
11:          $\mathcal{I}_{\mathcal{X}'}^i \leftarrow \mathsf{GenInt}_{\mathcal{X}'}(t_{j_2})$
12:       **end if**
13:       $\mathcal{I}^i \leftarrow \mathcal{I}_{\mathcal{X}}^i \cap \mathcal{I}_{\mathcal{X}'}^i$
14:       **if** $(\mathcal{I}^i == [0, 2^\ell))$ **then**
15:          $\mathcal{B}^i = \{1^\ell\}$
16:       **else**
17:          $\mathcal{B}^i \leftarrow \mathsf{IntEnc}(\mathcal{I}^i)$
18:       **end if**
19:     **end for**
20:     Denote $\mathcal{B}^i = (h_1^i, \ldots, h_{\sigma_i}^i)$, where $\sigma_i \leq 2\ell - 2$ for each i.
21:     **for all** $((q_1, \ldots, q_n) \in [\sigma_1] \times \cdots \times [\sigma_n])$ **do**
22:       $\mathcal{B} = \mathcal{B} \cup \{H_c(\|_{i=1}^n h_{q_i}^i)\}$
23:     **end for**
24: **end for**
25: **while** $(|\mathcal{B}| < \alpha)$ **do**
26:     $r \leftarrow_\$ \{0, 1\}^q$
27:     $\mathcal{B} = \mathcal{B} \cup \{r\}$
28: **end while**
29: **return** $\mathcal{B}$

Algorithm 6. Eval $(\mathcal{B}$ with $|\mathcal{B}| = 2^d(2\ell - 2)^n)$

Input: $(x_1, \ldots, x_n)$, ℓ, d
Output: 0 or 1
1: **for** $i = 1$ to n **do**
2:     $\mathcal{E}^i \leftarrow \{1^\ell\} \cup \{H(f_0(x_i)), \ldots, H(f_{\ell-1}(x_i))\}$
3: **end for**
4: $\mathcal{E}^i = \{h_1^i, \ldots, h_\sigma^i\}$, $\sigma \leq \ell + 1$
5: **for all** $(q_1, \ldots, q_i) \in [\sigma_1] \times \cdots \times [\sigma_n]$ **do**
6:     **if** $H_c(\|_{i \in [n]} h_{q_i}^i) \in \mathcal{B}$ **then**
7:         **return** 1
8:     **end if**
9: **end for**
10: **return** 0

---

is injective, $H_c(h_{k_1}^1 \| \ldots \| h_{k_n}^n)$ will not be contained in $\mathcal{B}$, except if it equals to one of its dummy entries. Since the number of possible encodings (in Eval) input to H_c is $2^{(\ell+1)n}$, and $w > 2\ell$, $q > 2wn$, the probability that Eval incorrectly accepts the input is given by $\frac{2^{(\ell+1)n}}{2^q} = \frac{1}{2^{3\ell n}}$. Finally, the probability that the α hash values output by $\mathcal{O}$ are good is given by $(1 - \frac{1}{2^{3\ell n}})^\alpha \approx 1 - negl(\lambda)$. □

Complexity Analysis. We now discuss the size complexity of the obfuscated decision tree. Let $\lambda \in \mathbb{N}$, and d, ℓ, n, q be polynomials in λ. Along an accepting path, the upper bound on the size of $\mathcal{B}^i$ is $2\ell - 2$ (see Lemma 5). There are n such set of encodings, hence the total number of possible encodings input to H_c is $(2\ell - 2)^n$. Since, the number of accepting paths is $O(2^d)$, the overall complexity of storing the obfuscated tree is $O(2^{d+n} \cdot \ell^n \cdot q)$, where q is the output size of H_c. Note that the upper bound $\alpha = 2^d(2\ell - 2)^n$ on the number of hashes is much larger than will be needed for most evasive decision trees, so in practice this parameter could be chosen a lot smaller to get a more compact obfuscated program.

Next, we analyze the time complexity of the evaluation procedure (Algorithm 6). Each node corresponds to set of encodings $\mathcal{E}^i$, where $|\mathcal{E}^i| = \ell + 1$. For n input attributes, the overall running time of the evaluation algorithm is of the order $O(\ell^n \log(\alpha))$ operations. Since the query response time is exponential in n, we restrict to decision trees of constant number of input elements.

We now prove polynomial slowdown only for some special cases.

Lemma 8 (Polynomial Slowdown). *Let* $\lambda \in \mathbb{N}$ *be the security parameter and* ℓ, n, d *be polynomials in* λ. *Define* $\mathcal{T}_\lambda$ *as a special family of evasive decision trees, where* $d = 5$, $n \leq 4$ *and* $\ell = \frac{\lambda}{4}$. *Then for every* $C \leftarrow \mathcal{T}_\lambda$, *there exists a polynomial* p *such that the running time of* $\mathcal{O}(C)$ *is bounded by* $p(|C|, \lambda)$.

Proof. Let $C \leftarrow \mathcal{T}_\lambda$ determines whether an input $(x_i)_{i=1}^n$ is contained in the decision region defined by $(c_i, c_i + w_i]$ with $w_i \in (0, w_{max})$. From Lemma 2, $w_{max} \leq 2^{\ell - \frac{\lambda}{n}}$, which specifies the maximum width of the intervals. For evasiveness, we require $\ell - \frac{\lambda}{n} \geq 0$, which gives $\ell n \geq \lambda$. Taking $\ell = \frac{\lambda}{4}$ and $n = 4$, we get $\ell n = \lambda$, which is a feasible condition for evasiveness. Since $d = 5$, we equate $d - 1 = n$. The cost of evaluating $\mathcal{O}$ is given by $\ell^n = (\frac{\lambda}{4})^4$. □

Parameters for Secure Construction. In the previous sections, we have discussed the choice of parameters that provide security to our decision tree obfuscator, i.e. for $\ell = \ell(\lambda)$, $d = d(\lambda)$, $n = n(\lambda)$, the conditions for evasiveness are given in Lemma 2. For the hardness of finding x_i that belongs to the decision region $(c_i, c_i + w_i]$, we require $w_{max}(\lambda) \leq 2^{\ell(\lambda) - \frac{\lambda}{n(\lambda)}}$ for every $i \in [n]$. In addition to that, we impose $d \leq n + 1$ (by construction). We now present some example parameters along with their bit-security in Table 1.

Table 1. Example parameter sets for an obfuscated decision tree with $w_{max} \leq 2^{\ell - \frac{\lambda}{n}}$. For $q = 512$ bits and *one* accepting path (q is the output size of hash function H_c), we calculate the size of the obfuscated program and the cost of the evaluation (Algorithms 5 and 6).

d	ℓ	n	λ	Program size	Evaluation cost
5	64	4	128	$127^4 \times 512$	$65^4 \times 512$
3	64	2	64	$127^2 \times 512$	$65^2 \times 512$

6   Proof of VBB Security

We prove VBB security in the random oracle model. For simplicity we restrict to decision trees with one accepting path, and let $\alpha = (2\ell - 2)^n$ be an upper bound on the number of hash values required for the obfuscated program.

We use cryptographic hash functions $H : \{0,1\}^* \rightarrow \{0,1\}^\omega$ and $H_c : \{0,1\}^* \rightarrow \{0,1\}^q$ in our construction, which we model as random oracles in our security proof. Our objective is to show that a PPT adversary having access to the obfuscated function has no advantage over a simulator having oracle access to the function. This is achieved by executing the adversary in a simulation, such that the adversary cannot distinguish between the simulated and real environment. We first give a brief intuition to our security proof.

Consider a simulator S who samples parameters ℓ, n following the conditions in Lemma 8, and sends them to adversary A. A now samples C and provides S oracle access to C. Since S does not know the program C, it simulates the obfuscated program and random oracles and provides answers to A's queries.

If the adversary never queries the circuit with an accepting input, everything is a correct simulation. However, if the adversary does query the circuit with an accepting input, then the security reduction immediately uses this clue to mount a model extraction attack, and hence learn the corresponding accepting path in the decision tree. The security reduction can then run the obfuscator correctly for that single accepting path, and program the random oracles to be consistent with the simulated O (the reduction does not learn anything about the other possible accepting paths). Hence, again everything is a correct simulation.

Theorem 1. *Let* $\lambda \in \mathbb{N}$ *be the security parameter and* $\ell = \ell(\lambda)$*,* $n = n(\lambda)$ *and* $\alpha = \alpha(\lambda)$ *satisfy the conditions required for Lemma 8. Let* $\mathcal{D}_\lambda$ *be a distribution of evasive decision trees. Then for random oracles* $H : \{0,1\}^* \rightarrow \{0,1\}^\omega$ *and* $H_c : \{0,1\}^* \rightarrow \{0,1\}^q$*, the decision tree obfuscator* $\mathcal{O}$ *is a VBB obfuscator.*

Proof. As evident from Lemma 6, $\mathcal{O}$ satisfies functionality preservation. Lemma 8 shows that $\mathcal{O}$ causes polynomial slowdown. Thus it suffices to show that there exists a (non-uniform) PPT simulator S for every (non-uniform) PPT adversary A, such that for an ensemble of decision tree evasive distributions $\mathcal{D}_\lambda$ (from Lemma 8), the following holds:

$$\left| \Pr_{C \leftarrow D_\lambda} [\mathcal{A}(\mathcal{O}(1^\lambda, C)) = 1] - \Pr_{C \leftarrow D_\lambda} [\mathcal{S}^C(1^\lambda) = 1] \right| \leq \mu(\lambda)$$

Every $C \leftarrow \mathcal{D}_\lambda$ identifies unique $(c_1, \ldots, c_n) \leftarrow X_n$ and $(w_1, \ldots, w_n) \leftarrow (0, w_{max})^n$, and on input $(x_1, \ldots, x_n)$ determines if $x_i \in (c_i, c_i + w_i]$ for all $i \in [n]$. Let $\mathcal{O}(1^\lambda, C) = \{h_1, \ldots, h_\alpha\}$ denote the correct obfuscation of C. Let $\mathcal{A}$ be a PPT adversary that takes as input $\mathcal{O}(1^\lambda, C)$. We use this adversary to design a PPT simulator $\mathcal{S}$ that simulates an execution of $\mathcal{A}$.

Since $\mathcal{A}$ expects the oracles H and H_c, $\mathcal{S}$ provides a simulation of both the oracles. In order to record the choices of the random oracles, $\mathcal{S}$ maintains two tables: $\mathcal{T}_1$ to record responses for queries to H, and $\mathcal{T}_2$ to record responses for queries to H_c.

Since $\mathcal{S}$ does not have access to $\mathcal{O}(1^\lambda, C)$, it prepares a purported obfuscation of C as follows: $\mathcal{S}$ samples α values uniformly at random from the co-domain of H_c, and sends $\{h'_1, \ldots, h'_\alpha\}$ to $\mathcal{A}$.

We assume that $\mathcal{A}$ makes polynomially many queries to both the random oracles. When $\mathcal{A}$ queries oracle H with $\mathbf{u}^*$, $\mathcal{S}$ looks for v such that $(\mathbf{u}^*, v) \in \mathcal{T}_1$ and returns it to the adversary. If no such v exists, then the simulation samples a distinct $v \in \{0, 1\}^\omega$ uniformly at random, adds $(\mathbf{u}^*, v)$ to $\mathcal{T}_1$, and returns v to $\mathcal{A}$.

When $\mathcal{A}$ makes a query $\mathbf{h}^*$ to the random oracle H_c, the simulator looks for a val such that $(\mathbf{h}^*, val) \in \mathcal{T}_2$ and returns it to $\mathcal{A}$. If there are no entries corresponding to $\mathbf{h}^*$, the simulator parses $\mathbf{h}^*$ as $(h_1^*, \ldots, h_n^*)$ and looks up table $\mathcal{T}_1$ to find an entry corresponding to each parsed string. If there does not exist such entry in $\mathcal{T}_1$, then a distinct $val \in \{0, 1\}^q$ is chosen uniformly at random, $(\mathbf{h}^*, val)$ is added to $\mathcal{T}_2$, and val is returned it $\mathcal{A}$.

If there exists a unique u such that $(u, h_i^*) \in \mathcal{T}_1$, then the simulator calculates $j \leftarrow \ell - |u|$, where $|u|$ denotes the bit length of u, and $x_i \leftarrow u \times 2^j$. Since u corresponds to μ_j (for a correct input), adding j zeroes yields an accepting input for C. Eventually, the simulator queries the oracle C with the $(x_i)_{i \in [n]}$. If C returns 1, $\mathcal{S}$ determines the c_i's and w_i's by doing standard model extraction attack for that *single* accepting path, calculates pairs (u, v) and registers the entries in $\mathcal{T}_1$.

Note that we do not learn anything about other possible accepting paths. Thereafter the simulator calculates the α input entries of $\mathcal{T}_2$, defines them to be α entries from the already published set $\{h'_1, \ldots, h'_\alpha\}$, registers the pairs in $\mathcal{T}_2$ and returns val to the adversary. If there are multiple entries in $\mathcal{T}_1$, the simulation halts.

We describe the simulation in form of pseudo code in Algorithm 7 and 8.

Algorithm 7. Oracle $H(\mathbf{u}^*)$

1: Find all v such that $(\mathbf{u}^*, v) \in \mathcal{T}_1$
2: **if** no such v exists **then**
3:     $v \leftarrow_\$ \{0, 1\}^\omega$
4:     $\mathcal{T}_1 \leftarrow \mathcal{T}_1 \cup (\mathbf{u}^*, v)$
5: **else**
6:     **if** $\exists w \, (\mathbf{u}^* \neq w)$, such that $(\mathbf{u}^*, v) \in \mathcal{T}_1$ and $(w, v) \in \mathcal{T}_1$ **then**
7:         **return** $\perp$
8:     **end if**
9: **end if**
10: **return** v

Algorithm 8. Oracle $H_c(\mathbf{h}^*)$

1: Find all val such that $(\mathbf{h}^*, val) \in \mathcal{T}_2$
2: **if** no such val exists **then**
3:     Parse $\mathbf{h}^* = (h_i^*)_{i \in [n]}$
4:     counter $= 0$
5:     **for** $i = 1$ to n **do**
6:         **if** $(u, h_i^*) \in \mathcal{T}_1$ **then**
7:             counter $\leftarrow$ counter $++$
8:             $j \leftarrow \ell - |u|$
9:             $x_i \leftarrow u \times 2^j$
10:        **end if**
11:    **end for**
12:    **if** (counter $== n$) **then**
13:        $b \leftarrow \mathcal{S}^C(x_1, \ldots, x_n)$
14:        **if** ($b == 1$) **then**
15:            Calculate $(c_i, w_i)_{i \in [n]}$ using model-extraction attack
16:            Run Algorithm 5 and calculate $(u, v), h$
17:            $\mathcal{T}_1 \leftarrow \mathcal{T}_1 \cup (u, v)$
18:            **if** $\exists u_1, u_2 \, (u_1 \neq u_2)$, such that $(u_1, v) \in \mathcal{T}_1$ and $(u_2, v) \in \mathcal{T}_1$ **then**
19:                **return** $\perp$
20:            **else**
21:                **for** $i = 1$ to α **do**
22:                    $val_i \leftarrow h_i'$
23:                    $\mathcal{T}_2 \leftarrow \mathcal{T}_2 \cup (h_i, val_i)$
24:                **end for**
25:            **end if**
26:        **end if**
27:        $val \leftarrow_\$ \{0, 1\}^q$
28:        $\mathcal{T}_2 \leftarrow \mathcal{T}_2 \cup (\mathbf{h}^*, val)$
29:    **end if**
30: **end if**
31: **return** val

The simulated view is distributed identically to the real view. Hence $\mathcal{A}$ cannot distinguish between the real and simulated obfuscation instances.   □

We now analyze the scenario where the simulator fails due to conflicts in table $\mathcal{T}_1$ and show that the probability of such conflicts is negligible in λ.

Lemma 9. *Let $\lambda \in \mathbb{N}$ be the security parameter and let ℓ, n, α be polynomials in λ. Let $\mathcal{D}_\lambda$ be an ensemble of evasive decision tree distributions, and let $\mathcal{O}(1^\lambda, C)$ be the obfuscation of $C \leftarrow \mathcal{D}_\lambda$. Consider Algorithms 7 and 8 and random oracles $H : \{0,1\}^* \rightarrow \{0,1\}^\omega$ and $H_c : \{0,1\}^* \rightarrow \{0,1\}^{q(\lambda)}$. Let $\eta = \eta(\lambda)$ be the number of entries in $\mathcal{T}_1$, then there exists a negligible function $\mu(\lambda)$ such that*

$$\Pr_{C \leftarrow \mathcal{D}_\lambda}[\mathcal{S}^C(1^\lambda) = \bot] \leq \mu(\lambda)$$

where $\mathcal{S}$ is a PPT algorithm with oracle access to C.

Proof. The simulation fails if and only if there is a conflict in table $\mathcal{T}_1$, and $\mathcal{S}$ halts. Conflicts may arise when $\mathcal{S}$ has already responded to an oracle query $\mathbf{u}^*$ to H with $v \leftarrow_\$ \{0,1\}^\omega$, and later on query $\mathbf{h}^*$ to H_c, it queries the oracle of C, and populates $\mathcal{T}_1$ with (w, v) such that $w \neq \mathbf{u}^*$. Let $\eta = \eta(\lambda)$ be the number of entries in $\mathcal{T}_1$. The probability that a conflict occurs in $\mathcal{T}_1$ is equal to the probability that a hash value is same as at least one of the η values in table $\mathcal{T}_1$. Since $H : \{0,1\}^* \rightarrow \{0,1\}^\omega$, there are 2^ω choices for the value. When there are no entries in $\mathcal{T}_1$, the collision probability is 0, when there is one entry in $\mathcal{T}_1$, the collision probability is $\frac{1}{2^\omega}$. Continuing the same way, when there are $\eta - 1$ entries in $\mathcal{T}_1$, the probability of collision is $\frac{(\eta-1)}{2^\omega}$. Assuming all the samples are independent, the final step is to draw a conclusion about $\mathcal{S}$'s probability of failure, given by

$$\Pr_{C \leftarrow \mathcal{D}_\lambda}[\mathcal{S}^C(1^\lambda) = \bot] = \frac{1 + \cdots + (\eta - 1)}{2^\omega}$$

$$= \frac{\eta^2 - \eta}{2^{\omega+1}}$$

$$\leq \mu(\lambda)$$

□

7   Conclusion

In this paper, we have introduced a new special-purpose virtual black-box obfuscator for evasive decision trees. While doing so, we have presented an encoder for hiding parameters in an interval-membership function. Our obfuscation construction blows up exponentially in the depth of the tree, hence an interesting problem would be to investigate solutions that work for more general class of evasive decision trees.

Acknowledgements. We thank Phillip Rogaway for discussions on methods for obfuscating inequalities. We thank the Marsden Fund of the Royal Society of New Zealand for supporting this research. We thank the reviewers of INDOCRYPT 2023 for their insightful comments.

References

1. Ateniese, G., Mancini, L.V., Spognardi, A., Villani, A., Vitali, D., Felici, G.: Hacking smart machines with smarter ones: how to extract meaningful data from machine learning classifiers. Int. J. Secure. Network. **10**(3), 137–150 (2015)
2. Banerjee, S., Galbraith, S.D., Khan, T., Castellanos, J.H., Russello, G.: Preventing reverse engineering of control programs in industrial control systems. In: Proceedings of the 9th ACM Cyber-Physical System Security Workshop, pp. 48–59 (2023)
3. Barak, B., Bitansky, N., Canetti, R., Kalai, Y.T., Paneth, O., Sahai, A.: Obfuscation for evasive functions. In: Lindell, Y. (ed.) TCC 2014. LNCS, vol. 8349, pp. 26–51. Springer, Heidelberg (2014). https://doi.org/10.1007/978-3-642-54242-8_2
4. Barak, B., et al.: On the (im)possibility of obfuscating programs. In: Kilian, J. (ed.) CRYPTO 2001. LNCS, vol. 2139, pp. 1–18. Springer, Heidelberg (2001). https://doi.org/10.1007/3-540-44647-8_1
5. Barni, M., Failla, P., Kolesnikov, V., Lazzeretti, R., Sadeghi, A.-R., Schneider, T.: Secure evaluation of private linear branching programs with medical applications. In: Backes, M., Ning, P. (eds.) ESORICS 2009. LNCS, vol. 5789, pp. 424–439. Springer, Heidelberg (2009). https://doi.org/10.1007/978-3-642-04444-1_26
6. Barni, M., et al.: Efficient privacy-preserving classification of ECG signals. In: 2009 First IEEE International Workshop on Information Forensics and Security (WIFS), pp. 91–95. IEEE (2009)
7. Bartusek, J., Lepoint, T., Ma, F., Zhandry, M.: New techniques for obfuscating conjunctions. In: Ishai, Y., Rijmen, V. (eds.) EUROCRYPT 2019. LNCS, vol. 11478, pp. 636–666. Springer, Cham (2019). https://doi.org/10.1007/978-3-030-17659-4_22
8. Bishop, A., Kowalczyk, L., Malkin, T., Pastro, V., Raykova, M., Shi, K.: A simple obfuscation scheme for pattern-matching with wildcards. In: Shacham, H., Boldyreva, A. (eds.) CRYPTO 2018. LNCS, vol. 10993, pp. 731–752. Springer, Cham (2018). https://doi.org/10.1007/978-3-319-96878-0_25
9. Blurock, E.S.: Automatic learning of chemical concepts: research octane number and molecular substructures. Comput. Chem. **19**(2), 91–99 (1995)
10. Bos, J.W., Lauter, K., Naehrig, M.: Private predictive analysis on encrypted medical data. J. Biomed. Inform. **50**, 234–243 (2014)
11. Bost, R., Popa, R.A., Tu, S., Goldwasser, S.: Machine learning classification over encrypted data. Cryptology ePrint Archive (2014)
12. Boyle, E., Ishai, Y., Meyer, P., Robere, R., Yehuda, G.: On low-end obfuscation and learning. In: 14th Innovations in Theoretical Computer Science Conference (ITCS 2023). Schloss Dagstuhl-Leibniz-Zentrum für Informatik (2023)
13. Brickell, J., Porter, D.E., Shmatikov, V., Witchel, E.: Privacy-preserving remote diagnostics. In: Proceedings of the 14th ACM Conference on Computer and Communications Security, pp. 498–507 (2007)
14. Canetti, R.: Towards realizing random oracles: hash functions that hide all partial information. In: Kaliski, B.S. (ed.) CRYPTO 1997. LNCS, vol. 1294, pp. 455–469. Springer, Heidelberg (1997). https://doi.org/10.1007/BFb0052255
15. Canetti, R., Rothblum, G.N., Varia, M.: Obfuscation of hyperplane membership. In: Micciancio, D. (ed.) TCC 2010. LNCS, vol. 5978, pp. 72–89. Springer, Heidelberg (2010). https://doi.org/10.1007/978-3-642-11799-2_5
16. Cong, K., Das, D., Park, J., Pereira, H.V.: SortingHat: efficient private decision tree evaluation via homomorphic encryption and transciphering. In: Proceedings of the 2022 ACM SIGSAC Conference on Computer and Communications Security, pp. 563–577 (2022)

17. Decaestecker, C., et al.: Methodological aspects of using decision trees to characterise leiomyomatous tumors. Cytometry J. Int. Soc. Anal. Cytol. **24**(1), 83–92 (1996)
18. Fredrikson, M., Jha, S., Ristenpart, T.: Model inversion attacks that exploit confidence information and basic countermeasures. In: Proceedings of the 22nd ACM SIGSAC Conference on Computer and Communications Security, pp. 1322–1333 (2015)
19. Galbraith, S.D., Zobernig, L.: Obfuscated fuzzy hamming distance and conjunctions from subset product problems. In: Hofheinz, D., Rosen, A. (eds.) TCC 2019. LNCS, vol. 11891, pp. 81–110. Springer, Cham (2019). https://doi.org/10.1007/978-3-030-36030-6_4
20. Garg, S., Gentry, C., Halevi, S., Raykova, M., Sahai, A., Waters, B.: Candidate indistinguishability obfuscation and functional encryption for all circuits. SIAM J. Comput. **45**(3), 882–929 (2016)
21. Goyal, R., Koppula, V., Waters, B.: Lockable obfuscation. In: 2017 IEEE 58th Annual Symposium on Foundations of Computer Science (FOCS), pp. 612–621. IEEE (2017)
22. Karnouskos, S.: Stuxnet worm impact on industrial cyber-physical system security. In: 37th Annual Conference of the IEEE Industrial Electronics Society, IECON 2011, pp. 4490–4494. IEEE (2011)
23. Kesarwani, M., Mukhoty, B., Arya, V., Mehta, S.: Model extraction warning in MLaaS paradigm. In: Proceedings of the 34th Annual Computer Security Applications Conference, pp. 371–380 (2018)
24. Lee, T., Edwards, B., Molloy, I., Su, D.: Defending against model stealing attacks using deceptive perturbations. arXiv preprint arXiv:1806.00054 (2018)
25. Tai, R.K.H., Ma, J.P.K., Zhao, Y., Chow, S.S.M.: Privacy-preserving decision trees evaluation via linear functions. In: Foley, S.N., Gollmann, D., Snekkenes, E. (eds.) ESORICS 2017. LNCS, vol. 10493, pp. 494–512. Springer, Cham (2017). https://doi.org/10.1007/978-3-319-66399-9_27
26. Pal, S., Gupta, Y., Shukla, A., Kanade, A., Shevade, S., Ganapathy, V.: A framework for the extraction of deep neural networks by leveraging public data. arXiv preprint arXiv:1905.09165 (2019)
27. Quiring, E., Arp, D., Rieck, K.: Forgotten siblings: unifying attacks on machine learning and digital watermarking. In: 2018 IEEE European Symposium on Security and Privacy (EuroS&P), pp. 488–502. IEEE (2018)
28. Silverstein, C., Shieber, S.M.: Predicting individual book use for off-site storage using decision trees. Libr. Q. **66**(3), 266–293 (1996)
29. Tramèr, F., Zhang, F., Juels, A., Reiter, M.K., Ristenpart, T.: Stealing machine learning models via prediction APIs. In: 25th USENIX Security Symposium (USENIX Security 2016), pp. 601–618 (2016)
30. Wee, H.: On obfuscating point functions. In: Proceedings of the Thirty-Seventh Annual ACM Symposium on Theory of Computing, pp. 523–532 (2005)
31. Wichs, D., Zirdelis, G.: Obfuscating compute-and-compare programs under LWE. In: 2017 IEEE 58th Annual Symposium on Foundations of Computer Science (FOCS), pp. 600–611. IEEE (2017)
32. Zheng, H., Ye, Q., Hu, H., Fang, C., Shi, J.: BDPL: a boundary differentially private layer against machine learning model extraction attacks. In: Sako, K., Schneider, S., Ryan, P.Y.A. (eds.) ESORICS 2019. LNCS, vol. 11735, pp. 66–83. Springer, Cham (2019). https://doi.org/10.1007/978-3-030-29959-0_4

Privacy-Preserving Plagiarism Checking

Nidhish Bhimrajka, Sujit Chakrabarti, Ashish Choudhury(✉),
and Supreeth Varadarajan

International Institute of Information Technology Bangalore, Bengaluru, India
{nidhish.bhimrajka,sujitkc,ashish.choudhury,
supreeth.varadarajan}@iiitb.ac.in

Abstract. Plagiarism is a pressing issue in academia and industry, where individuals often modify existing content without providing due credit to the original creator and unethically claiming the work to be their own. Existing plagiarism detection tools require code owners to share their code, raising privacy and intellectual property concerns. This paper presents a privacy-preserving protocol for plagiarism detection that eliminates the need for code disclosure during similarity computation. Our protocol builds upon the winnowing-based method for efficient and accurate structural similarity computation. It allows plagiarism detection while preserving code privacy, making it valuable for scenarios where code owners are reluctant to reveal their source code.

Keywords: Privacy-preserving plagiarism · cosine similarity · secret sharing · multiparty computation · shuffling

1 Introduction

The issue of code plagiarism, where individuals copy code segments without proper attribution, has become a significant concern. While several similarity-checking algorithms exist [6,10–12], they typically require the source code to be available in *clear text* for comparison. This poses a problem, particularly when proprietary software is accused of plagiarism. Extensive research has been conducted in the field of similarity checking, but the challenge of developing a *privacy-preserving* plagiarism checker that can compute similarity without accessing the clear source code remains unaddressed. Motivated by this, we pose the following question:

Is there a plagiarism-checking protocol that can securely compute the similarity between two source codes without revealing any additional information about the source codes?

A. Choudhury—This research is an outcome of the R&D work undertaken in the project under the Visvesvaraya PhD Scheme of the Ministry of Electronics & Information Technology, Government of India, being implemented by Digital India Corporation.

A. Chattopadhyay et al. (Eds.): INDOCRYPT 2023, LNCS 14460, pp. 105–125, 2024.
https://doi.org/10.1007/978-3-031-56235-8_6

We present the *first* protocol that fulfils the above requirement. The protocol involves two *clients* who wish to find the "similarity score" between their respective codes without revealing any additional information about their individual codes. In order to achieve this objective, we employ the method of secure multiparty computation in conjunction with a plagiarism-checking protocol to ensure both the accuracy of plagiarism scores and the privacy of the clients' code. Following [13], we use the method of *syntax-tree fingerprinting* to generate fingerprints for the clients' codes and then find the "cosine similarity" between these two vectors of fingerprints. The work of [13] generates the fingerprints for *three* different levels: *level 0*, which means the *entire* syntax tree, and *level 1* and *level 2*, which represent sub-trees of decreasing depth. Using these fingerprints, the cosine similarity is measured individually for the clients' codes at the three levels, and the weighted sum is considered. For simplicity, in our privacy-preserving protocol, we focus on the fingerprints generated at *level 0* and compute the cosine similarity between these fingerprints. Apart from the detailed theoretical analysis of our protocol, we also implement our protocol. Through experimental evaluations and analysis, we demonstrate the effectiveness and practical efficiency of our protocol.

Privacy-Preserving Plagiarism Checking via MPC. We deploy techniques from the domain of *multiparty computation* (MPC) [2,7,14] in our protocol. Informally, an MPC protocol allows a set of mutually distrusting parties to securely perform *any* computation over their private inputs, without revealing anything additional about their inputs. Ever since its inception, a plethora of work has investigated the MPC problem in various settings, and several interesting and fundamental results have been formulated regarding the possibility and feasibility of MPC. Indeed, a privacy-preserving protocol for plagiarism checking follows directly from the known results for MPC for any generic function [5,14]. However, the resultant protocol *may not* be practically efficient. This is because all *generic* MPC protocols abstract the underlying computation by some arithmetic circuit over some algebraic structure (which could be a ring or a field) and then securely evaluate each gate in the circuit. One *may not* know the most efficient arithmetic circuit representation for the cosine similarity function, and consequently, the derived protocol *may not* be the most efficient. We instead focus on getting a *practically-efficient* protocol by exploiting the "structure" of the cosine similarity function and deploying tailor-made privacy-preserving techniques, *customized* for the cosine similarity function.

1.1   Technical Overview

We follow the principle of *secure outsourced computation* via *server-aided computation* and envision a scenario, where apart from the clients C_1 and C_2, we have *three* "helper" servers S_0, S_1 and S_2. And the goal is to design a protocol that requires a "minimal" overhead for the clients, with "bulk" of the computation being performed by the servers. Each client holds one of the codes, and they outsource the computation of cosine similarity to the servers in a privacy-preserving way by sending necessary information. We assume an adversarial model where *at most one* client and one server can collude and get corrupted by a *passive* (semi-honest) adversary. To securely share information to the servers, we exploit the properties of the efficient ASTRA secure 3-Party computation (3PC) [4], which allows *three* mutually distrusting parties to securely compute any function, even if one of the servers get passively corrupt. In a nutshell, the clients secret-share their fingerprint vectors, generated using syntax tree fingerprinting, among three servers. This information is all the clients need to send to the servers. The servers then jointly execute a protocol to securely compute the cosine similarity score between the two secret-shared vectors without gaining any additional information about the fingerprint vectors. Finally, the cosine similarity score between the codes is returned to the clients.

In more detail, each C_i has input $\{(f_{i,\ell}, v_{i,\ell})\}_{\ell=1,...,L} \in \mathbb{F}^{2L}$, where $L \geq 1$ and is known *publicly* and where $\mathbb{F}$ is a finite *field*. Here $f_{i,\ell}$ denotes a fingerprint and $v_{i,\ell}$ denotes the corresponding "magnitude/value". Moreover, all values $f_{i,1}, \ldots, f_{i,L}$ are *distinct*. For $i = 1, 2$, let $L_i \overset{def}{=} \sqrt{v_{i,1}^2 + \ldots + v_{i,L}^2}$.[1] Furthermore, let $(f_{1,q_1}, f_{2,r_1}), \ldots, (f_{1,q_{\mathsf{Com}}}, f_{2,r_{\mathsf{Com}}})$ be the pairwise common fingerprints. Then the protocol outputs CS and Com for both the clients without revealing any additional information, where $0 \leq \mathsf{Com} \leq L$ and where

$$\mathsf{CS} \overset{def}{=} \frac{(v_{1,q_1} \cdot v_{2,r_1}) + \ldots + (v_{1,q_{\mathsf{Com}}} \cdot v_{2,r_{\mathsf{Com}}})}{\mathsf{L}_1 \cdot \mathsf{L}_2}.$$

The main challenge to achieving the above goal is to identify the common fingerprints in a privacy-preserving way without disclosing any additional information. The protocol proceeds as follows. Let $\vec{F}_i = (f_{i,1}, f_{i,2}, ..., f_{i,m})$ be the *fingerprint vector* and let $\vec{V}_i = (v_{i,1}, v_{i,2}, ..., v_{i,m})$ be the *value vector*. Each client C_i secret-shares $\vec{F}_i$ and $\vec{V}_i$ among the servers, along with the inverse of the magnitude of their value vectors IL_i; i.e. $\mathsf{IL}_i = \mathsf{L}_i^{-1}$. The servers then jointly compute AES-encryption of the elements of $\vec{F}_1$ and $\vec{F}_2$ in a secret-shared fashion to get the shares of encrypted vectors $\vec{\mathbf{F}}_1$ and $\vec{\mathbf{F}}_2$. The AES encryption is performed under a *random* key, which will be apriori secret-shared among the servers and will be available as part of a *trusted* one-time set-up. The key is assumed to be secret-shared in such a way that *no single* server will learn the key from its share of the key. The computation of secret-shared encrypted vectors $\vec{\mathbf{F}}_1$ and $\vec{\mathbf{F}}_2$ from $\vec{F}_1$ and $\vec{F}_2$ is performed using the existing ASTRA 3PC protocol. Note that since each component in $\vec{F}_i$ is distinct, consequently, each element of $\vec{\mathbf{F}}_i$ will also be distinct. Consequently, the servers *may* reconstruct the secret-shared encrypted vectors $\vec{\mathbf{F}}_1$ and $\vec{\mathbf{F}}_2$ and then identify the pairwise common components in the encrypted vectors. Unfortunately, doing so *will* reveal information about the client's fingerprint vectors, leading to a privacy breach. This is because collusion between

---

[1] For actual codes, both fingerprints as well as values will be integers, which will be encoded as field elements (since the ASTRA 3PC performs computation over a field/ring). Each client can first compute $\sqrt{v_{i,1}^2 + \ldots + v_{i,L}^2}$ over integers and then encode the resultant value as an element of $\mathbb{F}$ to get L_i.

a corrupt server and one of the clients will lead to the knowledge of the *exact* position of the common elements in the fingerprint vectors, which constitutes a leakage and a potential privacy breach.

To deal with the above leakage, instead of reconstructing the secret-shared encrypted vectors $\vec{F}_1$ and $\vec{F}_2$, the servers randomly *shuffle* the secret-shared vectors $\vec{F}_1$ and $\vec{F}_2$ and the corresponding secret-shared value vectors $\vec{V}_1$ and $\vec{V}_2$. The shuffling happens in a privacy-preserving fashion, with any potentially corrupt server learning no information about the underlying permutation. We present an efficient, secure 3PC protocol to enable the shuffling. Once the shuffling is done, the servers reconstruct the secret-shared shuffled encrypted fingerprint vectors and then identify the common components. Note that since the comparison happens over shuffled encrypted vectors, the potential leakage no longer happens, as a potentially corrupt server will *not* know the exact permutation applied as part of the shuffling. Once the common components are identified, the corresponding secret-shared elements from the shuffled value vectors are multiplied using the secure multiplication protocol of ASTRA 3PC. This is followed by further multiplying with the secret-shared IL_1 and IL_2, which leads to the servers learning secret-shared CS, after which it is reconstructed and revealed to the clients. The protocol is depicted pictorially in Fig. 1.

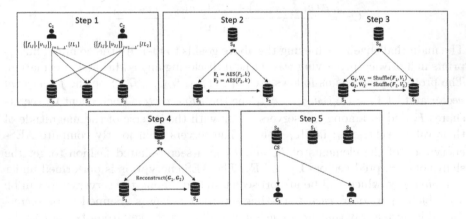

Fig. 1. Pictorial Depiction of the Steps of the Secure Cosine Similarity protocol

Implementation. In our secure cosine similarity algorithm, the fingerprint domain consists of a 128-bit binary ring, while the domain of the values corresponding to each fingerprint is set to the field $\mathscr{Z}_p$, where $p = 2^{61} - 1$. Once we obtain the similarity of the two codes using our protocol, we compare it with the similarity calculated over clear text. To measure the time complexity, we utilize the combined length of the fingerprints as a parameter.

We conduct tests on a customized dataset comprising Python codes of varying lengths to evaluate our work. This dataset is derived from the dataset used in the implementation of [13]. Looking ahead, we determine that the protocol's runtime is linear in the combined fingerprints generated from the two codes. The time required to securely compare two code pairs, each consisting of an average of 70 and 250 lines of code, respectively, is approximately 6 and 21 s respectively. It is worth noting that the percentage error between the plaintext similarity and the secure similarity measurement is approximately of the order 10^{-6}. This discrepancy, albeit negligible, arises due to the inherent errors introduced by secure floating-point computations.

2   Preliminaries and Definitions

Our protocols perform all computations over some finite field $(\mathbb{F}, +, \cdot)$. We assume the pairwise-secure channel model, where there exist two clients C_1 and C_2 and three helper servers S_0, S_1 and S_2, different from C_1 and C_2. Each client C_i has input $\{(f_{i,\ell}, v_{i,\ell})\}_{\ell=1,\ldots,L} \in \mathbb{F}^{2L}$, where $L \geq 1$ and is known publicly. Moreover, all values of $f_{i,1}, \ldots, f_{i,L}$ are *distinct*. The distrust in the system is modelled by a *computationally-bounded* (PPT) adversary Adv, who can corrupt *at most* one client and one server. That is, the Adv can pick any one subset from the *adversary structure* $\mathcal{Z}_{\mathsf{Adv}} = \{\emptyset, \{C_1\}, \{C_2\}, \{S_0\}, \{S_1\}, \{S_2\}, \{C_1, S_0\}, \{C_1, S_1\}, \{C_1, S_2\}, \{C_2, S_0\}, \{C_2, S_1\}, \{C_2, S_2\}\}$ for corruption. We consider semi-honest (passive) corruption, where the parties under the control of Adv *do not* deviate from the protocol instructions; however, the computation and communication of such parties can be observed by the adversary. We call the parties under the adversary's control as *corrupt* and the remaining parties as *honest*. We assume the existence of a pseudorandom permutation (PRP) $F : \{0, 1\}^{\ell_{key}} \times \mathbb{F} \to \mathbb{F}$ [9], which is emulated by AES in our implementation with appropriate domain and codomain.

2.1   ASTRA 3-Party Secret Sharing

We follow the secret-sharing semantics of ASTRA secure 3-*party computation* (3PC) [4], where all the values during the secure computation of cosine similarity are secret-shared among the servers S_0, S_1 and S_2, in such a way that to learn the underlying secret-shared value, we need the shares of *at least* two servers.

Definition 1 ($[\cdot]$-sharing [4]). *A value $s \in \mathbb{F}$ is said to be $[\cdot]$-shared among the servers S_0, S_1 and S_2, if there exist $\mathsf{m}_s, \lambda_s, \lambda_{s,1}, \lambda_{s,2} \in \mathbb{F}$ where $s = \mathsf{m}_s - \lambda_s$ and $\lambda_s = \lambda_{s,1} + \lambda_{s,2}$, such that the following holds.*

- S_0 *has* $(\lambda_{s,1}, \lambda_{s,2})$.
- S_1 *has* $(\mathsf{m}_s, \lambda_{s,1})$.
- S_2 *has* $(\mathsf{m}_s, \lambda_{s,2})$.

We use the notation $[s]_{S_0}, [s]_{S_1}$ and $[s]_{S_2}$ to denote the shares of s, corresponding to a $[\cdot]$-sharing of s, held by S_0, S_1 and S_2 respectively, where $[s]_{S_0} = (\lambda_{s,1}, \lambda_{s,2})$,

$[s]_{S_1} = (m_s, \lambda_{s,1})$ and $[s]_{S_2} = (m_s, \lambda_{s,2})$. A $[\cdot]$-sharing of s is called *random*, if the corresponding $\lambda_{s,1}$ and $\lambda_{s,2}$ are randomly chosen.[2]

A vector of values $\vec{S} = (S^{(1)}, \ldots, S^{(N)}) \in \mathbb{F}^N$ is said to be $[\cdot]$-shared, if each $S^{(i)} \in \vec{S}$ is $[\cdot]$-shared. Note that $[\cdot]$-sharing is *linear* and allows the servers to *non-interactively* compute any *linear* function over $[\cdot]$-shared inputs. In more detail, given $[a]$ and $[b]$ and publicly-known constants $c_1, c_2 \in \mathbb{F}$, then S_0, S_1 and S_2 can *locally* compute their shares, corresponding to $[c_1 \cdot a + c_2 \cdot b]$ from their shares corresponding to $[a]$ and $[b]$.

In general, let $g : \mathbb{F}^M \to \mathbb{F}^N$ be a publicly known linear function where $(y^{(1)}, \ldots, y^{(N)}) = g(x^{(1)}, \ldots, x^{(M)})$. Then given $[x^{(1)}], \ldots, [x^{(M)}]$, the servers S_0, S_1 and S_2 can non-interactively compute their shares corresponding to $[y^{(1)}], \ldots, [y^{(N)}]$, by applying the function g to their shares corresponding to $[x^{(1)}], \ldots, [x^{(M)}]$. In the rest of the paper, we will say that "*the servers locally compute* $([y^{(1)}], \ldots, [y^{(N)}]) = g([x^{(1)}], \ldots, [x^{(M)}])$" to mean the above.

Reconstructing a $[\cdot]$-Shared Value. Let s be a value which is $[\cdot]$-shared, with the servers S_0, S_1 and S_2 holding the shares $[s]_{S_0} = (\lambda_{s,1}, \lambda_{s,2})$, $[s]_{S_1} = (m_s, \lambda_{s,1})$ and $[s]_{S_2} = (m_s, \lambda_{s,2})$ respectively. The servers can then reconstruct s with a minimal communication overhead as follows: server S_0 sends $\lambda_{s,2}$ to S_1, server S_1 sends $\lambda_{s,1}$ to S_2 and server S_2 sends m_s to S_0. As a result, all the servers will have $m_s, \lambda_{s,1}$ and $\lambda_{s,2}$ and hence they can output $s = m_s - (\lambda_{s,1} + \lambda_{s,2})$. We denote this protocol as $\Pi_{\mathsf{RecPub}}([s])$, which takes one round of communication and incurs a communication of 3 elements from $\mathbb{F}$.

2.2   Security Definition of MPC Protocols

We prove the security of our protocols using the standard *simulation-paradigm* [8]. In this paradigm, we show that whatever Adv learns by participating in a protocol can be efficiently recreated/simulated solely based on the function output and the inputs of the corrupt parties. In our protocol for computing cosine similarity, we will use certain sub-protocols for specific subtasks. To prove the security of the overall protocol in a modular fashion, we will use the *sequential-composition theorem* of [3], which we will briefly discuss next.

The Sequential Modular Composition. Let Π_f be a protocol for securely computing some given function f and suppose Π_f uses a subprotocol Π_g for securely computing another function, say g. Then the composition theorem states that it is sufficient to consider the execution of Π_f in a *hybrid* model, where some trusted third party (TTP) is designated with the task of computing g, instead of the parties running the subprotocol Π_g. This model is denoted as the *g-hybrid model* since it involves both interaction among the parties, as well as the interaction between the parties and the TTP for computing g.

---

[2] This will automatically imply that m_s will be also a random element of $\mathbb{F}$.

2.3 Various Subprotocols Used in Our Protocol

For securely computing cosine similarity, we will use certain sub-protocols. Here, we describe the properties required from these sub-protocols. Looking ahead, we will prove the security of our cosine-similarity protocol using the modular composition theorem. To facilitate this, we will abstract out the underlying functions, which are securely computed by the sub-protocols.

Secure Shuffle Protocol

We assume the existence of protocol Π_{Shuffle}, with the following inputs and outputs.

- **Inputs:** The inputs to the protocol are vectors $\vec{A} = (A^{(1)}, \ldots, A^{(L)}) \in \mathbb{F}^L$ and $\vec{B} = (B^{(1)}, \ldots, B^{(L)}) \in \mathbb{F}^L$, both of which are $[\cdot]$-shared among S_0, S_1 and S_2.
- **Outputs:** Let $\sigma : \mathbb{F}^L \to \mathbb{F}^L$ be a *random* permutation and let $\vec{A} = \sigma(\vec{A})$ and $\vec{B} = \sigma(\vec{B})$. The protocol outputs a random $[\cdot]$-sharing of $\vec{A}$ and $\vec{B}$.

Informally, the security of Π_{Shuffle} means that Adv *does not* learn anything additional about $\vec{A}, \vec{B}, \vec{A}, \vec{B}$ and σ, except that $\vec{A} = \sigma(\vec{A})$ and $\vec{B} = \sigma(\vec{B})$. To formally capture this, we envision a function $\mathcal{F}_{\text{Shuffle}}$, where:

$$\mathcal{F}_{\text{Shuffle}}\left(\bot, \bot, ([\vec{A}]_{S_0}, [\vec{B}]_{S_0}), ([\vec{A}]_{S_1}, [\vec{B}]_{S_1}), ([\vec{A}]_{S_2}, [\vec{B}]_{S_2})\right)$$
$$= \left(\bot, \bot, ([\vec{A}]_{S_0}, [\vec{B}]_{S_0}), ([\vec{A}]_{S_1}, [\vec{B}]_{S_1}), ([\vec{A}]_{S_2}, [\vec{B}]_{S_2})\right),$$

where $[\vec{A}]$ and $[\vec{B}]$ denotes a random $[\cdot]$-sharing of $\vec{A}$ and $\vec{B}$. Namely, in the function $\mathcal{F}_{\text{Shuffle}}$, the clients have no input and output (denoted by $\bot$), while the servers participate with their respective shares of $[\vec{A}]$ and $[\vec{B}]$ as inputs and obtain their respective shares corresponding to a random $[\cdot]$-sharing of $\vec{A}$ and $\vec{B}$ as output. Formally, the security of the protocol Π_{Shuffle} means that the protocol Π_{Shuffle} is a secure protocol for computing $\mathcal{F}_{\text{Shuffle}}$, in the presence of an adversary Adv who can corrupt any one subset of parties from the adversary structure $\mathcal{Z}_{\text{Adv}} = \{\emptyset, \{C_1\}, \{C_2\}, \{S_0\}, \{S_1\}, \{S_2\}, \{C_1, S_0\}, \{C_1, S_1\}, \{C_1, S_2\}, \{C_2, S_0\}, \{C_2, S_1\}, \{C_2, S_2\}\}$. Even though one can obtain an instantiation of Π_{Shuffle} from the ASTRA 3PC protocol for securely computing any generic function over $[\cdot]$-shared inputs, looking ahead, we will give a *customized* instantiation of Π_{Shuffle}.

Securely Computing PRP Output

We assume the existence of protocol Π_{PRP} with the following inputs and outputs.

- **Inputs:** The input to the protocol is a pair of vectors $\vec{X} = (x^{(1)}, \ldots, x^{(L)}) \in \mathbb{F}^L$ and $\vec{Y} = (y^{(1)}, \ldots, y^{(L)}) \in \mathbb{F}^L$, where each $x^{(\ell)}$ and $y^{(\ell)}$ is $[\cdot]$-shared.
- **Outputs:** Let $\vec{X} = (\mathbf{x}^{(1)}, \ldots, \mathbf{x}^{(L)})$ and $\vec{Y} = (\mathbf{y}^{(1)}, \ldots, \mathbf{y}^{(L)})$, where for $\ell = 1, \ldots, L$, the value $\mathbf{x}^{(\ell)} \stackrel{def}{=} F_k(x^{(\ell)})$ and $\mathbf{y}^{(\ell)} \stackrel{def}{=} F_k(y^{(\ell)})$. Here $F : \{0, 1\}^{\ell_{key}} \times$

$\mathbb{F} \to \mathbb{F}$ is a publicly known PRP and k is a randomly chosen key from the set $\{0,1\}^{\ell_{key}}$, which is randomly $[\cdot]$-shared among the servers. The output of the protocol is a *random* $[\cdot]$-sharing of each element $\mathbf{x}^{(\ell)}$ and $\mathbf{y}^{(\ell)}$.

Informally, the security of Π_{PRP} means that Adv *does not* learn anything additional about $\vec{X}, \vec{Y}, \vec{\mathbf{X}}, \vec{\mathbf{Y}}$ and k, except that each $\mathbf{x}^{(\ell)} \stackrel{def}{=} F_k(x^{(\ell)})$ and $\mathbf{y}^{(\ell)} \stackrel{def}{=} F_k(y^{(\ell)})$. To formally capture the security requirements of the protocol Π_{PRP}, we envision a function $\mathcal{F}_{\mathsf{PRP}}$, where:

$$\mathcal{F}_{\mathsf{PRP}}\left(\bot, \bot, \{[x^{(\ell)}]_{\mathsf{S}_0}, [y^{(\ell)}]_{\mathsf{S}_0}\}_{\ell=1,\ldots,L}, \{[x^{(\ell)}]_{\mathsf{S}_1}, [y^{(\ell)}]_{\mathsf{S}_1}\}_{\ell=1,\ldots,L}, \{[x^{(\ell)}]_{\mathsf{S}_2}, [y^{(\ell)}]_{\mathsf{S}_2}\}_{\ell=1,\ldots,L}\right)$$
$$= \left(\bot, \bot, \{[\mathbf{x}^{(\ell)}]_{\mathsf{S}_0}, [\mathbf{y}^{(\ell)}]_{\mathsf{S}_0}\}_{\ell=1,\ldots,L}, \{[\mathbf{x}^{(\ell)}]_{\mathsf{S}_1}, [\mathbf{y}^{(\ell)}]_{\mathsf{S}_1}\}_{\ell=1,\ldots,L}, \{[\mathbf{x}^{(\ell)}]_{\mathsf{S}_2}, [\mathbf{y}^{(\ell)}]_{\mathsf{S}_2}\}_{\ell=1,\ldots,L}\right).$$

Namely, the clients have no input and output, while the servers participate with their shares of $[x^{(\ell)}]$ and $[y^{(\ell)}]$ as input and obtain their shares corresponding to a random $[\cdot]$-sharing of $\mathbf{x}^{(\ell)}$ and a random $[\cdot]$-sharing of $\mathbf{y}^{(\ell)}$ as output. The security of Π_{PRP} means that Π_{PRP} is a secure protocol for computing $\mathcal{F}_{\mathsf{PRP}}$, in the presence of an adversary Adv who can corrupt any one subset of parties from the adversary structure $\mathcal{Z}_{\mathsf{Adv}} = \{\emptyset, \{\mathsf{C}_1\}, \{\mathsf{C}_2\}, \{\mathsf{S}_0\}, \{\mathsf{S}_1\}, \{\mathsf{S}_2\}, \{\mathsf{C}_1, \mathsf{S}_0\}, \{\mathsf{C}_1, \mathsf{S}_1\}, \{\mathsf{C}_1, \mathsf{S}_2\}, \{\mathsf{C}_2, \mathsf{S}_0\}, \{\mathsf{C}_2, \mathsf{S}_1\}, \{\mathsf{C}_2, \mathsf{S}_2\}\}$. Our instantiation of Π_{PRP} is derived from the ASTRA 3PC protocol, which allows us to securely compute any generic function over $[\cdot]$-shared inputs.

Securely Multiplying Secret-Shared Values

We assume the existence of protocol Π_{Mult} with the following inputs and outputs.

- **Input:** The input to the protocol is a pair of elements (a, b), each of which is $[\cdot]$-shared.
- **Output:** Let $c = a \cdot b$. The output of the protocol is a random $[\cdot]$-sharing of c.

Informally, the security of Π_{Mult} means that Adv *does not* learn anything additional about the triplet (a, b, c), except that $c = a \cdot b$. To formally capture the security requirements of the protocol Π_{Mult}, we envision a function $\mathcal{F}_{\mathsf{Mult}}$, where:

$$\mathcal{F}_{\mathsf{Mult}}\left(\bot, \bot, \{[a]_{\mathsf{S}_0}, [b]_{\mathsf{S}_0}\}, \{[a]_{\mathsf{S}_1}, [b]_{\mathsf{S}_1}\}, \{[a]_{\mathsf{S}_2}, [b]_{\mathsf{S}_2}\}\right) = \left(\bot, \bot, [c]_{\mathsf{S}_0}, [c]_{\mathsf{S}_1}, [c]_{\mathsf{S}_2}\right).$$

Namely, the clients have no input and output, while the servers participate with their respective shares of $[a], [b]$ as input and obtain their respective shares corresponding to a random $[\cdot]$-sharing of c as output, where $c = a \cdot b$. The security of Π_{Mult} means that it is a secure protocol for computing $\mathcal{F}_{\mathsf{Mult}}$, in the presence of an adversary Adv who can corrupt any one subset of parties from the adversary structure $\mathcal{Z}_{\mathsf{Adv}} = \{\emptyset, \{\mathsf{C}_1\}, \{\mathsf{C}_2\}, \{\mathsf{S}_0\}, \{\mathsf{S}_1\}, \{\mathsf{S}_2\}, \{\mathsf{C}_1, \mathsf{S}_0\}, \{\mathsf{C}_1, \mathsf{S}_1\}, \{\mathsf{C}_1, \mathsf{S}_2\}, \{\mathsf{C}_2, \mathsf{S}_0\}, \{\mathsf{C}_2, \mathsf{S}_1\}, \{\mathsf{C}_2, \mathsf{S}_2\}\}$. We use the instantiation of Π_{Mult} from the ASTRA 3PC protocol, involving 2 rounds of communication, and 3 elements from $\mathbb{F}$ are communicated.

3   Computing Cosine Similarity Securely

In this section, we will present our protocol Π_{CS} (Fig. 2) for securely computing the cosine similarity between the vectors $\vec{F}_1$ and $\vec{F}_2$ held by C_1 and C_2 respectively. The inputs and outputs of the protocol are as follows.

- **Inputs:** For $i = 1, 2$, client C_i has the input $\{(f_{i,\ell}, v_{i,\ell})\}_{\ell=1,...,L}$, where $f_{i,1}, \ldots, f_{i,L}$ are distinct elements from $\mathbb{F}$. The servers have *no* input.
- **Outputs:** For $i = 1, 2$, let $L_i \overset{def}{=} \sqrt{v_{i,1}^2 + \ldots + v_{i,L}^2}$. Moreover, let $(f_{1,q_1}, f_{2,r_1}), \ldots, (f_{1,q_{Com}}, f_{2,r_{Com}})$ be the pairwise *common* fingerprints. The protocol outputs CS and Com for both the clients, where $0 \leq$ Com $\leq L$ and

$$CS \overset{def}{=} \frac{(v_{1,q_1} \cdot v_{2,r_1}) + \ldots + (v_{1,q_{Com}} \cdot v_{2,r_{Com}})}{L_1 \cdot L_2}.$$

To capture the security of Π_{CS}, we envision a function $\mathcal{F}_{CS}$, where

$$\mathcal{F}_{CS}\Big(\{(f_{1,\ell}, v_{1,\ell})\}_{\ell=1,...,L}, \{(f_{2,\ell}, v_{2,\ell})\}_{\ell=1,...,L}, \bot, \bot, \bot\Big) = \Big((CS, Com), (CS, Com), \bot, \bot, \bot\Big).$$

The high-level idea behind Π_{CS} and its pictorial illustration has already been discussed in Sect. 1.1. We present Π_{CS} in the *hybrid* model, assuming TTPs for securely computing the functions $\mathcal{F}_{Shuffle}, \mathcal{F}_{PRP}$ and $\mathcal{F}_{Mult}$. We will abuse the notations and denote these TTPs as $\mathcal{F}_{Shuffle}, \mathcal{F}_{PRP}$ and $\mathcal{F}_{Mult}$, respectively.

The proof of Theorem 1 will be available in the full version of the paper.

Theorem 1. *Let* $\mathcal{F}_{CS}\Big(\{(f_{1,\ell}, v_{1,\ell})\}_{\ell=1,...,L}, \{(f_{2,\ell}, v_{2,\ell})\}_{\ell=1,...,L}, \bot, \bot, \bot\Big) =$ $\Big((CS, Com), (CS, Com), \bot, \bot, \bot\Big)$, *where* C_i *has the input* $\{(f_{i,\ell}, v_{i,\ell})\}_{\ell=1,...,L}$, *with* $f_{i,1}, \ldots, f_{i,L}$ *being distinct elements from* $\mathbb{F}$ *and where*

$$CS \overset{def}{=} \frac{(v_{1,q_1} \cdot v_{2,r_1}) + \ldots + (v_{1,q_{Com}} \cdot v_{2,r_{Com}})}{L_1 \cdot L_2}, \quad L_i \overset{def}{=} \sqrt{v_{i,1}^2 + \ldots + v_{i,L}^2},$$

such that $(f_{1,q_1}, f_{2,r_1}), \ldots, (f_{1,q_{Com}}, f_{2,r_{Com}})$ *are pairwise common fingerprints. Moreover, let* Adv *be an adversary, who can corrupt any one subset from the adversary structure* $\mathcal{Z}_{Adv} = \{\emptyset, \{C_1\}, \{C_2\}, \{S_0\}, \{S_1\}, \{S_2\}, \{C_1, S_0\}, \{C_1, S_1\}, \{C_1, S_2\}, \{C_2, S_0\}, \{C_2, S_1\}, \{C_2, S_2\}\}$. *Then protocol* Π_{CS} *is a secure protocol for computing* $\mathcal{F}_{CS}$ *in the* $(\mathcal{F}_{Shuffle}, \mathcal{F}_{PRP}, \mathcal{F}_{Mult})$-*hybrid model.*

The protocol makes 2 *calls to* $\mathcal{F}_{Shuffle}$, 1 *call to* $\mathcal{F}_{PRP}$ *and* Com+2 *calls to* $\mathcal{F}_{Mult}$, *where* Com $= \mathcal{O}(L)$. *Additionally,* $17 + 24L$ *elements from* $\mathbb{F}$ *are communicated. The protocol involves* 3 *rounds of communication.*

Protocol $\Pi_{\mathsf{CS}}(\mathsf{C}_1, \mathsf{C}_2, \mathsf{S}_0, \mathsf{S}_1, \mathsf{S}_2, \{(f_{1,\ell}, v_{1,\ell})\}_{\ell=1,\ldots,L}, \{(f_{2,\ell}, v_{2,\ell})\}_{\ell=1,\ldots,L})$

- **Step 1: Secret-sharing the inputs among the servers:** For $i = 1, 2$, client C_i does the following, on having the input $\{(f_{i,\ell}, v_{i,\ell})\}_{\ell=1,\ldots,L}$.
  - Compute $\mathsf{L}_i \stackrel{def}{=} \sqrt{v_{i,1}^2 + \ldots + v_{i,L}^2}$ and $\mathsf{IL}_i \stackrel{def}{=} \mathsf{L}_i^{-1}$. Randomly select $\lambda_{\mathsf{IL}_i,1}, \lambda_{\mathsf{IL}_i,2} \in \mathbb{F}$ and compute $\mathsf{m}_{\mathsf{IL}_i} = \mathsf{IL}_i + \lambda_{\mathsf{IL}_i,1} + \lambda_{\mathsf{IL}_i,2}$. Set $[\mathsf{IL}_i]_{\mathsf{S}_0} = (\lambda_{\mathsf{IL}_i,1}, \lambda_{\mathsf{IL}_i,2})$, $[\mathsf{IL}_i]_{\mathsf{S}_1} = (\mathsf{m}_{\mathsf{IL}_i}, \lambda_{\mathsf{IL}_i,1})$ and $[\mathsf{IL}_i]_{\mathsf{S}_2} = (\mathsf{m}_{\mathsf{IL}_i}, \lambda_{\mathsf{IL}_i,2})$.
  - For $\ell = 1, \ldots, L$, randomly select $\lambda_{f_{i,\ell},1}, \lambda_{f_{i,\ell},2} \in \mathbb{F}$ and compute $\mathsf{m}_{f_{i,\ell}} = f_{i,\ell} + \lambda_{f_{i,\ell},1} + \lambda_{f_{i,\ell},2}$. Set $[f_{i,\ell}]_{\mathsf{S}_0} = (\lambda_{f_{i,\ell},1}, \lambda_{f_{i,\ell},2})$, $[f_{i,\ell}]_{\mathsf{S}_1} = (\mathsf{m}_{f_{i,\ell}}, \lambda_{f_{i,\ell},1})$ and $[f_{i,\ell}]_{\mathsf{S}_2} = (\mathsf{m}_{f_{i,\ell}}, \lambda_{f_{i,\ell},2})$.
  - For $\ell = 1, \ldots, L$, randomly select $\lambda_{v_{i,\ell},1}, \lambda_{v_{i,\ell},2} \in \mathbb{F}$ and compute $\mathsf{m}_{v_{i,\ell}} = v_{i,\ell} + \lambda_{v_{i,\ell},1} + \lambda_{v_{i,\ell},2}$. Set $[v_{i,\ell}]_{\mathsf{S}_0} = (\lambda_{v_{i,\ell},1}, \lambda_{v_{i,\ell},2})$, $[v_{i,\ell}]_{\mathsf{S}_1} = (\mathsf{m}_{v_{i,\ell}}, \lambda_{v_{i,\ell},1})$ and $[v_{i,\ell}]_{\mathsf{S}_2} = (\mathsf{m}_{v_{i,\ell}}, \lambda_{v_{i,\ell},2})$.
  - Generate $[\mathsf{IL}_i]$, $\{[f_{i,\ell}]\}_{\ell=1,\ldots,L}$ and $\{[v_{i,\ell}]\}_{\ell=1,\ldots,L}$, by sending the following information to the different servers.
    - Send $[\mathsf{IL}_i]_{\mathsf{S}_0}, \{[f_{i,\ell}]_{\mathsf{S}_0}\}_{\ell=1,\ldots,L}$ and $\{[v_{i,\ell}]_{\mathsf{S}_0}\}_{\ell=1,\ldots,L}$ to S_0.
    - Send $[\mathsf{IL}_i]_{\mathsf{S}_1}, \{[f_{i,\ell}]_{\mathsf{S}_1}\}_{\ell=1,\ldots,L}$ and $\{[v_{i,\ell}]_{\mathsf{S}_1}\}_{\ell=1,\ldots,L}$ to S_1.
    - Send $[\mathsf{IL}_i]_{\mathsf{S}_2}, \{[f_{i,\ell}]_{\mathsf{S}_2}\}_{\ell=1,\ldots,L}$ and $\{[v_{i,\ell}]_{\mathsf{S}_2}\}_{\ell=1,\ldots,L}$ to S_2.
- **Step 2: Generating the shares of the output of the PRP on fingerprints:** The servers $\mathsf{S}_0, \mathsf{S}_1$ and S_2 call $\mathcal{F}_{\mathsf{PRP}}$ with inputs $\{[f_{1,\ell}]_{\mathsf{S}_0}, [f_{2,\ell}]_{\mathsf{S}_0}\}_{\ell=1,\ldots,L}, \{[f_{1,\ell}]_{\mathsf{S}_1}, [f_{2,\ell}]_{\mathsf{S}_1}\}_{\ell=1,\ldots,L}$ and $\{[f_{1,\ell}]_{\mathsf{S}_2}, [f_{2,\ell}]_{\mathsf{S}_2}\}_{\ell=1,\ldots,L}$ and obtain the outputs $\{[\mathsf{f}_{1,\ell}]_{\mathsf{S}_0}, [\mathsf{f}_{2,\ell}]_{\mathsf{S}_0}\}_{\ell=1,\ldots,L}, \{[\mathsf{f}_{1,\ell}]_{\mathsf{S}_1}, [\mathsf{f}_{2,\ell}]_{\mathsf{S}_1}\}_{\ell=1,\ldots,L}$ and $\{[\mathsf{f}_{1,\ell}]_{\mathsf{S}_2}, [\mathsf{f}_{2,\ell}]_{\mathsf{S}_2}\}_{\ell=1,\ldots,L}$ respectively.[a]
- **Step 3: Shuffling the PRP outputs:** Let $\vec{\mathsf{F}}_i \stackrel{def}{=} (\mathsf{f}_{i,1}, \ldots, \mathsf{f}_{i,L})$ and $\vec{V}_i \stackrel{def}{=} (v_{i,1}, \ldots, v_{i,L})$, for $i = 1, 2$. For $i = 1, 2$, the servers $\mathsf{S}_0, \mathsf{S}_1$ and S_2 call $\mathcal{F}_{\mathsf{Shuffle}}$ with inputs $([\vec{\mathsf{F}}_i]_{\mathsf{S}_0}, [\vec{V}_i]_{\mathsf{S}_0})$, $([\vec{\mathsf{F}}_i]_{\mathsf{S}_1}, [\vec{V}_i]_{\mathsf{S}_1})$ and $([\vec{\mathsf{F}}_i]_{\mathsf{S}_2}, [\vec{V}_i]_{\mathsf{S}_2})$ and obtain the outputs $([\vec{\mathsf{G}}_i]_{\mathsf{S}_0}, [\vec{W}_i]_{\mathsf{S}_0})$, $([\vec{\mathsf{G}}_i]_{\mathsf{S}_1}, [\vec{W}_i]_{\mathsf{S}_1})$ and $([\vec{\mathsf{G}}_i]_{\mathsf{S}_2}, [\vec{W}_i]_{\mathsf{S}_2})$ respectively.[b]
- **Step 4: Reconstructing the shuffled PRP outputs:** For $i = 1, 2$, let $\vec{\mathsf{G}}_i \stackrel{def}{=} (\mathsf{g}_{i,1}, \ldots, \mathsf{g}_{i,L})$. For $\ell = 1, \ldots, L$, the servers participate in instances $\Pi_{\mathsf{RecPub}}([\mathsf{g}_{i,\ell}])$ of Π_{RecPub}, to reconstruct $\mathsf{g}_{i,\ell}$.
- **Step 5: Computing the cosine similarity:** The servers do the following.
  - Compare the elements of $\vec{\mathsf{G}}_1$ and $\vec{\mathsf{G}}_2$ to find the common elements in $\vec{\mathsf{G}}_1$ and $\vec{\mathsf{G}}_2$. Let Com be the number of common elements, where $0 \leq \mathsf{Com} \leq L$. Moreover, let $(i_1, j_1), \ldots, (i_{\mathsf{Com}}, j_{\mathsf{Com}})$ be the indices of the common elements such that $\mathsf{g}_{1,i_k} = \mathsf{g}_{2,j_k}$, for $k = 1, \ldots, \mathsf{Com}$.
  - For $i = 1, 2$, let $\vec{W}_i \stackrel{def}{=} (w_{i,1}, \ldots, w_{i,L})$. For $k = 1, \ldots, \mathsf{Com}$, the servers $\mathsf{S}_0, \mathsf{S}_1$ and S_2 call $\mathcal{F}_{\mathsf{Mult}}$ with inputs $\{[w_{1,i_k}]_{\mathsf{S}_0}, [w_{2,j_k}]_{\mathsf{S}_0}\}, \{[w_{1,i_k}]_{\mathsf{S}_1}, [w_{2,j_k}]_{\mathsf{S}_1}\}$ and $\{[w_{1,i_k}]_{\mathsf{S}_2}, [w_{2,j_k}]_{\mathsf{S}_2}\}$ and obtain the outputs $[\alpha_k]_{\mathsf{S}_0}, [\alpha_k]_{\mathsf{S}_1}$ and $[\alpha_k]_{\mathsf{S}_2}$ respectively, where $\alpha_k = w_{1,i_k} \cdot w_{2,j_k}$.[c]
  - The servers locally compute $[\alpha] = [\alpha_1] + \ldots + [\alpha_{\mathsf{Com}}]$.
  - The servers call $\mathcal{F}_{\mathsf{Mult}}$ with inputs $\{[\mathsf{IL}_2]_{\mathsf{S}_0}, [\mathsf{IL}_1]_{\mathsf{S}_0}\}, \{[\mathsf{IL}_2]_{\mathsf{S}_1}, [\mathsf{IL}_1]_{\mathsf{S}_1}\}$ and $\{[\mathsf{IL}_2]_{\mathsf{S}_2}, [\mathsf{IL}_1]_{\mathsf{S}_2}\}$ and obtain the outputs $[\beta]_{\mathsf{S}_0}, [\beta]_{\mathsf{S}_1}$ and $[\beta]_{\mathsf{S}_2}$ respectively, where $\beta = \mathsf{IL}_2 \cdot \mathsf{IL}_1 = (\mathsf{L}_1 \cdot \mathsf{L}_2)^{-1}$.
  - The servers call $\mathcal{F}_{\mathsf{Mult}}$ with inputs $\{[\alpha]_{\mathsf{S}_0}, [\beta]_{\mathsf{S}_0}\}, \{[\alpha]_{\mathsf{S}_1}, [\beta]_{\mathsf{S}_1}\}$ and $\{[\alpha]_{\mathsf{S}_2}, [\beta]_{\mathsf{S}_2}\}$ and obtain the outputs $[\mathsf{CS}]_{\mathsf{S}_0}, [\mathsf{CS}]_{\mathsf{S}_1}$ and $[\mathsf{CS}]_{\mathsf{S}_2}$ respectively, where $\mathsf{CS} = \alpha \cdot \beta$.
  - The servers execute the protocol $\Pi_{\mathsf{RecPub}}([\mathsf{CS}])$ and publicly reconstruct CS. Server S_0 then sends $(\mathsf{CS}, \mathsf{Com})$ to C_1 and C_2, who output $(\mathsf{CS}, \mathsf{Com})$.

---

[a] Recall that the clients C_1 and C_2 have *no* inputs and outputs for $\mathcal{F}_{\mathsf{PRP}}$.
[b] Recall that the clients C_1 and C_2 have *no* inputs and outputs for $\mathcal{F}_{\mathsf{Shuffle}}$.
[c] Recall that the clients C_1 and C_2 have *no* inputs and outputs for $\mathcal{F}_{\mathsf{Mult}}$.

Fig. 2. Protocol for securely computing $\mathcal{F}_{\mathsf{CS}}$ in the $(\mathcal{F}_{\mathsf{Shuffle}}, \mathcal{F}_{\mathsf{PRP}}, \mathcal{F}_{\mathsf{Mult}})$-hybrid model

4   Secure Shuffle Protocol

In this section, we present our instantiation of the protocol Π_{Shuffle} for securely computing the function $\mathcal{F}_{\text{Shuffle}}$ (recall $\mathcal{F}_{\text{Shuffle}}$ from Sect. 2.3). The protocol Π_{Shuffle} takes input vectors $\vec{A}, \vec{B} \in \mathbb{F}^L$, which are $[\cdot]$-shared. The output of Π_{Shuffle} consists of a random $[\cdot]$-sharing of the vectors $\sigma(\vec{A})$ and $\sigma(\vec{B})$, where $\sigma : \mathbb{F}^L \to \mathbb{F}^L$ represents a *random* permutation *unknown* to the servers. The idea behind Π_{Shuffle} is as follows: let σ_{01}, σ_{12} and σ_{02} denote random permutations over $\mathbb{F}^L$, known to the *pair* of servers $(S_0, S_1), (S_1, S_2)$ and (S_0, S_2) respectively. Notice that *each* server will know *exactly* two out of the three permutations. Let σ be the *composition* of the permutations σ_{02}, σ_{12} and σ_{01}. That is:

$$\sigma \stackrel{def}{=} (\sigma_{02} \, o \, \sigma_{12} \, o \, \sigma_{01})$$

It then follows that σ will be a *random* permutation over $\mathbb{F}^L$, with Adv being unaware of σ, since a corrupt server will *not* know the random permutation σ_{ij}, known only to the *honest* servers S_i and S_j. The task of computing a random $[\cdot]$-sharing of $\sigma(\vec{A})$ and $\sigma(\vec{B})$ then reduces to the problem of letting a *pair* of servers S_i and S_j to securely apply a random permutation σ_{ij} known *only* to S_i and S_j on secret-shared vectors and generate a random secret-sharing of the resultant permuted vectors. Motivated by this, we next design a protocol Π_{Helper}, which precisely does this. Looking ahead, in protocol Π_{Shuffle}, we will call the protocol Π_{Helper} *three* times.

4.1   Protocol Π_{Helper}

Protocol Π_{Helper} (Fig 3) has the following inputs and outputs.

- **Inputs:** Indices $i, j \in \{0, 1, 2\}$, such that $i \neq j, i < j$, and a permutation $\sigma_{ij} : \mathbb{F}^L \to \mathbb{F}^L$ which is available with S_i and S_j, and a vector $\vec{U} = (U^{(1)}, \ldots, U^{(L)}) \in \mathbb{F}^L$ which is $[\cdot]$-shared among S_0, S_1 and S_2.
- **Outputs:** Let $\vec{V} = \sigma_{ij}(\vec{U})$. The protocol outputs a random $[\cdot]$-sharing of $\vec{V}$.

Informally, the security of Π_{Helper} means that the adversary *does not* learn any additional information about $\vec{U}, \vec{V}$ and σ_{ij}, except that $\vec{V} = \sigma_{ij}(\vec{U})$. To formally capture this, we envision a function $\mathcal{F}_{\text{Helper}}$, where:

$$\mathcal{F}_{\text{Helper}}\Big(\bot, \bot, ([\vec{U}]_{S_0}, \sigma_{ij}, \text{ if } S_0 \in \{S_i, S_j\}), ([\vec{U}]_{S_1}, \sigma_{ij}, \text{ if } S_1 \in \{S_i, S_j\}),$$

$$([\vec{U}]_{S_2}, \sigma_{ij}, \text{ if } S_2 \in \{S_i, S_j\})\Big) = \Big(\bot, \bot, [\vec{V}]_{S_0}, [\vec{V}]_{S_1}, [\vec{V}]_{S_2}\Big),$$

where $[\vec{V}]$ denotes a random $[\cdot]$-sharing of $\vec{V}$. The high-level idea behind the protocol is as follows: to generate a random $[\cdot]$-sharing of $\vec{V}$, server S_0 can pick *random* "mask-shares" $\vec{\lambda}_{V,1}, \vec{\lambda}_{V,2} \in \mathbb{F}^L$ and give $\vec{\lambda}_{V,1}$ and $\vec{\lambda}_{V,2}$ to S_1 and S_2 respectively, which will complete "half" the generation of random $[\cdot]$-sharing of

$\vec{V}$. To complete the sharing, the goal will be to enable *both* S_1 and S_2 securely compute $\vec{m}_V = \vec{V} + \vec{\lambda}_{V,1} + \vec{\lambda}_{V,2}$. To achieve this remaining goal, the protocol proceeds as follows: let $[\vec{U}]_{S_0} = (\vec{\lambda}_{U,1}, \vec{\lambda}_{U,2})$, $[\vec{U}]_{S_1} = (\vec{m}_U, \vec{\lambda}_{U,1})$ and $[\vec{U}]_{S_2} = (\vec{m}_U, \vec{\lambda}_{U,2})$, where $\vec{\lambda}_{U,1}, \vec{\lambda}_{U,2}, \vec{m}_U \in \mathbb{F}^L$, such that:

$$\vec{U} = \vec{m}_U - \vec{\lambda}_{U,1} - \vec{\lambda}_{U,2}.$$

It then follows that the following holds:

$$\vec{V} = \sigma_{ij}(\vec{U}) = \sigma_{ij}(\vec{m}_U - \vec{\lambda}_{U,1} - \vec{\lambda}_{U,2})$$
$$= \sigma_{ij}(\vec{m}_U - \vec{\lambda}_{U,1}) - \sigma_{ij}(\vec{\lambda}_{U,2}).$$

Consequently, the following holds:

$$\vec{m}_V = \vec{V} + \vec{\lambda}_{V,1} + \vec{\lambda}_{V,2}$$
$$= \left(\sigma_{ij}(\vec{m}_U - \vec{\lambda}_{U,1}) + \vec{\lambda}_{V,1}\right) + \left(\vec{\lambda}_{V,2} - \sigma_{ij}(\vec{\lambda}_{U,2})\right)$$
$$= \vec{m}_{V,1} + \vec{m}_{V,2},$$

where $\vec{m}_{V,1} \overset{def}{=} \left(\sigma_{ij}(\vec{m}_U - \vec{\lambda}_{U,1}) + \vec{\lambda}_{V,1}\right)$ and $\vec{m}_{V,2} \overset{def}{=} \left(\vec{\lambda}_{V,2} - \sigma_{ij}(\vec{\lambda}_{U,2})\right)$. Thus to enable the servers S_1 and S_2 learn $\vec{m}_V$, it is sufficient to let S_1 and S_2 learn $\vec{m}_{V,1}$ and $\vec{m}_{V,2}$ respectively, after which they can exchange $\vec{m}_{V,1}$ and $\vec{m}_{V,2}$ with each other and compute $\vec{m}_V$. To enable S_1 and S_2 learn $\vec{m}_{V,1}$ and $\vec{m}_{V,2}$ *securely*, we need to follow *different* set of steps, depending upon the identity of the servers S_i and S_j in the protocol Π_{Helper}, who hold the permutation σ_{ij}. There are *three* different possibilities, depending upon whether S_i and S_j are the pair of servers (S_1, S_2), (S_0, S_1) and (S_0, S_2) respectively.

The simplest case is when (S_1, S_2) is the pair of servers (S_i, S_j). In this case, S_1 and S_2 can directly compute $\vec{m}_{V,1}$ and $\vec{m}_{V,2}$ respectively, since they have all the required information. It is easy to see that *privacy* holds. Namely, if S_0 is the *corrupt* server, then it *does not* learn anything additional, since it does not learn about $\vec{m}_{V,1}$ and $\vec{m}_{V,2}$ and hence $\vec{m}_V$. On the other hand, if S_1 (resp. S_2) is *corrupt*, then the random mask-share $\vec{\lambda}_{V,2}$ (resp. $\vec{\lambda}_{V,1}$) maintains the privacy.

Next consider the case when (S_0, S_1) constitutes the pair of servers (S_i, S_j). In this case, S_1 still *has* all the necessary information to compute $\vec{m}_{V,1}$. However, S_2 *does not* have σ_{ij} to compute $\vec{m}_{V,2}$. Fortunately, S_0 *has* the "capability" to compute $\vec{m}_{V,2}$, which it *can* send to S_2. Unfortunately, doing so may *reveal* the permutation σ_{ij} to a potentially *corrupt* S_2 which already knows $\vec{\lambda}_{V,2}$ and $\vec{\lambda}_{U,2}$. To mitigate this leakage, S_0 *randomizes* the vector $\vec{m}_{V,2}$ before sending it to S_2. In more detail, S_0 randomly picks $\vec{\gamma_1}, \vec{\gamma_2}$, subject to the condition that $\vec{\gamma_1} + \vec{\gamma_2} = 0^L$ holds, and sends $\vec{\gamma_1}$ to S_1. We then observe that the following holds:

$$\vec{m}_V = \vec{m}_V + 0^L$$

$$= \left(\vec{V} + \vec{\lambda}_{V,1} + \vec{\lambda}_{V,2}\right) + \left(\vec{\gamma_1} + \vec{\gamma_2}\right)$$

$$= \left(\sigma_{ij}(\vec{m}_U - \vec{\lambda}_{U,1}) + \vec{\lambda}_{V,1} + \vec{\gamma_1}\right) + \left(\vec{\lambda}_{V,2} - \sigma_{ij}(\vec{\lambda}_{U,2}) + \vec{\gamma_2}\right)$$

$$= (\vec{m}_{V,1} + \vec{\gamma_1}) + (\vec{m}_{V,2} + \vec{\gamma_2}),$$

Now, while S_1 will have all the information to compute $(\vec{m}_{V,1} + \vec{\gamma_1})$, server S_0 can compute $(\vec{m}_{V,2} + \vec{\gamma_2})$ and send it to S_2. The randomness of the "zero-shares" $\vec{\gamma_1}$ and $\vec{\gamma_2}$ maintains the privacy of σ_{ij} from a potentially *corrupt* S_2. On the other hand, if S_0 is the *corrupt* server, then it *does not* learn anything additional, since it does not learn about $\vec{m}_{V,1}$ and hence $\vec{m}_V$. Moreover, if S_1 is *corrupt*, then the random mask-share $\vec{\lambda}_{V,2}$ maintains the privacy.

Finally, consider the case when (S_0, S_2) constitutes the pair of servers (S_i, S_j). In this case, while S_2 can directly compute $\vec{m}_{V,2}$, server S_1 *does not* have σ_{ij} to compute $\vec{m}_{V,1}$. Unfortunately, *unlike* the previous case, S_0 also *cannot* compute $\vec{m}_{V,1}$, since it *does not* have $\vec{m}_U$. The way out in this case is that we do a *different* "interpretation" for the terms $\vec{m}_{V,1}$ and $\vec{m}_{V,2}$. Namely, we observe that the following holds:

$$\vec{m}_V = \vec{m}_V$$

$$= \vec{V} + \vec{\lambda}_{V,1} + \vec{\lambda}_{V,2}$$

$$= \left(\vec{\lambda}_{V,1} - \sigma_{ij}(\vec{\lambda}_{U,1})\right) + \left(\sigma_{ij}(\vec{m}_U - \vec{\lambda}_{U,2}) + \vec{\lambda}_{V,2}\right)$$

$$= \vec{m}_{V,1} + \vec{m}_{V,2},$$

where $\vec{m}_{V,1}$ and $\vec{m}_{V,2}$ are now defined as $\left(\vec{\lambda}_{V,1} - \sigma_{ij}(\vec{\lambda}_{U,1})\right)$ and $\left(\sigma_{ij}(\vec{m}_U - \vec{\lambda}_{U,2}) + \vec{\lambda}_{V,2}\right)$ respectively. Now while S_2 can compute $\vec{m}_{V,2}$, server S_0 can compute $\vec{m}_{V,1}$ and hand it over to S_1. However, doing so, may *leak* information about the permutation σ_{ij} to a potentially *corrupt* S_1. Hence, as done in the previous case, S_0 and S_2 *randomize* $\vec{m}_{V,1}$ and $\vec{m}_{V,2}$ respectively, by adding *random* "zero-shares" $\vec{\gamma_1}$ and $\vec{\gamma_2}$ to $\vec{m}_{V,2}$ and $\vec{m}_{V,1}$ respectively, where $\vec{\gamma_1} + \vec{\gamma_2} = 0^L$.

Protocol $\Pi_{\mathsf{Helper}}(S_0, S_1, S_2, i, j, [\vec{U}]_{S_0}, [\vec{U}]_{S_1}, [\vec{U}]_{S_2}, \sigma_{ij})$

Input: S_0, S_1 and S_2 has inputs $[\vec{U}]_{S_0} = (\vec{\lambda}_{U,1}, \vec{\lambda}_{U,2})$, $[\vec{U}]_{S_1} = (\vec{m}_U, \vec{\lambda}_{U,1})$ and $[\vec{U}]_{S_2} = (\vec{m}_U, \vec{\lambda}_{U,2})$ respectively, where $\vec{\lambda}_{U,1}, \vec{\lambda}_{U,2}, \vec{m}_U \in \mathbb{F}^L$, such that $\vec{U} = \vec{m}_U - \vec{\lambda}_{U,1} - \vec{\lambda}_{U,2}$. Server S_i and S_j has an additional input, namely a permutation $\sigma_{ij} : \mathbb{F}^L \to \mathbb{F}^L$.

- **Step 1: Distributing Mask and Pad-shares:** Server S_0 does the following.
  - Randomly pick $\vec{\lambda}_{V,1}, \vec{\lambda}_{V,2} \in \mathbb{F}^L$ and send $\vec{\lambda}_{V,1}$ and $\vec{\lambda}_{V,2}$ to S_1 and S_2 respectively.
  - If $S_i = S_0$ and $S_j \in \{S_1, S_2\}$, then randomly pick $\vec{\gamma_1}, \vec{\gamma_2} \in \mathbb{F}^L$ such that $\vec{\gamma_1} + \vec{\gamma_2} = 0^L$. Send $\vec{\gamma_1}$ to server S_j.
- **Step 2: Computing Masked Output Vectors:** The servers do the following, depending upon the identity of S_i and S_j.
  - If $S_i = S_1$ and $S_j = S_2$
    - S_1 computes $\vec{m}_{V,1} = \left(\sigma_{ij}(\vec{m}_U - \vec{\lambda}_{U,1}) + \vec{\lambda}_{V,1} \right)$.
    - S_2 computes $\vec{m}_{V,2} = \left(\vec{\lambda}_{V,2} - \sigma_{ij}(\vec{\lambda}_{U,2}) \right)$.
    - S_1 and S_2 exchange $\vec{m}_{V,1}$ and $\vec{m}_{V,2}$ to reconstruct $\vec{m}_V = \vec{m}_{V,1} + \vec{m}_{V,2}$.
  - If $S_i = S_0$ and $S_j = S_1$
    - S_1 computes $\vec{m}_{V,1} = \left(\sigma_{ij}(\vec{m}_U - \vec{\lambda}_{U,1}) + \vec{\lambda}_{V,1} + \vec{\gamma_1} \right)$.
    - S_0 computes $\vec{m}_{V,2} = \left(\vec{\lambda}_{V,2} - \sigma_{ij}(\vec{\lambda}_{U,2}) + \vec{\gamma_2} \right)$ and sends it to S_2.
    - S_1 and S_2 exchange $\vec{m}_{V,1}$ and $\vec{m}_{V,2}$ to reconstruct $\vec{m}_V = \vec{m}_{V,1} + \vec{m}_{V,2}$.
  - If $S_i = S_0$ and $S_j = S_2$
    - S_0 computes $\vec{m}_{V,1} = \left(\vec{\lambda}_{V,1} - \sigma_{ij}(\vec{\lambda}_{U,1}) + \vec{\gamma_2} \right)$ and sends it to S_1.
    - S_2 computes $\vec{m}_{V,2} = \left(\sigma_{ij}(\vec{m}_U - \vec{\lambda}_{U,2}) + \vec{\lambda}_{V,2} + \vec{\gamma_1} \right)$.
    - S_1 and S_2 exchange $\vec{m}_{V,1}$ and $\vec{m}_{V,2}$ to reconstruct $\vec{m}_V = \vec{m}_{V,1} + \vec{m}_{V,2}$.
- **Output Computation:** The servers compute their output as follows.
  - S_0 outputs $[\vec{V}]_{S_0} = (\vec{\lambda}_{V,1}, \vec{\lambda}_{V,2})$.
  - S_1 outputs $[\vec{V}]_{S_1} = (\vec{m}_V, \vec{\lambda}_{V,1})$.
  - S_2 outputs $[\vec{V}]_{S_2} = (\vec{m}_V, \vec{\lambda}_{V,2})$.

Fig. 3. Helper Protocol for Π_{Shuffle}

The proof of Theorem 2 will be available in the full version of the paper.

Theorem 2. *Let* $\mathcal{F}_{\mathsf{Helper}}\left(\perp, \perp, ([\vec{U}]_{S_0}, \sigma_{ij}, \text{ if } S_0 \in \{S_i, S_j\}), ([\vec{U}]_{S_1}, \sigma_{ij}, \text{ if } S_1 \in \{S_i, S_j\}), ([\vec{U}]_{S_2}, \sigma_{ij}, \text{ if } S_2 \in \{S_i, S_j\}) \right) = \left(\perp, \perp, [\vec{V}]_{S_0}, [\vec{V}]_{S_1}, [\vec{V}]_{S_2} \right)$, *where* $\vec{V} \stackrel{def}{=} \sigma_{ij}(\vec{U})$ *and where* $[\vec{V}]$ *denotes a random* $[\cdot]$*-sharing of* $\vec{V}$. *Moreover, let* Adv *be a PPT semi-honest adversary, who can corrupt any one subset of parties from the adversary structure* $\mathcal{Z}_{\mathsf{Adv}} = \{\emptyset, \{C_1\}, \{C_2\}, \{S_0\}, \{S_1\}, \{S_2\}, \{C_1, S_0\}, \{C_1, S_1\}, \{C_1, S_2\}, \{C_2, S_0\}, \{C_2, S_1\}, \{C_2, S_2\}\}$. *Then protocol* Π_{Helper} *is a secure protocol for computing* $\mathcal{F}_{\mathsf{Helper}}$.

The protocol involves communicating $4L$ *elements from* $\mathbb{F}$ *and 3 rounds of communication in the worst case.*

4.2   The Shuffle Protocol

Protocol Π_{Shuffle} is formally present in Fig. 4. The high-level idea behind Π_{Shuffle} has already been discussed in Sect. 4. We present and analyze protocol Π_{Shuffle} in the hybrid model, where we assume the existence of a TTP for securely computing the function $\mathcal{F}_{\text{Helper}}$. We will abuse the notation and denote the TTP as $\mathcal{F}_{\text{Helper}}$. Hence, we will present the protocol Π_{Shuffle} in the $\mathcal{F}_{\text{Helper}}$-hybrid model.

Protocol $\Pi_{\text{Shuffle}}(S_0, S_1, S_2, [\vec{A}]_{S_0}, [\vec{A}]_{S_1}, [\vec{A}]_{S_2}, [\vec{B}]_{S_0}, [\vec{B}]_{S_1}, [\vec{B}]_{S_2})$

- **Step 1: Generating pair-wise random permutations:** The servers do the following.
  - S_0 picks two random permutations $\sigma_{01} : \mathbb{F}^L \rightarrow \mathbb{F}^L$ and $\sigma_{02} : \mathbb{F}^L \rightarrow \mathbb{F}^L$ and sends σ_{01} and σ_{02} to S_1 and S_2 respectively.
  - S_1 picks a random permutation $\sigma_{12} : \mathbb{F}^L \rightarrow \mathbb{F}^L$ and sends it to S_2.
- **Step 2: Applying the permutations:** The servers do the following.
  - **(Applying the Permutation σ_{01}):**
    - Servers S_0, S_1 and S_2 call $\mathcal{F}_{\text{Helper}}$ with inputs $\left([\vec{A}]_{S_0}, \sigma_{01}\right), \left([\vec{A}]_{S_1}, \sigma_{01}\right)$ and $\left([\vec{A}]_{S_2}\right)$ and obtain the output $[\mathbf{\vec{A}}]_{S_0}, [\mathbf{\vec{A}}]_{S_1}$ and $[\mathbf{\vec{A}}]_{S_2}$ respectively, where $\mathbf{\vec{A}} = \sigma_{01}(\vec{A})$.
    - Servers S_0, S_1 and S_2 call $\mathcal{F}_{\text{Helper}}$ with inputs $\left([\vec{B}]_{S_0}, \sigma_{01}\right), \left([\vec{B}]_{S_1}, \sigma_{01}\right)$ and $\left([\vec{B}]_{S_2}\right)$ and obtain the output $[\mathbf{\vec{B}}]_{S_0}, [\mathbf{\vec{B}}]_{S_1}$ and $[\mathbf{\vec{B}}]_{S_2}$ respectively, where $\mathbf{\vec{B}} = \sigma_{01}(\vec{B})$.
  - **(Applying the Permutation σ_{12}):**
    - Servers S_0, S_1 and S_2 call $\mathcal{F}_{\text{Helper}}$ with inputs $\left([\mathbf{\vec{A}}]_{S_0}\right), \left([\mathbf{\vec{A}}]_{S_1}, \sigma_{12}\right)$ and $\left([\mathbf{\vec{A}}]_{S_2}, \sigma_{12}\right)$ and obtain the output $[\mathbb{\vec{A}}]_{S_0}, [\mathbb{\vec{A}}]_{S_1}$ and $[\mathbb{\vec{A}}]_{S_2}$ respectively, where $\mathbb{\vec{A}} = \sigma_{12}(\mathbf{\vec{A}})$.
    - Servers S_0, S_1 and S_2 call $\mathcal{F}_{\text{Helper}}$ with inputs $\left([\mathbf{\vec{B}}]_{S_0}\right), \left([\mathbf{\vec{B}}]_{S_1}, \sigma_{12}\right)$ and $\left([\mathbf{\vec{B}}]_{S_2}, \sigma_{12}\right)$ and obtain the output $[\mathbb{\vec{B}}]_{S_0}, [\mathbb{\vec{B}}]_{S_1}$ and $[\mathbb{\vec{B}}]_{S_2}$ respectively, where $\mathbb{\vec{B}} = \sigma_{12}(\mathbf{\vec{B}})$.
  - **(Applying the Permutation σ_{02}):**
    - Servers S_0, S_1 and S_2 call $\mathcal{F}_{\text{Helper}}$ with inputs $\left([\mathbb{\vec{A}}]_{S_0}, \sigma_{02}\right), \left([\mathbb{\vec{A}}]_{S_1}\right)$ and $\left([\mathbb{\vec{A}}]_{S_2}, \sigma_{02}\right)$ and obtain the output $[\vec{\mathbb{A}}]_{S_0}, [\vec{\mathbb{A}}]_{S_1}$ and $[\vec{\mathbb{A}}]_{S_2}$ respectively, where $\vec{\mathbb{A}} = \sigma_{02}(\mathbb{\vec{A}})$.
    - Servers S_0, S_1 and S_2 call $\mathcal{F}_{\text{Helper}}$ with inputs $\left([\mathbb{\vec{B}}]_{S_0}, \sigma_{02}\right), \left([\mathbb{\vec{B}}]_{S_1}\right)$ and $\left([\mathbb{\vec{B}}]_{S_2}, \sigma_{02}\right)$ and obtain the output $[\vec{\mathbb{B}}]_{S_0}, [\vec{\mathbb{B}}]_{S_1}$ and $[\vec{\mathbb{B}}]_{S_2}$ respectively, where $\vec{\mathbb{B}} = \sigma_{02}(\mathbb{\vec{B}})$.
- **Computing the Output:** S_0, S_1 and S_2 output $\left([\vec{\mathbb{A}}]_{S_0}, [\vec{\mathbb{B}}]_{S_0}\right), \left([\vec{\mathbb{A}}]_{S_1}, [\vec{\mathbb{B}}]_{S_1}\right)$ and $\left([\vec{\mathbb{A}}]_{S_2}, [\vec{\mathbb{B}}]_{S_2}\right)$ respectively.

Fig. 4. Secure Shuffle Protocol

The proof of Theorem 3 will be available in the full version of the paper.

Theorem 3. *Let* $\mathcal{F}_{\mathsf{Shuffle}}\left(\bot, \bot, ([\vec{A}]_{\mathsf{S}_0}, [\vec{B}]_{\mathsf{S}_0}), ([\vec{A}]_{\mathsf{S}_1}, [\vec{B}]_{\mathsf{S}_1}), ([\vec{A}]_{\mathsf{S}_2}, [\vec{B}]_{\mathsf{S}_2})\right) =$
$\left(\bot, \bot, ([\vec{A}]_{\mathsf{S}_0}, [\vec{B}]_{\mathsf{S}_0}), ([\vec{A}]_{\mathsf{S}_1}, [\vec{B}]_{\mathsf{S}_1}), ([\vec{A}]_{\mathsf{S}_2}, [\vec{B}]_{\mathsf{S}_2})\right)$, *where* $\vec{A} \stackrel{def}{=} \sigma(\vec{A})$ *and* $\vec{B} \stackrel{def}{=}$
$\sigma(\vec{B})$ *and* $\sigma : \mathbb{F}^L \to \mathbb{F}^L$ *is a random permutation and where* $[\vec{A}]$ *and*
$[\vec{B}]$ *denotes a random* $[\cdot]$*-sharing of* $\vec{A}$ *and* $\vec{B}$ *respectively. Moreover, let* Adv
be a semi-honest adversary, who can corrupt any one subset from $\mathcal{Z}_{\mathsf{Adv}} =$
$\{\emptyset, \{\mathsf{C}_1\}, \{\mathsf{C}_2\}, \{\mathsf{S}_0\}, \{\mathsf{S}_1\}, \{\mathsf{S}_2\}, \{\mathsf{C}_1, \mathsf{S}_0\}, \{\mathsf{C}_1, \mathsf{S}_1\}, \{\mathsf{C}_1, \mathsf{S}_2\}, \{\mathsf{C}_2, \mathsf{S}_0\}, \{\mathsf{C}_2, \mathsf{S}_1\},$
$\{\mathsf{C}_2, \mathsf{S}_2\}\}$. *Then protocol* Π_{Shuffle} *is a secure protocol for computing* $\mathcal{F}_{\mathsf{Shuffle}}$ *in*
the $\mathcal{F}_{\mathsf{Helper}}$*-hybrid model. The protocol makes 6 calls to* $\mathcal{F}_{\mathsf{Helper}}$ *and requires 1*
round. Additionally, 3L elements from $\mathbb{F}$ *are communicated.*

5   Implementation and Experiments

5.1   Setting

Hardware. Our experiments are conducted on a 12th Gen Intel(R) Core(TM) i7-12700H x64-based processor with a clock speed of 2.30 GHz. The experiments utilize 14 CPU cores and 16 GB of memory.

Implementation. The protocol is implemented in Python 3.11.1 and is run on a 64-bit Windows operating system. The implementation utilized a *fast* implementation of AES-128 and other low-level cryptographic operations. For generating random numbers, we used the random module of Python. To emulate the operation of servers, we employ the *multiprocessing* module in Python. This allows us to simulate multiple servers running concurrently, with each process representing the execution of an individual server. It is important to note that the multiprocessing framework, while utilized, is not explicitly exploited for runtime optimization. This design decision aligns with our commitment to fairly comparing the protocol's performance.

Domain of Input/Output Shares. The fingerprint shares are represented as elements of the ring $(\{0, 1\}^{128}, \oplus, *)$, where $\oplus$ denotes the bitwise XOR operation and $*$ denotes the bitwise AND operation.[3] The value shares, on the other hand, exist in the finite field $\mathbb{Z}_p$, where $p = 2^{61} - 1$. This specific choice of p, a *Mersenne prime*, optimizes the computation of the inverse operation. For situations involving floating-point computations, a strategic choice is made to represent value shares as shares of floating-point numbers. The floating point operations are

---

[3] It should be noted that in this context, a ring is being utilized instead of a field due to the convenience it offers in implementing AES-128 (PRP). The decision to use rings does not compromise the *privacy* and *correctness* of the protocol, as the computation involving fingerprints does not entail any multiplicative inverse operations in the protocol or the implementation.

performed using the techniques outlined in [1]. This resource-intensive approach is judiciously employed in select protocol steps where floating-point results are inevitable, with the actual usage of floating-point multiplication limited to just two instances.

Cryptographic Security Parameter. The protocol operates within the context of a cryptographic security parameter κ. This parameter correlates with the size of the AES-128 key, serving as a *critical* factor in our design considerations.

5.2   Experimental Results and Analysis

Performance Evaluation: We present an analysis of the performance of our secure plagiarism-checking protocol based on the varying lengths of the combined fingerprint vectors of both clients. The duration required to calculate similarity utilizing our approach will *significantly surpass* the time expended when calculating similarity in plain text. This discrepancy arises from the involvement of client-server or server-server communication at every stage of the computation within our method for providing privacy, which constitutes a substantial portion of the overall execution time. In Table 1, we showcase the execution times for our protocol when applied to source codes containing lines of code ranging from 10 to 250, corresponding to which the combined fingerprint lengths range from 0 to 2600. The reported run time encapsulates the entire process, assuming that the servers have received the necessary shares from the clients. This includes the time taken to execute the Pseudo-random Permutation (PRP) operation (utilizing AES-128) on each fingerprint, performing shuffling operations on the fingerprint and value vectors, reconstructing the shuffled fingerprint output, and ultimately calculating the cosine similarity on the shared values. Our evaluation methodology encompasses 25 sets, each comprising 20 code pairs with code length in the $[10 \cdot i, 10 \cdot i + 10]$ range, where $i = 0, \ldots, 24$. The protocol is executed for each code pair within a set, and the reported run time is averaged over all runs within the respective fingerprint range because the length of the fingerprints generated for the test codes varies with the size and depth of these codes. Fig 5 exhibits a clear linear correlation between the protocol's average run time and the fingerprint vectors' combined lengths. Additionally, for larger files containing around 1500 lines of code where the combined lengths of the fingerprint vectors averaged at 4070, the average run time of the protocol came up to 43.17 s.

Table 1. Average run time in seconds for various parts of the protocol with varying combined length of the fingerprint vectors of the two programs. Columns include the combined lengths of the fingerprint vectors, the total run time averaged over all runs within the set, the average run times of PRP, shuffle, reconstruction, output computation and the percentage error.

Fing. Len.	Run Time	PRP	Shuffle	Rec.	Output Comp.	Error
0−100	3.39	0.21	0.17	0.04	2.03	6.58×10^{-6}
100−200	3.69	0.33	0.27	0.06	2.08	7.45×10^{-6}
200−300	4.43	0.55	0.5	0.1	2.22	8.25×10^{-6}
300−400	5.07	0.72	0.65	0.12	2.42	8.76×10^{-6}
400−500	5.79	0.93	0.8	0.16	2.67	8.38×10^{-6}
500−600	6.68	1.17	1.05	0.2	2.92	8.72×10^{-6}
600−700	7.37	1.33	1.21	0.24	3.17	8.4×10^{-6}
700−800	8.17	1.54	1.4	0.27	3.39	7.87×10^{-6}
800−900	8.94	1.73	1.6	0.3	3.63	7.77×10^{-6}
900−1000	9.55	1.93	1.71	0.33	3.78	8.2×10^{-6}
1000−1100	10.22	2.12	1.86	0.37	3.99	7.91×10^{-6}
1100−1200	11.17	2.38	2.12	0.42	4.26	8.08×10^{-6}
1200−1300	11.93	2.58	2.33	0.46	4.47	9.0×10^{-6}
1300−1400	12.53	2.75	2.46	0.49	4.64	9.32×10^{-6}
1400−1500	13.23	2.95	2.58	0.54	4.82	8.44×10^{-6}
1500−1600	13.61	3.08	2.73	0.56	4.94	9.18×10^{-6}
1600−1700	14.63	3.33	2.91	0.59	5.27	8.78×10^{-6}
1700−1800	15.41	3.54	3.1	0.63	5.49	8.34×10^{-6}
1800−1900	15.91	3.66	3.21	0.66	5.62	8.49×10^{-6}
1900 − 2000	16.66	3.85	3.38	0.69	5.83	9.01×10^{-6}
2000−2100	17.31	4.03	3.52	0.72	6.0	8.6×10^{-6}
2100−2200	18.13	4.24	3.69	0.75	6.2	8.32×10^{-6}
2200−2300	18.69	4.42	3.85	0.78	6.29	9.81×10^{-6}
2300−2400	19.6	4.57	4.1	0.9	6.52	7.36×10^{-6}
2400−2500	20.73	4.78	4.58	1.0	6.76	8.61×10^{-6}
2500−2600	21.37	4.88	4.87	1.12	6.81	4.74×10^{-6}

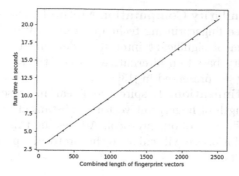

Fig. 5. A scatter plot showing the linearity of the average runtimes compared to the combined length of fingerprint vectors.

Optimization Potential: A notable observation from our experiments is that a substantial portion of the run time is attributed to the final phase of our implementation, influenced by the techniques introduced in [1]. While this phase accounts for a significant proportion of the execution time, it is worth highlighting that further advancements in state-of-the-art floating-point protocols could potentially offer opportunities for optimization. For instance, when assessing the cosine similarity between two codes, each of length 140, featuring an average fingerprint vector size of 800, our protocol takes approximately 9 s. Notably, a considerable portion, around 3.5 s, is consumed by the floating-point computations in this context.

Protocol Correctness and Error Analysis: We verify the correctness of our protocol against the benchmark set by [13]. Our evaluation reveals an error margin of only 10^{-6}%, attributed to the inherent limitations of floating-point computations. Considering the overall precision requirements of the problem being considered, this error is negligible and does not compromise our protocol's results' overall integrity and validity.

6   Conclusion and Open Problems

This paper introduces the *first* privacy-preserving plagiarism checker capable of computing the similarity between source codes while maintaining strict privacy guarantees. There are several interesting research directions to pursue in this domain. Here, we list a few of them.

- **Byzantine Adversaries.** While our current work focuses on passive adversaries, a natural extension is to investigate the resilience of our protocol against Byzantine (malicious) adversaries. Considering adversaries capable of deviating arbitrarily from protocol steps and propagating erroneous messages or values offers fertile ground for further exploration.

- **Alternative Similarity Computation Methods.** Our protocol capitalizes on the syntax-tree fingerprinting technique from [13] to compute similarity. An open problem of significant interest is devising techniques to securely compute similarity based on alternative state-of-the-art methods, including machine learning, as presented in [13].
- **Run time Optimization.** Despite the linear increase in run time with the combined length of fingerprint vectors, substantial potential remains for enhancing the efficiency of our protocol. Addressing this challenge and refining run-time performance will enhance the protocol's practical viability and adoption prospects.

References

1. Aliasgari, M., Blanton, M., Zhang, Y., Steele, A.: Secure computation on floating point numbers (2012). https://eprint.iacr.org/2012/405
2. Ben-Or, M., Goldwasser, S., Wigderson, A.: Completeness theorems for non-cryptographic fault-tolerant distributed computation (extended abstract), pp. 1–10 (1988)
3. Canetti, R.: Universally composable security: a new paradigm for cryptographic protocols. In: Proceedings 42nd IEEE Symposium on Foundations of Computer Science, pp. 136–145 (2001)
4. Chaudhari, H., Choudhury, A., Patra, A., Suresh, A.: ASTRA: high throughput 3PC over rings with application to secure prediction. In: Proceedings of the 2019 ACM SIGSAC Conference on Cloud Computing Security Workshop, CCSW 2019, pp. 81–92. Association for Computing Machinery, New York (2019)
5. Choudhury, A., Patra, A.: Secure Multi-party Computation Against Passive Adversaries. Synthesis Lectures on Distributed Computing Theory. Springer, Cham (2023). https://doi.org/10.1007/978-3-031-12164-7
6. Gitchell, D., Tran, N.: SIM: a utility for detecting similarity in computer programs. In: The Proceedings of the Thirtieth SIGCSE Technical Symposium on Computer Science Education, SIGCSE 1999, pp. 266–270. Association for Computing Machinery, New York (1999)
7. Goldreich, O., Micali, S, Wigderson, A.: How to play any mental game or a completeness theorem for protocols with honest majority. In: Aho, A.V. (ed.) Proceedings of the 19th Annual ACM Symposium on Theory of Computing, 1987, New York, New York, USA, pp. 218–229. ACM (1987)
8. Goldreich, O.: Foundations of Cryptography - Volume 2: Basic Applications. Cambridge University Press, Cambridge (2004)
9. Katz, J., Lindell, Y.: Introduction to Modern Cryptography. Chapman and Hall/CRC (2014)
10. Liu, C., Chen, C., Han, J., Yu, P.S.: GPLAG: detection of software plagiarism by program dependence graph analysis. In: Proceedings of the 12th ACM SIGKDD International Conference on Knowledge Discovery and Data Mining, KDD 2006, pp. 872–881. Association for Computing Machinery, New York (2006)
11. Prechelt, L., Malpohl, G.: Finding plagiarisms among a set of programs with JPlag. J. Univ. Comput. Sci. **8**, 03 (2003)
12. Schleimer, S., Wilkerson, D.S., Aiken, A.: Winnowing: local algorithms for document fingerprinting. In: SIGMOD 2003, pp. 76–85. Association for Computing Machinery, New York (2003)

13. Verma, A., Udhayanan, P., Shankar, R.M., Nikhila, K.N., Chakrabarti, S.K.: Source-code similarity measurement: syntax tree fingerprinting for automated evaluation. In: Proceedings of the First International Conference on AI-ML Systems, AIMLSystems 2021. Association for Computing Machinery, New York (2021)
14. Yao, A.C.: Protocols for secure computations (extended abstract). In: FOCS, pp. 160–164. IEEE Computer Society (1982)

PURED: A Unified Framework
for Resource-Hard Functions

Alex Biryukov[1,2] and Marius Lombard-Platet[1(✉)]

[1] DCS, University of Luxembourg, Esch-sur-Alzette, Luxembourg
{alex.biryukov,marius.lombard-platet}@uni.lu
[2] SnT, University of Luxembourg, Esch-sur-Alzette, Luxembourg

Abstract. Algorithm hardness can be described by 5 categories: hardness in computation, in sequential computation, in memory, in energy consumption (or bandwidth), in code size. Similarly, hardness can be a concern for solving or for verifying, depending on the context, and can depend on a secret trapdoor or be universally hard. Two main lines of research investigated such problems: cryptographic puzzles, that gained popularity thanks to blockchain consensus systems (where solving must be moderately hard, and verification either public or private), and white box cryptography (where solving must be hard without knowledge of the secret key). In this work, we improve upon the classification framework proposed by Biryukov and Perrin in Asiacypt 2017 and offer a united hardness framework, PURED, that can be used for measuring all these kinds of hardness, both in solving and verifying. We also propose three new constructions that fill gaps previously uncovered by the literature (namely, trapdoor proof of CMC, trapdoor proof of code, and a hard challenge in sequential time trapdoored in verification), and analyse their hardness in the PURED framework.

Keywords: puzzle cryptography · white-box cryptography · memory hardness · VDF · trapdoor problems

1 Introduction

In some specific situations, it might be preferable that algorithms are inefficient: for instance, RSA assumes that factoring is hard in computation. Hard problems can be used to slow down attackers or honest parties (the problems are then usually called puzzles). Such examples include code protection [30], resistance against password cracking [43], or blockchain consensus [11,33]. A brief summary of hardnesses in the literature can be found in Table 1. A more exhaustive survey has been carried in [3], but did not tackle how to measure the puzzle hardness.

Because of the variety of such applications and the resources considered, research on the general topic is quite scattered and not always consistent with each other. Our goal is twofold: expand existing literature on a generic hardness framework, and give a unified measure of hardness. We also along the way illustrate our framework with new constructions that fill gaps in the literature, and can be of individual interest.

A. Chattopadhyay et al. (Eds.): INDOCRYPT 2023, LNCS 14460, pp. 126–149, 2024.
https://doi.org/10.1007/978-3-031-56235-8_7

Table 1. Panorama of existing resource-hard problems. Our constructions and reductions are in grey cells. We also discuss about easy verification in Sect. 3.4

Solving	Verification	SeqTime	CPU	Mem	BW	Code
Moderate/Hard	Moderate/Hard	See Section 3.2				
	Trapdoored	Iterated modular squaring in group of unknown order	Any non-NP problem	Argon [14], BalloonHash [23], scrypt [40]	Argon [14], scrypt [40]	Lookup table of random numbers
		SeqTime challenge (see Section 6)	Encryption of NP-complete witness, of proof of work	[unknown]	[unknown]	[unknown]
	Easy	Proof of sequential work [28,39], trapdoorless VDF or in trustless group of unknown order [22,31] or in MPC-generated RSA groups [26,49]	NP-complete problems; Proof-of-Work [32]; witness or PoW encryption[35, 38]	Proof-of-Space [33]; Equihash, ethash [16,34]; MTP/Itsuku [15,27]	[unknown]	Proof of storage [37]
Trapdoored	Moderate/Hard	See Section 3.1				
	Trapdoored	RSA time-lock [45]	Encrypted PoK of a secret key or of factoring, Skipper [17]	Diodon [17]	[unknown]	SPACE, ASASA, lookup on a large trapdoored S-box [13,20,21]
	Easy	Trapdoor VDF [29,41,51]	EUF-CMA signatures, Schnorr protocol [47], proof of factoring [42], signatures of knowledge [25]	Trapdoor proof of CMC (See Section 5), memory lock (see Section 3.3)	Bandwidth lock (see Section 3.3)	HSig-BigLUT (see Section 4)
Easy	Moderate/Hard	See Section 3.1				
	Trapdoored	[unknown]	Encrypt 0 (with IND-CPA scheme)	[unknown]	[unknown]	[unknown]
	Easy	Return 0	Return 0	Return 0	Return 0	Return 0

Our Contributions. We present the following results:

- A formal hardness framework, PURED, uniting hardness over different kinds of resources, such as CPU, memory, code, sequential time, as well as over different issues: solving and verification, and with or without trapdoors.
- Generic reductions between different hardness classes.
- We propose three constructions, with hardness proved in our PURED model:
  - a trapdoored (in solving) proof of code, that cannot be implemented by an algorithm much smaller than the honest one, unless a secret trapdoor is known. On the other hand, verification only requires a few lines of code;
  - a trapdoor (in solving) proof of CMC that requires a significant amount of memory, unless a secret trapdoor is known, while verification is always easy;
  - a trapdoored (in verification) challenge of sequential time. This problem necessarily takes time to solve, but can be fast to verify if one knows the secret trapdoor.

Outline. In Sect. 2, we introduce the problem and define our PURED framework. Then, in Sect. 3 we offer reductions between different hardness classes. Sections 4 to 6 propose three new constructions, respectively a trapdoor proof of CMC, a trapdoor solving proof of code, and a SeqTime challenge, along with their proofs of hardness in the PURED framework. We conclude in Sect. 7.

2   General Resource-Hardness Framework

2.1   Resources

We here give a quick panorama of the resources we consider in this paper.

CPU. Computation complexity is defined by the number of operations (which can be counted as clock cycles, multiplications, exponentiations...) required to evaluate a program. CPU does not account for parallelization.

SeqTime is defined by the minimal sequential time (measured in cycles, multiplications...) for solving a problem. While CPU can be linked to circuit size, SeqTime can be linked to circuit depth and measures the minimal number of consecutive steps to compute a function SeqTime problems are non-parallelisable and do not depend on the number of processors, and have been introduced notably for time lock puzzles [45], then for verifiable delay functions [22,41,51]. Typical examples include iterated hashing and iterated squaring.

Mem. In many cases, a memory intensive algorithm can be rewritten to use very little memory, at the expense of an increased computation time. Time-memory product [40], then cumulative memory complexity (CMC, [8]) addressed this issue. CMC (further refined in [5]) counts the sum of memory requirements at each step of the execution, and does not depend on parallelization. In this paper, we use the CMC memory-hardness framework, which is notably the framework in which scrypt hardness has been proven [6].

Code. Code complexity is the minimum size of a binary implementing a given algorithm. Code hardness has been used notably for whitebox cryptography [13,20,21,30], and applies to programs that cannot be efficiently compressed. Code complexity usually relies on lookup tables given to the user at generation time. If the solver/verifier can generate these tables on the fly, then the problem becomes Mem hard rather than code hard.

BW. Bandwidth complexity has been introduced to count the number of off-chip memory accesses, which is invariant between CPUs and ASICs, thus offering a hardware-agnostic complexity. Energy complexity [4,7], then bandwidth complexity [18,44], lower-bound the energy cost. Bandwidth complexity increases with the square root of the CMC [18] , and does not vary with parallelization.

Other Complexities. On top of the aforementioned complexities, other complexities have been proposed [1,2,9,48,50]. However, either they have been superseded by one of the previous complexities, or either their usage remains minimal.

2.2   Resource-Hardness Game

Problem Class. The high level idea is that a problem class is specifically crafted so that it requires hardness in resource R, i.e. it requires at least u units of R to solve (resp. verify) an instance (resp. a solution), for any u.

This gives us the following definition of a problem class.

Definition 1. *A problem class instance is a tuple* $(\mathcal{P}_u^\lambda, \text{Solve}, \text{Verify})$, *with* Solve : $\mathcal{P}_u^\lambda \times \{0,1\}^* \to \{0,1\}^*$, Verify : $\{0,1\}^* \to \{0,1\}$ *the reference algorithm implementations for solving and verifying the problem, and* λ *is a security parameter. Solve is a (potentially) nondeterministic algorithm that takes as input a problem instance* $P \in \mathcal{P}_u^\lambda$, *and a random tape[1]* z: Solve(P, z).

Furthermore, Solve *and* Verify *implementations must be sound, i.e. any output of* Solve *is valid for* Verify: $\forall P \in \mathcal{P}_u^\lambda, \forall z \in \{0,1\}^*$, Verify$(P, \text{Solve}(P, z)) = 1$.

A problem class $\mathfrak{C}^*$ *is an algorithm* Generate(u, λ) *that is easy to evaluate and generates a problem class instance* $\mathfrak{C}$ *where* $\mathfrak{C} = (\mathcal{P}_u^\lambda, \text{Solve}, \text{Verify})$. *Furthermore, in case of a solving (resp. verification) trapdoor, it also privately discloses the trapdoor* t_S *(resp.* t_V*) to privileged parties.*

Note that generation is trustless if and only if no secret trapdoor are created. We also assume that solving (or verifying) a problem from a problem class instance does not yield any advantage for solving (or verifying) another problem from another instance of the same problem class[2].

Finally, while we model the probabilistic nature of solving algorithm, probabilistic verification algorithms are left out of scope of this work.

---

[1] In this paper, we often omit the random tape and write $s \xleftarrow{\$} \text{Solve}(P)$ when information on z is irrelevant, and $s \leftarrow \text{Solve}(P, z)$ otherwise.

[2] For instance, consider the problem of inverting a hash for the hash function $H(\cdot, r)$ where r is a random string for each problem class instance. For a given r, inverting two hashes is cheaper than double the cost of inverting one hash. However, in the case of inverting one hash of $H(\cdot, r)$ and one hash of $H(\cdot, r')$, this is not true.

Hardness Game. The hardness (either in solving or verifying) of a problem class is evaluated through the following protocol. For Verify, since using the probability of being correct on a random input yields the 0 function as an overwhelmingly perfect verification algorithm for hard to solve problems, we rather use a variant of Youden's J statistic [52], where 0 indicates no better performance than random, and ± 1 a perfectly good (or bad) performance.

Definition 2 (Algorithm advantage). *The advantage* Adv^s *of an algorithm* $\mathcal{A}$ *on solving a problem class* $\mathfrak{C}^*$ *is* $\mathsf{Adv}^s(\mathcal{A}) = Pr_{P,z,\mathfrak{C}}(\mathsf{Verify}(P, \mathcal{A}(P, z)) = 1)$, *where the randomness is taken over* $P \in \mathcal{P}_u^\lambda$, *the random tape* z *of* $\mathcal{A}$ *and the generation of* $\mathfrak{C} = (\mathcal{P}_u^\lambda, \mathsf{Solve}, \mathsf{Verify}) \xleftarrow{\$} \mathfrak{C}^*.Generate(u, \lambda)$.

The advantage Adv_D^v *of an algorithm* $\mathcal{A}$ *on verifying a problem class* $\mathfrak{C}^*$ *over a domain* D *is*

$$\mathsf{Adv}_D^v(\mathcal{A}) = Pr_{P,s,\mathfrak{C}}(\mathcal{A}(P, s) = 1 | \mathsf{Verify}(P, s) = 1) + Pr_{P,s,\mathfrak{C}}(\mathcal{A}(P, s) = 0 | \mathsf{Verify}(P, s) = 0) - 1$$

where the randomness is taken over $P \in \mathcal{P}_u^\lambda$, $s \in D$, *and the generation of* $\mathfrak{C} = (\mathcal{P}_u^\lambda, \mathsf{Solve}, \mathsf{Verify}) \xleftarrow{\$} \mathfrak{C}^*.Generate(u, \lambda)$.

Remark 1. For the verifying game, the advantage is domain dependent as there might be strings that are obvious non solutions. Depending on the problem class, determining if the challenge belongs to the set of possible solutions might be a difficult problem in itself, thus the requirement to specify the domain of the verification advantage.

Hard Problem Class. While Solve and Verify are already defined in the previous section, they might not be the algorithms chosen by an attacker, who might want to compromise accuracy in exchange of a more efficient resource usage.

We note the consumption of resource R with the function $\mathsf{Res}(\cdot)$: an algorithm $\mathcal{A}$ using u units of resource R on entry P is noted $\mathsf{Res}_R(\mathcal{A}(P)) = u$. The resource R will be omitted if the context allows.

Our definition relies on the hardness game defined earlier, however we use probabilities rather than an interactive game. Our definition extends the hardness framework used in [6, 28], and generalizes other similar approaches [12, 17]. A high-level understanding of our definition is: *If a problem class is hard, then for any algorithm solving (or verifying) a problem instance, the probability this algorithm uses less than some amount of resource is minimal.*

Definition 3 (PURED resource hardness). *Let* u *be a nonnegative number,* λ *be a security parameter, and* $\mathfrak{C} = (\mathcal{P}_u^\lambda, \mathsf{Solve}, \mathsf{Verify})$ *a problem class.*

- *We say that the problem class* $\mathfrak{C}^*$ *is a* $(p, u, R, \delta, \epsilon)$-*solving hard problem (with* ε *a function of* p, u *and* δ), *if*

$$\forall \mathcal{A}, \mathsf{Adv}^s(\mathcal{A}) \geq p \Rightarrow Pr_{P,z,\mathfrak{C}}[\mathsf{Res}(\mathcal{A}(P, z)) < \delta u] < \epsilon + \mathsf{negl}(\lambda)$$

where the probability is taken over the sampling of $P \in \mathcal{P}_u^\lambda$, *the random tape* z, *and the generation of* $\mathfrak{C} = (\mathcal{P}_u^\lambda, \mathsf{Solve}, \mathsf{Verify}) \xleftarrow{\$} \mathfrak{C}^*.Generate(u, \lambda)$, *and* negl *denotes a function negligible in* λ.

– *Similarly, $\mathfrak{C}^*$ is $(p, u, R, \delta, \epsilon)$ verifying hard problem over D if:*

$$\forall \mathcal{A}, \mathsf{Adv}_D^v(\mathcal{A}) \geq p \Rightarrow Pr_{P,s,\mathfrak{C}} \left[Res(\mathcal{A}(P,s)) < \delta u \right] < \epsilon + \mathsf{negl}(\lambda)$$

Where the probability is taken over the sampling of $P \in \mathcal{P}_u^\lambda, s \in D$ and the generation of $\mathfrak{C} = (\mathcal{P}_u^\lambda, \mathsf{Solve}, \mathsf{Verify}) \xleftarrow{\$} \mathfrak{C}^*.\mathsf{Generate}(u, \lambda)$.

Definition 4 (PURED for trapdoor hardness). *Similarly, a problem class is trapdoored if it is at most $(p, \log u, R, \delta, \epsilon)$ solving hard for people knowing the trapdoor, but $(p, u, R, \delta, \epsilon)$ hard otherwise. More precisely,*

– *The $(p, u, R, \delta, \epsilon)$ solving hard problem class $\mathfrak{C}^*$ is said to be trapdoored if there exists $\mathcal{A}$ such that*

$$\mathsf{Adv}^s(\mathcal{A}(t_{\mathfrak{C}}, \cdot)) \geq p \wedge Pr_{P,z,\mathfrak{C}} \left[Res(\mathcal{A}(t_{\mathfrak{C}}, P, z)) < \delta \log u \right] \geq \epsilon + \mathsf{negl}(\lambda)$$

With $t_{\mathfrak{C}}$ the solving trapdoor generated by $\mathfrak{C}^*.\mathsf{Generate}$, see Definition 1.
– *The $(p, u, R, \delta, \epsilon)$ verifying hard over a domain D problem class $\mathfrak{C}^*$ is said to be trapdoored if there exists $\mathcal{A}$ such that*

$$\mathsf{Adv}_D^v(\mathcal{A}(t_{\mathfrak{C}}, \cdot)) \geq p \wedge Pr_{P,s,\mathfrak{C}} \left[Res(\mathcal{A}(t_{\mathfrak{C}}, P, s)) < \delta \log u \right] \geq \epsilon + \mathsf{negl}(\lambda)$$

A moderately hard problem is one that is hard, for some parameters deemed acceptable. An easy problem can be solved within a small amount of resource usage.

Remark 2. Following [28], u and δ are two different parameters instead of just one, as u indicates the estimated hardness of the problem, with a typical $\epsilon - \delta$ approach.

One can show that proof of work of difficulty d (i.e. given s find x so that $H(x\|s)$ starts with d zeroes) is $(p, p2^d, \mathsf{CPU}, \delta, \delta)$ solving hard in the ROM.

Similarly, according to [6] the function scrypt [40] is $(p, \frac{M^2 n}{25}, \mathsf{Mem}, 1 - \frac{100 \log M}{n}, 1 - p + 0.08 M^6 2^{-n} + 2^{-M/30})$ solving hard in the parallel random oracle model, where M and n are (integer) parameters.

Finally, in the literature, for CPU related problems, the notion of hardness is sometimes defined by (t, p) security, which means that any attacker running in time t cannot succeed with probability more than p. Thus, in our framework a (t, p) secure problem class is $(p, t, \mathsf{CPU}, \delta, \mathbb{1}_{\delta \geq 1})$ hard, where $\mathbb{1}$ is the indicator function.

We assume that ϵ is increasing in δ, and decreasing in p.

As noted previously, this definition is only hard *on average*. We only define systematic hardness on problems relying on a function that can be approached as a random oracle.

Definition 5 (systematic hardness). *Let u be a nonnegative number, λ be a security parameter, and $\mathfrak{C}_h^*$ a problem class that relies on a function h, modeled as a random variable.*

We say that the problem class $\mathfrak{C}_h^*$ is a systematic $(p, u, R, \delta, \epsilon)$ solving hard problem, if

$$\forall \mathcal{A}, \forall P \in \mathcal{P}_u^\lambda, Pr_h\left[\text{Verify}(P, s) = 1\right] \geq p \Rightarrow Pr_h\left[\text{Res}(\mathcal{A}(P)) < \delta u\right] < \epsilon + \text{negl}(\lambda)$$

Where randomness is taken over the choice of h in the random oracle model. The notion of systematic verifying hardness is similarly defined.

2.3 Bounded Adversaries

An important remark is that the current definition of our PURED framework leaves the possibility of an unbounded adversary in Code and SeqTime problems, which obviously cause issues if the problem relies on an underlying problem, assumed computationally intractable (e.g. factoring a RSA modulus for Wesolowki VDF [51]). Hence, we additionally request, for Code and SeqTime, to specify the adversary computational bound. Note that for CPU and Mem, this bound is already natively included in our framework.

For instance, we can say that a problem class is $(p, u, R, \delta, \varepsilon, s)$ solving hard if it is $(p, u, R, \delta, \varepsilon)$ solving hard, for any PPT adversary bounded by 2^s operations (which include oracle calls or table lookups).

Our Constructions in the PURED Framework. We give the hardness of the constructions we present below in Sects. 4 to 6. The different parameters are described in the associated sections.

- HSig-BigLUT$_u^{\lambda,\gamma}$ (Sect. 4)
    • Systematically $(p, u, \text{Code}, \delta, \mathbb{1}_{p=0}\mathbb{1}_{\delta<1})$ trapdoor solving hard against a 2^λ bounded adversary, with $u = \gamma L - \lambda \approx \lambda(L-1)$
    • Verification complexity of $O(\lambda)$ in Code
- Trapdoor proof of CMC TdPoCMC$_{M,L,k}^\lambda$ (Sect. 5)
    • $((pq)^k, \frac{qML\lambda}{25}, \text{Mem}, 1-\frac{100\log M}{\lambda}, 1-p+\text{negl}(M)+\text{negl}(\lambda))$ trapdoor solving hard
    • Verification complexity at most $O\left(k^2\lambda \log \frac{L}{k}(\lambda + \log L)\right)$ in Mem
- SeqTime Challenge$_{T,w,t}^\lambda$ (Sect. 6), against an adversary that is 2^λ bounded, and makes at most 2^q calls to the oracle:
    • Systematically $(p, 2T-1, \text{SeqTime}, \delta, \mathbb{1}_{\delta\geq1})$ solving hard
    • Systematically $(p, 2T-1, \text{SeqTime}, \delta, \mathbb{1}_{\delta\geq\left(1-p-\frac{\log_2(T)w}{2^w-2^q-1}\right)^{1/n}})$ solving hard
      if the solver knows the trapdoor
    • $(p, 2T-1, \text{SeqTime}, \delta, \mathbb{1}_{\delta\geq1})$ trapdoored verifying hard over $\{0,1\}^w$

3 Problem Class Reductions

In this section, we show generic reductions from one difficulty class to another. In all these theorems, we do not specify the computational bound of the adversary (when needed), since it always transfers from the old class to the new one (even though better bounds might exist in the new class), thus can be omitted for simplicity.

3.1 Leveraging Trapdoored Solving Hard into Verifying Hard

We show how to obtain a problem class hard in verification from a problem class that is trapdoored in solving. For our proof, we need what we call a *deterministic* problem class, which means that for any problem instance P, there is only one solution s, which implies that Solve is deterministic.

Theorem 1. *Let* $\mathfrak{C}^*, \mathfrak{D}^*$ *be problem classes, let* H *be a hash function of codomain* $\{0,1\}^n$. *If* $\mathfrak{C}^*$ *is* $(p, u, R, \delta, \epsilon)$ *trapdoored solving hard and deterministic, then there exists a problem class* $\mathfrak{E}^*$ *that is* $(p - \frac{1-p}{2^n}, u, R, \delta, \epsilon)$ *verifying hard in the random oracle model on the domain* $\{\mathfrak{C}, P, c : \mathfrak{C} = (\mathcal{P}_u^\lambda, \_, \_) \xleftarrow{\$}$ $\mathfrak{C}^*.Generate(u, \lambda); P \xleftarrow{\$} \mathcal{P}_u^\lambda; c \in \{0,1\}^n\}$.

The case when $\mathfrak{D}^*$ is not deterministic is left for future work.

Proof. The idea of this proof is that the solver will generate on the fly their own trapdoor problem class, then encode the solution inside that class. For instance, for a CPU hard problem, instead of encrypting the result with the verifier's public key, the solver XORs the solution with a random message encrypted using a public key generated on the spot. We now give a formal proof of this idea for any kind of resource hardness.

Let $\mathfrak{D}^*$ be a problem class of unspecified hardness, and $\mathfrak{D} = (\mathcal{Q}_u^\lambda, \text{Solve}_\mathfrak{D}, \text{Verify}_\mathfrak{D})$ be an instance of $\mathfrak{D}^*$. Let $\mathfrak{C}^*$ be a $(p, u, R, \delta, \epsilon)$ trapdoor solving problem class.

Our goal is to create a problem class that is $(p, u, R, \delta, \epsilon)$ verifying hard.

Let $\mathfrak{E}^*$ be a problem class such that $Generate(u, \lambda)$ returns a tuple $\mathfrak{E} = (\mathcal{Q}_u^\lambda, \text{Solve}_\mathfrak{E}, \text{Verify}_\mathfrak{E})$, with the following properties.

- Upon a problem entry $Q \in \mathcal{Q}_u^\lambda$, $\text{Solve}_\mathfrak{E}(Q)$ will run $s \leftarrow \text{Solve}_\mathfrak{D}(Q)$, then generate an instance $\mathfrak{C} = (\mathcal{P}_u^\lambda, \text{Solve}_\mathfrak{C}, \text{Verify}_\mathfrak{C})$ of $\mathfrak{C}^*$ with solving trapdoor $t_\mathfrak{C}$ (note that $\text{Solve}_\mathfrak{E}(Q)$ knows $t_\mathfrak{C}$). Then, they sample a random P from $\mathcal{P}_u^\lambda$ and run the trapdoored function $s' \leftarrow \text{Solve}_\mathfrak{C}(t_\mathfrak{C}, P)$. Finally, $\text{Solve}_\mathfrak{E}(Q)$ returns $(\mathfrak{C}, P, s \oplus H(s'))$, where H is a preimage resistant hash function of codomain size higher than the size of s.
- Upon a possible solution $(\mathfrak{C}, P, c)$, $\text{Verify}_\mathfrak{E}$ is defined by $\text{Verify}_\mathfrak{E}(Q, (\mathfrak{C}, P, c)) = \text{Verify}_\mathfrak{D}(Q, c \oplus H(\text{Solve}_\mathfrak{C}(P)))$.

We now prove that $\mathfrak{C}^*$ is verifying hard over the set $\{\mathfrak{C}, P, c : \mathfrak{C} = (\mathcal{P}_u^\lambda, \_, \_) \xleftarrow{\$}$ $\mathfrak{C}^*.Generate(u, \lambda); P \xleftarrow{\$} \mathcal{P}_u^\lambda; c \in \{0,1\}^n\}$.

Let $\mathcal{V}$ be a PPT algorithm, with a non-null verifying advantage p.

In order to run $\mathcal{V}$ on input $(\mathfrak{C}, P, c)$, the adversary can either find a preimage of c in the random oracle model), or treat it as a random variable (assuming the random oracle model).

- If the $\mathcal{V}$ treats c as a random variable, the distributions $\mathcal{V}(Q, (\mathfrak{C}, P, c)))$ and $\mathcal{V}(Q, (\mathfrak{C}, P, r))$ are indistinguishable, so no matter the input, $\mathcal{V}$ will answer randomly, with a given probability q. Thus, $\mathcal{V}$ verification advantage is $\mathsf{Adv}_s(\mathcal{V}) = q + (1 - q) - 1 = 0$, which contradicts the hypothesis that $\mathcal{V}$ has advantage $p > 0$.

- If $\mathcal{V}$ computes a preimage of c, given that we operate in the random oracle model, this means that $\mathcal{V}$ actually computes the value $\mathsf{Solve}_{\mathfrak{C}}(P)$, so the probability for $\mathcal{V}$ to use less than δu resources is less than ϵ. We still must assess $\mathcal{V}$ advantage. Let $(\mathfrak{C}, P, c)$ be a valid answer. If $\mathsf{Solve}_{\mathfrak{C}}(P)$ answers correctly with probability p, then $\mathcal{V}$ validates the input correctly with probability p in the best case[3]. Now, if $(\mathfrak{C}, P, c)$ is not a valid answer, there is the possibility that $\mathsf{Solve}_{\mathfrak{C}}$ does not answer correctly (with probability $1 - p$) and that the wrong answer of Solve actually gives a valid on c (with probability $\frac{1}{2^n}$). Hence, on an invalid answer, $\mathcal{V}$ will answer correctly with probability $1 - \frac{1-p}{2^n}$, hence $\mathcal{V}$ has advantage $p - \frac{1-p}{2^n}$, hence the result.

Furthermore, we observe that $\mathfrak{C}^*$ is not much harder to solve than $\mathfrak{D}^*$.

Note that the same result does not immediately transfer for creating easy verification problems: adding the solving trapdoor in the public information at generation might be problematic if it is equal to the solving trapdoor, as is the case for the RSA time-lock [45].

3.2 Leveraging Solving Hard to Verification Hard

We process similarly here. As noted in our table, there exist problems which are assumed to be hard both in solving and verification, mostly because the verification algorithm consists of the solving algorithm. However, these problems might lack a formal proof on the verification hardness, given that verification was not considered during their design.

Let $\mathfrak{C}^*$ be a deterministic problem class $(p, u, R, \delta, \epsilon)$ hard in solving. We show how to create a new problem class that is close to $\mathfrak{C}^*$, safe for the fact that it is now also hard in verification.

Theorem 2. *Let $\mathfrak{C}^*$ be a problem class. If $\mathfrak{C}^*$ is $(p, u, R, \delta, \epsilon)$ solving hard, then there exists a problem class $\mathfrak{C}^*$ that is $(p - \frac{1-p}{2^n}, u, R, \delta, \epsilon)$ verifying hard on a specific domain. Furthermore,*

- *If $R = \mathsf{CPU}$, then $\mathfrak{C}^*$ is $(p, u, R, \delta, \epsilon')$ solving hard, with $\epsilon'(p, \delta) = \epsilon(\sqrt{p}, \delta)$.*
- *If $R = \mathsf{SeqTime}$, $R = \mathsf{Mem}$, $R = \mathsf{Code}$ or $R = \mathsf{BW}$, then $\mathfrak{C}^*$ is $(p, 2u, R, \delta, \epsilon'')$ solving hard, with $\epsilon''(p, \delta) = \max\limits_{\substack{0 \le p_1 \le p \\ 0 \le \delta_1 \le 2\delta}} \epsilon(p_1, \delta_1)\epsilon(p/p_1, 2\delta - \delta)$.*

Proof. We use essentially the same construction as in Theorem 1, and thus will reuse the same notations, but this time there is no trapdoor involved hence we can take $\mathfrak{D}^* = \mathfrak{C}^*$ (we name $\mathfrak{C}_1$ the instance from $\mathfrak{C}^*$, and $\mathfrak{C}_2$ the instance from $\mathfrak{D}^*$), and thus verifying a problem in $\mathfrak{C}^*$ requires solving one problem in $\mathfrak{C}^*$

---

[3] This is the best case since the only bits of information $\mathcal{V}$ disposes of are from $\mathsf{Solve}_{\mathfrak{C}}(P)$. As we saw previously answering randomly gives a null advantage.

(which is $(p, u, R, \delta, \epsilon)$ solving hard), and verifying one problem in $\mathfrak{C}^*$. Thus $\mathfrak{C}^*$ is at least $(p, u, R, \delta, \epsilon)$ verification hard.

For solving hardness, in order to solve a problem in $\mathfrak{C}^*$, an algorithm $\mathcal{A}$ must solve two instances P_1, P_2 of $\mathfrak{C}_1, \mathfrak{C}_2$, respectively.

- If $R = \mathsf{SeqTime}$, the two instances can be solved in parallel.
- If $R = \mathsf{CPU}$ or $R = \mathsf{Code}$, because the two problem class instances are different, solving one does not offer any advantage in solving the other (see the remarks below Definition 1) so there is no other possibility than dedicate twice the amount of computation (or code).
- Similarly, if $R = \mathsf{Mem}$ (resp. $R = \mathsf{BW}$), the two problems can be solved one after the other, or in parallel , and both methods lead to a doubled cost.

Let us now consider an algorithm $\mathcal{A}$ that solves instances $P \in \mathcal{P}$ for the problem class instance $\mathfrak{C}$ with probability q. In order to solve P, $\mathcal{A}$ must solve two problems from $\mathfrak{C}^*$ (but of different instances), namely P_1 and P_2. On average, let us assume that $\mathcal{A}$ solves P_1 (resp. P_2) with probability p_1 (resp. p_2), with $p_1 p_2 = q$.

Case where $R = \mathsf{CPU}$. Both problems P_1 and P_2 can be solved in parallel. Thus, because $\mathfrak{C}^*$ is $(p, u, R, \delta, \epsilon)$ solving hard, we get that $\Pr[\mathsf{Res}(\mathcal{A}_1(P_1)) < \delta u] < \epsilon(p_1, \delta)$ where $\mathcal{A}_1$ is the part of $\mathcal{A}$ dedicated to solving P_1. Similar formula applies to P_2.

Because solving P uses less than δu resources of R if and only if both solving P_1 and solving P_2 require less than δu of resources each, we get $\Pr[\mathsf{Res}(\mathcal{A}(P)) < \delta u] = \Pr[\mathsf{Res}(\mathcal{A}_1(P_1)) < \delta u] \Pr[\mathsf{Res}(\mathcal{A}_2(P_2)) < \delta u] < \epsilon(p_1, \delta)\epsilon(q/p_1, \delta)$.

Because ϵ is decreasing in p, the right hand side is lower than $1\epsilon(\sqrt{q}, \delta)$, no matter the values of p_1 and p_2. Hence, $\mathfrak{C}^*$ is $(p, u, R, \delta, \epsilon')$ solving hard, with $\epsilon'(p, \delta) = \epsilon(\sqrt{p}, \delta)$.

Case where $R = \mathsf{SeqTime}$ or $R = \mathsf{Mem}$ or $R = \mathsf{Code}$. In this case, the algorithm $\mathcal{A}$ must not allot more than $2\delta u$ units of R *cumulated* to $\mathcal{A}_1$ and $\mathcal{A}_2$.

Let $\mathcal{A}$ be an algorithm solving $\mathfrak{C}$, with probability q. Using the same notations as before, let us assume that $\mathcal{A}_1$ solves P_1 with probability p_1, and $\mathcal{A}_2$ solves P_2 with probability $p_2 = q/p_1$. If $\delta_1 u$ units are devoted to solving P_1, then $\mathcal{A}$ can devote up to $(2\delta - \delta_1)u$ units to P_2, so that the total amount of resources devoted does not exceed δ.

Hence, $\mathcal{A}$ succeeds solving P with probability $p_1 p_2 = q$, and $\Pr[\mathsf{Res}(\mathcal{A}(P)) < 2\delta u] < \epsilon(p_1, \delta_1)\epsilon(q/p_1, 2\delta - \delta_1)$, from which we derive our upper bound.

3.3 Leveraging Trapdoored Solving Hard and Trapdoored Verification to Easy Verification

From a deterministic problem class that is trapdoored solving, it is trivial to create a problem class that is easy in verifying: it suffices, at generation, to publish the solving trapdoor. However, this also breaks the solving complexity, thus reducing the interest of doing so.

Algorithm 1. Resource-lock Solve and Verify algorithm

Inputs: $\mathfrak{C}$ is a trapdoored solving hard class problem of trapdoor t, H is a second preimage resistant hash function whose outputs are more than $|t|$ bits, $c = H(\mathfrak{C}.\mathsf{Solve}(m)) \oplus t$, $m \in \mathfrak{C}.\mathcal{P}$

function SOLVE(m, c, $\mathfrak{C}$, H)	**function** VERIFY$((m, c, \mathfrak{C}, H), v)$
$s \leftarrow \mathfrak{C}.\mathsf{Solve}(m)$	▷ $(m, c, \mathfrak{C}, H)$ is a problem instance,
return $H(s) \oplus c$	v is a response
end function	**return** $H(\mathfrak{C}.\mathsf{Solve}(m, v)) \oplus c \stackrel{?}{=} v$
	end function

However, we can design one-time problems with problems that are deterministic (i.e. only one possible solution) and trapdoored in solving. As a matter of fact, this approach had been used by Rivest for his time capsule challenge [46]. In this section, we simply generalize the concept with Algorithm 1, and give its security in our framework.

With this construction, we get the following result.

Theorem 3. *Let $\mathfrak{C}$ be a problem class that is $(p, u, R, \delta, \epsilon)$ solving trapdoored hard, of solving trapdoor t. Then the resource-lock problem class of Algorithm 1 defined on top of $\mathfrak{C}$ is publicly verifiable and $(p, u, R, \delta, \epsilon)$ solving hard for at most one instance.*

Proof. We simply show that the problem class $\mathcal{D}$ defined with Algorithm 1 on top of $\mathfrak{C}$ offers the claimed security, since the soundness is immediate. It is also immediate that verification is easy, and that once one instance has been solved and published (hence once the solving trapdoor has been published), the solving hardness disappears.

Let us explore the solving strategies for a solver of the resource-lock problem class that does not dispose of the solving trapdoor. For Verify to accept a solution, either the solver finds the trapdoor, either a preimage on H, or the solution to the instance of $\mathfrak{C}$. Finding the trapdoor is, by hypothesis, assumed to be infeasible with probability higher than $\mathsf{negl}(\lambda)$. Similarly, finding a preimage of $H(\mathfrak{C}.\mathsf{Solve}(m, v))$ is equally as hard as finding the trapdoor (since the hash codomain has the same size as the trapdoor), and hence negligible in λ. Finally, solving $\mathfrak{C}$ without the trapdoor is $(p, u, R, \delta, \epsilon)$ hard, which concludes the proof.

3.4   Leveraging Any Problem Class to Easy Verification

In this section, we briefly describe how proofs of computation allow to transform problem classes into similar problem classes, but with easy verification.

Proofs of computation have been pioneered in [10], and much refined since then. Proofs of computation are often derived from zero-knowledge proofs [36].

Because of the generality of the construction, one can build succinct zero-knowledge proofs for any language in NP. Because of usual requirements on zk-proofs, checking a proof cannot take much CPU (or SeqTime), and because of the succinctness, Mem usage is low as well. Furthermore, the code of these constructions is also relatively small.

4   HSig-BigLUT: **Code**, Systematic Trapdoored-Hard Solving, Easy Verification Problem Class

4.1   Primer on Homomorphic Signature and the BFKW Scheme

It is obvious that a homomorphic signature scheme cannot be forgery resistant, but the definition can be adapted as follows

Definition 6 (Homomorphic EUF-KMA security). *Let Sig = (Gen, Sign, Verify) be a signature scheme over a vector space* $\mathbb{F}$. *Sig is Homomorphic EUF-KMA secure if the advantage of any PPT adversary* $\mathcal{A}$, *who knows the signatures of some messages* $\{\mathbf{m}\}$, *in forging a valid signature (outside of the span of* $\{\mathbf{m}\}$*) is negligible:*

$$
Pr\left[\begin{array}{l} \mathbf{m}^* \notin \mathsf{Span}(\{\mathbf{m}\}) \wedge \\ \mathsf{Verify}(\mathsf{pk}, \mathbf{m}^*, \sigma^*) = 1 \end{array}\middle|\begin{array}{l} (\mathsf{pk}, \mathsf{sk}) \xleftarrow{\$} \mathsf{Gen}(1^\lambda); \{\mathbf{m}\} \xleftarrow{\$} \mathbb{F}; \mathsf{Span}(\{\mathbf{m}\}) \subsetneq \mathbb{F} \\ (\mathbf{m}^*, \sigma^*) \xleftarrow{\$} \mathcal{A}\left(\mathsf{pk}, (m, \mathsf{Sign}(\mathsf{sk}, m))_{m \in \{\mathbf{m}\}}\right) \end{array}\right] < \mathsf{negl}(\lambda)
$$

Where $\mathsf{Sign}(\mathsf{sk}, \{\mathbf{m}\})$ *is the set of signatures of the messages of* $\{\mathbf{m}\}$, *and* $\mathsf{Span}(\{\mathbf{m}\})$ *the vector space generated by the vectors of* $\{\mathbf{m}\}$.

In this paper, we rely on the BFKW scheme [24], which allows for signing m vectors of $\mathbb{F}_p^n$, with m known in advance. The BFWK scheme is proven to be homomorphically EUF-KMA resistant under co-CDH in the ROM [24, Theorem 6], assuming that $\frac{1}{p} = \mathsf{negl}(\lambda)$.

For a keypair $(\mathsf{pk}, \mathsf{sk})$, two scalars β_1, β_2, two vectors $\mathbf{v}_1, \mathbf{v}_2$ and their respective signatures σ_1, σ_2, we have $\mathsf{BFKW}_m.\mathsf{Verify}(\mathsf{pk}, \beta_1 \mathbf{v}_1 + \beta_2 \mathbf{v}_2, \sigma_1^{\beta_1} \sigma_2^{\beta_2}) = 1$. The scheme relies on $m + 1$ public parameters (generators of a finite group), which can be stored compactly by storing the seed that generated them. For more details, please refer to Appendix A.2 or [24].

4.2   HSig-BigLUT Construction

We now show how to leverage the BFKW signature scheme in a code-hard solving problem with public verification.

Let us consider a lookup table LUT consisting with L entries. Each entry $LUT[i]$ is of the form $LUT[i] = \mathsf{BFKW}_L.\mathsf{Sign}(\mathbf{e}_i)$, where $\mathbf{e}_i$ is the i-th element of the canonical base of $\mathbb{F}_p^L$.

Our problem instances are of the form $x \in \{0,1\}^*$, which is hashed via a hash function H of codomain $(\mathbb{F}_p^*)^L$. The solver has to return a linear combination of all the signatures from the LUT, where the coefficients are taken from $H(x)$. In other words the solver must return $\mathsf{BFKW}_L.\mathsf{Sign}\left(\sum_i H(x)[i] \cdot \mathbf{e}_i\right)$. Verification of the challenge is a simple signature verification.

We thus formalize the algorithms in Algorithm 2.

Algorithm 2. HSig-BigLUT$_u^\lambda$ Generate, Solve and Verify algorithms
Parameter: h is a hash function such that $h(x)$ returns L values of $\mathbb{F}_p^*$

---

function GENERATE(u, λ)
    $p \leftarrow$ prime of λ bits
    $L \leftarrow \lfloor \frac{u}{\lambda} \rfloor$
    pk, sk $\leftarrow$ instance of BFKW with a
    random id id, and for signing over the
    canonical basis $(\mathbf{e}_i)_{1 \leq i \leq L}$ of $\mathbb{F}_p^L$
    $LUT \leftarrow$ empty table of L entries
    for $i \in \{1, \ldots, L\}$ **do**
        $LUT[i] \leftarrow$ BFKW$_\mathsf{L}$.Sign(sk, $\mathbf{e}_i$)
    end for
    $\mathfrak{C} \leftarrow (\{0,1\}^*, \mathsf{Solve}^{LUT}, \mathsf{Verify})$

    return $(\mathfrak{C}, \mathsf{sk}, \perp)$
                 ▷ solving trapdoor is sk
end function

function SOLVE$^{LUT}(x)$
$$s \leftarrow \prod_{i=1}^{L} LUT[i]^{h(x)[i]}$$
    return s
end function

function VERIFY(x, s)
$$\mathbf{v} \leftarrow \sum_{i=1}^{L} h(x)[i] \cdot \mathbf{e}_i$$
    return BFKW$_\mathsf{L}$.Verify(pk, $\mathbf{v}$, s)
end function

Theorem 4. *Under the co-CDH in the ROM and assuming $1/p = \mathsf{negl}(\lambda)$, HSig-BigLUT consisting of L entries of size γ is systematically $(p, \gamma L - \lambda, \mathsf{Code}, \delta, \mathbb{1}_{p=0}\mathbb{1}_{\delta<1})$-trapdoor solving hard against an 2^λ bounded adversary, and verification can be done with probability 1 using $O(\lambda)$ resources in Code, where γ is the compressed size of the BFKW scheme (i.e., the size of signatures once compressed)*[4].

Proof. Verification hardness. This part is trivial: for verification, one only needs the simple code written in Algorithm 2, which does not make any call to any lookup table. The solver only needs to store the public parameters, which essentially consist of $L + 1$ generators of a finite group (and one bilinear map), which can be compactly stored by storing the random seed that generated them.

Solving hardness First, a solver with trapdoor solves in $O(\lambda)$ by signing the message with the private key. Let us now focus on the case where the solver does not have the trapdoor.

We first see that an attacker can safely remove λ/L bits to each entry and yet compute the solution to any input in 2^λ steps. Thus, we consider the base u to be $u = \gamma L - \lambda$. Let us now consider an attacker that removes even more bits. Since, for any input, the adversary must reconstruct all signatures of the LUT, then bruteforcing the solution to a problem instance will cost them more than 2^λ operations, which is impossible. Thus a success probability of $p > 0$ indicates that the adversary has at least u bits of Code complexity, thus the result.

---

[4] Since BFKW relies on elliptic curves, γ is expected to be close from λ.

5 Trapdoor Proof of CMC: **Mem**, Trapdoored Solving, Easy Verification Problem Class

5.1 A Primer on Diodon [17]

Diodon (see Algorithm 3) is a memory trapdoored hard solving problem. [17] shows that without trapdoor, one must operate $\eta \times M$ squarings while storing $M \times \log_2(N)$ bits, while a user in possession of the trapdoor can instead use $L \times \log_2(N)$ squaring operations and $2\log_2(N)$ bits of memory. Its complexity is estimated around $\Omega(LM\log_2(N))$ in the CMC model [17, Appendix A].

There is no known efficient trapdoorless verification of Diodon.

Algorithm 3. Diodon Solve algorithm [17]
Parameters: public key $N = pq$ of λ bits, $\eta \in \mathbb{N}^*$, memory requirement M, length requirement L, problem instance $x \in \mathbb{Z}/N\mathbb{Z}$, H a hash function of λ bits

$V_0 \leftarrow x$ ▷ Expansion phase
for $i \in \{1,\ldots,M\}$ **do**
  $V_i \leftarrow V_{i-1}^{2^\eta} \mod N$
end for
$S_0 \leftarrow H(V_M)$ ▷ Hash chaining phase
for $i \in \{1,\ldots,L\}$ **do**
  $k_i \leftarrow S_{i-1} \mod M$
  $S_i \leftarrow H(S_{i-1}, V_{k_i})$
end for
return S_L

5.2 A Primer on VDFs

Wesoloski VDF (Verifiable Delay Function, [51], also described in Appendix A.1) relies on iterated squaring in RSA groups, and is $(p, T, \mathsf{SeqTime}, \delta, \mathbb{1}_{\delta \geq 1})$ trapdoor solving hard (assuming that factorization of the RSA modulus is computationally intractable), with easy verification. More details are given in [51]. In this paper, for (y, π) a solution to a Wesolowski VDF problem instance, we often refer to π as the *Wesolowski proof of y*.

5.3 Trapdoor Proof of CMC: The General Idea

The main idea of our proof of CMC is to adapt Diodon so that the verifier can validate a proposed solution without spending a significant amount of memory. This is done by, at each step of the hash chaining phase, adding a VDF proving the validity of the value. These VDFs are then aggregated in a Merkle tree.

A high-level view of the protocol is described in Fig. 1, and the full algorithm is described in Algorithm 4.

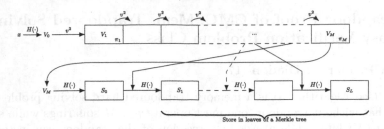

Fig. 1. A high-level view of the first steps (expansion and hash chaining phases) of the trapdoor proof of CMC. π_i is the Wesolowsky proof of the VDF computation of V_i. Then, we form a Merkle tree from the hash chain.

5.4 Trapdoored Proof of CMC Protocol

Summing it up, we get the Solve and Verify algorithms of our trapdoored proof of CMC in Algorithm 4.

Algorithm 4. Trapdoor proof of CMC TdPoCMC$_{M,L,k}^{\lambda}$ Solve and Verify algorithms
Parameters: N a RSA module of λ bits, M a memory requirement, L a hashing chain length, k a number of openings

---

function SOLVE(x) ▷ $x \in \mathcal{P}_u^{\lambda} = \mathbb{Z}/N\mathbb{Z}$
  $V_0 = x$
  for $\in \{1, \ldots, M\}$ **do**
    $V_i \leftarrow V_{i-1}^2 \mod N$
  end for
  $s \leftarrow V_L$
  $S_0 \leftarrow H(V_L) \mod L$
  for $i \in \{1, \ldots, L\}$ **do**
    $k_i \leftarrow S_{i-1} \mod L$
    $\pi_i \leftarrow$ the Wesolowski proof of
computation of V_{k_i}
    $S_i \leftarrow H(i, S_{i-1}, V_{k_i})$
  end for
  $r \leftarrow$ the root of a Merkle tree where the leaves are the tuples $\ell_i = (V_{k_i}, S_{i-1}, \pi_i)$
  $p \leftarrow$ Open k paths using the Fiat Shamir heuristic. For each opening that leads to leaf ℓ_i, also include ℓ_{i+1}
  return r, p
end function

function VERIFY($x, (r, p)$)
  for each opening o in p **do**
    $\ell := (V, S, \pi) \leftarrow$ the opening of the leaf, contained in o
    $i \leftarrow$ the index of ℓ in the tree
    $\ell' := (V', S', \pi') \leftarrow$ the following leaf in the tree, contained in o
    Check that (V, π) corresponds to a valid Wesolowski proof of the computation of $x^{2^{S \mod L}} \mod N$
    Check that the path from ℓ is valid and leads to the root r
    $S'' \leftarrow H(i, S, V)$
    Check that S'' is equal to S'
  end for
  If at least one of the check fails, return 0, otherwise return 1
end function

Theorem 5. *In a group where iterated squaring is $(p, T, SeqTime, \delta, \mathbb{1}_{\delta \geq 1})$ solving hard, the trapdoor proof of CMC problem class* TdPoCMC$_{M,L,k}^{\lambda}$ *is $((pq)^k, \frac{qML\lambda}{25}, Mem, 1 - \frac{100 \log M}{\lambda}, 1 - p + \mathsf{negl}(M) + \mathsf{negl}(\lambda))$ solving hard in the parallel ROM.*

Proof. Let us consider a solver that wishes to minimize their CMC, while having a good probability of passing verification. Looking at the security proof of scrypt, we see that the CMC of creating a (continuous) hash chain of size L/E, following an expansion phase of size M, with words of size λ, is around $\Omega(\frac{LM\lambda}{E})$. Because the hashes are index dependent, a chain cannot be reused at another place without immediately introducing a discontinuity.

Hence, computing the hash chain (with E discontinuities) will cost $\Omega(LM\lambda)$ to an attacker no matter what.

Furthermore, as discussed previously, the verifier, when asking for opening of leaf i, not only asks for S_i, V_i but also S_{i+1}, V_{i+1} so they can verify the chain has no discontinuity at step i. Let q be the proportion of steps correctly computed by the solver (without loss of generality, we can assume that other steps are computed at cost 0, and are always incorrect). Given results on scrypt, and notably the fact that scrypt is $(p, \frac{M^2\lambda}{25}, \mathsf{Mem}, 1 - \frac{100\log M}{\lambda}, 1 - p + 0.08M^6 2^{-\lambda} + 2^{-M/20})$ solving hard (scrypt uses $L = M$), we can assume that the generation of a 2-chain is $(p, \frac{2M\lambda}{25}, \mathsf{Mem}, 1 - \frac{100\log M}{\lambda}, 1 - p + \mathsf{negl}(M) + \mathsf{negl}(\lambda))$ hard. The probability that k openings do not reveal a cheater is then of $(pq)^k$. Thus, our protocol with k openings is at least $((pq)^k, \frac{ML\lambda}{25}, \mathsf{Mem}, 1 - \frac{100\log M}{\lambda}, 1 - p + \mathsf{negl}(M) + \mathsf{negl}(\lambda))$ hard, hence the claim.

We now give the verification complexity of the trapdoor proof of CMC. Most importantly, the verification cost does not depend on M, which allows verification to be much smaller than solving when $L \ll M$.

Theorem 6. *The trapdoor proof of CMC problem class* $\mathsf{TdPoCMC}_{M,L,k}^{\lambda}$ *can be verified in a CMC of at most* $O\left(k^2\lambda \log \frac{L}{k}(\lambda + \log L)\right)$.

Proof. A verifier receives k openings of the Merkle tree (plus one root value that can be neglected). An opening consists of:

- One path down the Merkle tree ($\log L$ nodes of size λ each)
- Two leaves, the leaf ℓ_i and the following leaf ℓ_{i+1}. We have $\ell_i = (V_{k_i}, S_{i-1}, \pi_i)$ hence one leaf is $\lambda + \lambda + 2\lambda = 4\lambda$ bits.

Verifying the VDF takes $O(\lambda)$ operations, verifying the tree opening takes $O(\log L)$ time, hence k verifications take $O(k(\lambda + \log L))$ time.

An opening consists of one node each of depth $1, 2, \ldots, L - 1$, and two nodes of depth L. Thus, with k openings, there are at most 2^j nodes of depth $j + 1$ for $j + 1 \leq \log(k)$. On the other hand, we cannot assume that there will be any collision on the nodes of depth higher than $\log k$. In total, we thus have k nodes of depth lower than $\log k$, and $k(\log L - \log k)$ nodes of higher depth, thus a higher bound of $k(\log L + 1 - \log k)$ nodes. Summing it, the proof uses $k(\log L + 1 - \log k)$ nodes of size λ and $2k$ leaves of size 4λ each, hence a memory requirement of at most $O\left(k\lambda \log \frac{L}{k}\right)$.

Multiplying the time requirements with the memory requirements, we get a CMC of at most $O\left(k^2\lambda \log \frac{L}{k}(\lambda + \log L)\right)$.

6 SeqTime Challenge: SeqTime Systematic Hard Solving and Trapdoored Hard Verifying Problem Class

6.1 A Primer on Proofs of Sequential Work

Proofs of sequential work can be seen as VDFs for which there exist cheating strategies, hence while it can be proved that the cheater must spend at least some time to solve the problem, the solution is not unique.

While other constructions exist (see for instance [39]) we hereby rely on the proof of sequential work $POSW_{n,w,t}$ described in [28], which is proven to be systematically $(p, 2^{n+1} - 1, \mathsf{SeqTime}, \delta, \mathbb{1}_{\delta \geq \left(1 - p - \frac{nw}{2^{w}-2q-1}\right)^{1/t}})$ solving hard in the parallel ROM against a 2^q bounded adversary (in oracle calls), where n, w, t are various parameters. See [28] for more details, or Appendix A.3.

6.2 Our Construction

The construction we propose in this section, SeqTime challenge, consists of solving two puzzles, one being trapdoored hard and the other one being trapdoorless in SeqTime. Both puzzles are solved in parallel, then we use a construction similar to the lock puzzles (see Sect. 3.3).

Let x be the solution to the iterated squaring problem, and y, z the solution to the POSW instance, along with its proof. We require that the hash size is bigger than y, i.e. $|h(x)| \geq |y|$. The algorithms for Solve and Verify are summarized in Algorithm 5.

Algorithm 5. SeqTime Challenge$^{\lambda}_{T,w,q,t}$ Solve and Verify algorithm
Parameters: A semiprime N of λ bits, a time parameter T (being a power of 2), a hash function H_x of codomain size 2^x

---

function SOLVE(a, b)
  $s, o \leftarrow POSW_{\log_2 T, w, t}.\mathsf{Solve}(a)$
  $y \leftarrow b^{2^{2T-1}} \mod N \triangleright s$ and y
are computed in parallel
  $h \leftarrow H_{|o|}(y)$
  return $(s, o \oplus h)$
end function

function VERIFY$((a, b), (s, c))$
  $y \leftarrow b^{2^{2T-1}} \mod N$
  $o \leftarrow H_{|c|}(y) \oplus c$
  return $POSW_{\log_2 T, w, t}.\mathsf{Verify}(a, (s, o))$
end function

Theorem 7. *Let $T, w \in \mathbb{N}, H$ be a hash function of codomain $\{0, 1\}^w$.*

Let us use a group where, against a 2^λ bounded adversary, iterated squaring is systematically $(p, T, \mathsf{SeqTime}, \delta, \mathbb{1}_{\delta \geq 1})$. Against a 2^λ bounded adversary with at most 2^q oracle calls, the SeqTime challenge is systematically $(p, 2T - 1, \mathsf{SeqTime}, \delta, \mathbb{1}_{\delta \geq 1})$ solving hard in the parallel ROM. If the solver knows the trapdoor, then the problem class is systematically $(p, 2T - 1, \mathsf{SeqTime}, \delta, \mathbb{1}_{\delta \geq \left(1 - p - \frac{\log_2(T)w}{2^{w}-2q-1}\right)^{1/n}})$ solving hard in the random oracle model.

The *SeqTime* challenge is also $(p, 2T - 1, SeqTime, \delta, \mathbb{1}_{\delta \geq 1})$ trapdoored verifying hard on $\{0, 1\}^w$.

Proof. **Solving hardness.** On instance (a, b), a honest solver will run $\mathrm{POSW}_{\log_2 T, w, t}.\mathsf{Solve}(a)$ and compute $b^{2^{2T-1}} \bmod N$ in parallel. Computing $b^{2^{2T-1}} \bmod N$ requires at least T consecutive steps to have a nonnegligible chance of success, so there is no incentive for the solver to find a solution to the POSW in less than T consecutive steps, which then gives a success probability of 1.

Moreover, in order to get a valid solution from an instance a, b, the solver has to output (s, c) such that $c \oplus (b^{2^{2T-1}} \bmod N)$ is a valid opening for the POSW DAG tree of challenge a. Let us assume that the solver does not compute the value $b^{2^{2T-1}} \bmod N$. Because at least one bit of y defined as $y \leftarrow b^{2^{2T-1}} \bmod N$ can be seen as random (see [19]), the value $H_{|c|}(y)$ is random in the random oracle model, hence $c \oplus H_{|c|}(y)$ is random. Because the string $c \oplus H_{|c|}(y)$ is supposed to be a valid opening, it must contain at least the two hashes that lead to the root label, i.e. contain the labels l_1, l_2 such that $H(x, \varepsilon, l_1, l_2) = s$. Hence the random string $c \oplus H_{|c|}(y)$ must contain a preimage of s, which is negligible in the random oracle model. Hence, we conclude that the only strategy leading to a valid solution with nonnegligible probability must include a valid computation of $b^{2^{2T-1}} \bmod N$, at which point the adversary has no interest in deviating from the honest execution of the protocol.

However, we note that the solver in possession of the trapdoor can compute $b^{2^{2T-1}} \bmod N$ using the trapdoor, and is only limited by the solving complexity of the POSW.

Verification hardness. Let us assume that there exists an algorithm $\mathcal{A}$ such that there exists $p > 0$ with $\mathsf{Adv}_v(\mathcal{A}) > p$. If $\mathcal{A}$ takes less than $2T - 1$ steps, they cannot compute $b^{2^{2T-1}} \bmod N$, hence, using the same argument as above, must verify the POSW with nothing but a random string. We conclude that if $p > 0$ (i.e. if $\mathcal{A}$ has any advantage better than random), then it is impossible to succeed in less than $2T - 1$ consecutive steps, hence the complexity.

7   Conclusion and Future Work

In this paper, we presented the PURED framework, uniting the hardness models over different resources, and explored its properties. We also introduced three new problem classes, with a proven hardness.

Another interesting consideration to add would be to see how one can embed trustless problem generation in the framework.

Finally, we observe that in the literature, many schemes that are trapdoored hard to solve and verify use the same trapdoor for solving and verifying. It would be interesting to investigate on schemes relying on different trapdoors for both solving and verifying, and explore how this particularity could be embedded into the PURED framework.

Acknowledgements. The authors would like to thank Aleksei Udovenko for his suggestions on how to improve the HSig-BigLUT performance. The work was partly supported by the Luxembourg National Research Fund's (FNR) project CryptoFin C22/IS/17415825.

A   Related Constructions

A.1   Wesolowski's VDF [51]

The description of Wesolowski algorithm is summarised in Algorithm 6.

Algorithm 6. Wesolowski VDF Solve and Verify algorithm [51]

Parameters : G a group of unknown order, $g \in G$, $T \in \mathbb{N}$, H_{prime} hashes to a prime of 2λ bits, $\mathtt{bin}$ outputs a binary representation

function SOLVE(g, T)
   $y \leftarrow g^{2^T} \mod N$
   $\pi \leftarrow g^{\lfloor 2^T/\ell \rfloor} \mod N$
end function

function VERIFY$((g, T), (y, \pi))$
   $\ell \leftarrow H_{\text{prime}}(\mathtt{bin}(g) \| \mathtt{bin}(y))$
   $r \leftarrow$ remainder of 2^T divided by ℓ
   return $\pi^\ell g^r \overset{?}{=} y$
end function

A.2   BFKW Scheme [24]

Given a security parameter λ, a number m of vectors from $\mathbb{F}_p^n$ with $p > 2^\lambda$, the BFKW schemes generates the following (public) parameters:

Two groups $\mathbb{G}, \mathbb{G}_T$ of prime order p, a bilinear map $e : \mathbb{G} \times \mathbb{G} \to \mathbb{G}_T$, random generators[5] $g_1, \ldots, g_n, h$ of $\mathbb{G}$, a hash function H that maps to $\mathbb{G}$, and id is a public nonce to prevent signature reuse in a different setup[6], as well as vectors $\mathbf{v}_1, \ldots, \mathbf{v}_m$ to be signed.

The secret key is sk $\overset{\$}{\leftarrow} \mathbb{Z}_p$. The public key is pk $\leftarrow h^{\text{sk}}$. BFKW$_m$.Sign and BFKW$_m$.Verify algorithms are described in Algorithm 7. It is assumed that the decomposition in the base $(\mathbf{v}_1, \ldots, \mathbf{v}_m)$ is an easy task. Furthermore, for any coefficients β_1, β_2, and any two signature pairs $(\mathbf{m}_1, \sigma_1), (\mathbf{m}_2, \sigma_2)$, the signature of the message $\beta_1 \mathbf{m}_1 + \beta_2 \mathbf{m}_2$ is BFKW.Combine(pk, $(\{\beta_1, \sigma_1\}, \{\beta_2, \sigma_2\})) = \sigma_1^{\beta_1} \sigma_2^{\beta_2}$.

A.3   Proofs of Successive Work [28]

Let $n \in \mathbb{N}$, we create a complete binary tree (V, E) of depth n. G_n^{POSW} is defined as follows: $G_n^{\text{POSW}} = (V, E \cup E')$, where $(u, v) \in E'$ if and only if v is a leaf and u is a left sibling of a node belonging to the shortest path from v to the root. We also identify each node of depth n with a string, from 0^n to 1^n, in

---

[5] For code compactness, a user might prefer to store the random seed leading to these generators.

[6] In our context, the nonce prevents a cheating solver from using twice the same lookup table for two different code-hard instances.

Algorithm 7. BFKW_m Sign and Verify algorithms over the m-dimensional vector space $\text{Span}(\mathfrak{B}) \subset \mathbb{F}_p^n$, with $\mathfrak{B} = (\mathbf{v}_1, \ldots, \mathbf{v}_m)$ [24]

---

function $\text{BFKW}_m.\text{SIGN}(\text{sk}, \mathbf{w})$

Decompose w in $\mathfrak{B}$: $w = \sum_{i=1}^{m} \alpha_i \mathbf{v}_i$

return $\left(\prod_{i=1}^{m} H(id,i)^{\alpha_i} \prod_{j=1}^{n} g_j^{w_j} \right)^{\text{sk}}$

▷ w_j is the j-th coordinate of $\mathbf{w}$

end function

function $\text{BFKW}_m.\text{VERIFY}(\text{pk}, \mathbf{w}, \sigma)$

Decompose w in $\mathfrak{B}$: $w = \sum_{i=1}^{m} \alpha_i \mathbf{v}_i$

return $e(h, \sigma) \overset{?}{=}$

$e\left(\text{pk}, \prod_{i=1}^{m} H(\text{id},i)^{\alpha_i} \prod_{j=1}^{n} g_j^{w_j} \right)$

end function

---

the lexicographic order from left to right (the root is identified with ε). As an example, we give G_4^{POSW} in Fig. 2.

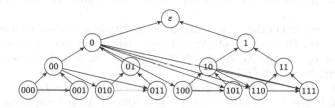

Fig. 2. Representation of the DAG G_4^{POSW}

The proof, for a challenge x, in computing the label of the root of the DAG, where the label of a node u having parents[7] $p_1, \ldots, p_k$ each of respective label $l_{p_1}, \ldots, l_{p_k}$ is given by $l_u = H(x, u, l_{p_1}, \ldots, l_{p_k})$, with H a hash function.

The POSW solving consists in labelling the root of the DAG. A node label depends on all its childrens nodes labels, thus a intuitively a solver must iteratively label each node. The solution to the problem instance consists of the root of the tree, along with a random opening of the tree à la Merkle: for a challenge leaf node u, the solver sends the labels of u along with the labels of all the siblings of the nodes on the path from u to the root ε. Then, the verifier checks that the labels do lead to the proposed solution.

Theorem 8 (from [28]). *The problem class $POSW_n$, using the DAG G_n^{POSW}, the hash function H of codomain $\{0,1\}^w$, is systematically $(p, \mathsf{SeqTime}, 2^{n+1} - 1, \delta, \mathbb{1}_{\delta \geq \left(1 - p - \frac{2nwq^2}{2^w}\right)^{1/n}})$ solving hard in the random oracle model, where q is the total number of (not necessarily sequential) queries to H.*

Moreover, there exists a verification algorithm that takes no more than n steps in $\mathsf{SeqTime}$.

---

[7] For an oriented graph $G = (V, E)$, u is a parent of v if $(u, v) \in E$.

References

1. Abadi, M., Burrows, M., Manasse, M., Wobber, T.: Moderately hard, memory-bound functions. ACM Trans. Internet Technol. **5**(2), 299–327 (2005)
2. Abliz, M., Znati, T.: A guided tour puzzle for denial of service prevention. In: 2009 Annual Computer Security Applications Conference, pp. 279–288 (2009)
3. Ali, I.M., Caprolu, M., Pietro, R.D.: Foundations, properties, and security applications of puzzles: a survey. ACM Comput. Surv. **53**(4), 72:1–72:38 (2020)
4. Alwen, J., Blocki, J.: Efficiently computing data-independent memory-hard functions. In: Robshaw, M., Katz, J. (eds.) CRYPTO 2016. LNCS, vol. 9815, pp. 241–271. Springer, Heidelberg (2016). https://doi.org/10.1007/978-3-662-53008-5_9
5. Alwen, J., Blocki, J., Pietrzak, K.: Sustained space complexity. In: Nielsen, J.B., Rijmen, V. (eds.) EUROCRYPT 2018. LNCS, vol. 10821, pp. 99–130. Springer, Cham (2018). https://doi.org/10.1007/978-3-319-78375-8_4
6. Alwen, J., Chen, B., Pietrzak, K., Reyzin, L., Tessaro, S.: Scrypt is maximally memory-hard. In: Coron, J.-S., Nielsen, J.B. (eds.) EUROCRYPT 2017. LNCS, vol. 10212, pp. 33–62. Springer, Cham (2017). https://doi.org/10.1007/978-3-319-56617-7_2
7. Alwen, J., et al.: On the memory-hardness of data-independent password-hashing functions. In: Proceedings of the 2018 on Asia Conference on Computer and Communications Security, ASIACCS 2018, pp. 51–65. Association for Computing Machinery, New York (2018)
8. Alwen, J., Serbinenko, V.: High parallel complexity graphs and memory-hard functions. In: Proceedings of the Forty-Seventh Annual ACM Symposium on Theory of Computing, STOC 2015, pp. 595–603. Association for Computing Machinery, New York (2015). https://doi.org/10.1145/2746539.2746622
9. Ateniese, G., Chen, L., Francati, D., Papadopoulos, D., Tang, Q.: Verifiable capacity-bound functions: a new primitive from Kolmogorov complexity (revisiting space-based security in the adaptive setting). Cryptology ePrint Archive, Paper 2021/162 (2021). https://eprint.iacr.org/2021/162,
10. Babai, L., Fortnow, L., Levin, L.A., Szegedy, M.: Checking computations in poly-logarithmic time. In: Koutsougeras, C., Vitter, J.S. (eds.) Proceedings of the 23rd Annual ACM Symposium on Theory of Computing, New Orleans, Louisiana, USA, 5–8 May 1991, pp. 21–31. ACM (1991). https://doi.org/10.1145/103418.103428
11. Back, A.: Hashcash - a denial of service counter-measure (2002)
12. Baldimtsi, F., Kiayias, A., Zacharias, T., Zhang, B.: Indistinguishable proofs of work or knowledge. In: Cheon, J.H., Takagi, T. (eds.) ASIACRYPT 2016. LNCS, vol. 10032, pp. 902–933. Springer, Heidelberg (2016). https://doi.org/10.1007/978-3-662-53890-6_30
13. Biryukov, A., Bouillaguet, C., Khovratovich, D.: Cryptographic schemes based on the ASASA structure: black-box, white-box, and public-key (extended abstract). In: Sarkar, P., Iwata, T. (eds.) ASIACRYPT 2014. LNCS, vol. 8873, pp. 63–84. Springer, Heidelberg (2014). https://doi.org/10.1007/978-3-662-45611-8_4
14. Biryukov, A., Dinu, D., Khovratovich, D.: Argon2: new generation of memory-hard functions for password hashing and other applications. In: 2016 IEEE European Symposium on Security and Privacy (EuroS&P), pp. 292–302 (2016). https://doi.org/10.1109/EuroSP.2016.31
15. Biryukov, A., Khovratovich, D.: Egalitarian computing (MTP 1.2). CoRR abs/1606.03588 (2016). http://arxiv.org/abs/1606.03588

16. Biryukov, A., Khovratovich, D.: Equihash: asymmetric proof-of-work based on the generalized birthday problem. Ledger **2**, 1–30 (2017). https://doi.org/10.5195/ledger.2017.48. https://ledger.pitt.edu/ojs/ledger/article/view/48
17. Biryukov, A., Perrin, L.: Symmetrically and asymmetrically hard cryptography. In: Takagi, T., Peyrin, T. (eds.) ASIACRYPT 2017. LNCS, vol. 10626, pp. 417–445. Springer, Cham (2017). https://doi.org/10.1007/978-3-319-70700-6_15
18. Blocki, J., Ren, L., Zhou, S.: Bandwidth-hard functions: reductions and lower bounds. In: Proceedings of the 2018 ACM SIGSAC Conference on Computer and Communications Security, CCS 2018, pp. 1820–1836. Association for Computing Machinery, New York (2018)
19. Blum, L., Blum, M., Shub, M.: Comparison of two pseudo-random number generators. In: Chaum, D., Rivest, R.L., Sherman, A.T. (eds.) Advances in Cryptology, pp. 61–78. Springer, Boston (1983). https://doi.org/10.1007/978-1-4757-0602-4_6
20. Bogdanov, A., Isobe, T.: White-box cryptography revisited: space-hard ciphers. In: Proceedings of the 22nd ACM SIGSAC Conference on Computer and Communications Security, CCS 2015, pp. 1058–1069. Association for Computing Machinery, New York (2015). https://doi.org/10.1145/2810103.2813699
21. Bogdanov, A., Isobe, T., Tischhauser, E.: Towards practical whitebox cryptography: optimizing efficiency and space hardness. In: Cheon, J.H., Takagi, T. (eds.) ASIACRYPT 2016. LNCS, vol. 10031, pp. 126–158. Springer, Heidelberg (2016). https://doi.org/10.1007/978-3-662-53887-6_5
22. Boneh, D., Bonneau, J., Bünz, B., Fisch, B.: Verifiable delay functions. In: Shacham, H., Boldyreva, A. (eds.) CRYPTO 2018. LNCS, vol. 10991, pp. 757–788. Springer, Cham (2018). https://doi.org/10.1007/978-3-319-96884-1_25
23. Boneh, D., Corrigan-Gibbs, H., Schechter, S.: Balloon hashing: a memory-hard function providing provable protection against sequential attacks. In: Cheon, J.H., Takagi, T. (eds.) ASIACRYPT 2016. LNCS, vol. 10031, pp. 220–248. Springer, Heidelberg (2016). https://doi.org/10.1007/978-3-662-53887-6_8
24. Boneh, D., Freeman, D., Katz, J., Waters, B.: Signing a linear subspace: signature schemes for network coding. In: Jarecki, S., Tsudik, G. (eds.) PKC 2009. LNCS, vol. 5443, pp. 68–87. Springer, Heidelberg (2009). https://doi.org/10.1007/978-3-642-00468-1_5
25. Chase, M., Lysyanskaya, A.: On signatures of knowledge. In: Dwork, C. (ed.) CRYPTO 2006. LNCS, vol. 4117, pp. 78–96. Springer, Heidelberg (2006). https://doi.org/10.1007/11818175_5
26. Chen, M., et al.: Multiparty generation of an RSA modulus. J. Cryptol. **35**(2), 12 (2022)
27. Coelho, F., Larroche, A., Colin, B.: Itsuku: a memory-hardened proof-of-work scheme. IACR Cryptology ePrint Archive, p. 1168 (2017). http://eprint.iacr.org/2017/1168
28. Cohen, B., Pietrzak, K.: Simple proofs of sequential work. In: Nielsen, J.B., Rijmen, V. (eds.) EUROCRYPT 2018. LNCS, vol. 10821, pp. 451–467. Springer, Cham (2018). https://doi.org/10.1007/978-3-319-78375-8_15
29. De Feo, L., Masson, S., Petit, C., Sanso, A.: Verifiable delay functions from supersingular isogenies and pairings. In: Galbraith, S.D., Moriai, S. (eds.) ASIACRYPT 2019. LNCS, vol. 11921, pp. 248–277. Springer, Cham (2019). https://doi.org/10.1007/978-3-030-34578-5_10
30. Delerablée, C., Lepoint, T., Paillier, P., Rivain, M.: White-box security notions for symmetric encryption schemes. In: Lange, T., Lauter, K., Lisoněk, P. (eds.) SAC 2013. LNCS, vol. 8282, pp. 247–264. Springer, Heidelberg (2014). https://doi.org/10.1007/978-3-662-43414-7_13

31. Dobson, S., Galbraith, S.D.: Trustless groups of unknown order with hyperelliptic curves. IACR Cryptology ePrint Archive, p. 196 (2020). https://eprint.iacr.org/2020/196
32. Dwork, C., Naor, M.: Pricing via processing or combatting junk mail. In: Brickell, E.F. (ed.) CRYPTO 1992. LNCS, vol. 740, pp. 139–147. Springer, Heidelberg (1993). https://doi.org/10.1007/3-540-48071-4_10
33. Dziembowski, S., Faust, S., Kolmogorov, V., Pietrzak, K.: Proofs of space. In: Gennaro, R., Robshaw, M. (eds.) CRYPTO 2015. LNCS, vol. 9216, pp. 585–605. Springer, Heidelberg (2015). https://doi.org/10.1007/978-3-662-48000-7_29
34. Ethereum Community: ethash—ethereum wiki. https://eth.wiki/en/concepts/ethash/ethash
35. Garg, S., Gentry, C., Sahai, A., Waters, B.: Witness encryption and its applications. In: Proceedings of the Forty-Fifth Annual ACM Symposium on Theory of Computing, STOC 2013, pp. 467–476. Association for Computing Machinery, New York (2013). https://doi.org/10.1145/2488608.2488667
36. Goldwasser, S., Micali, S., Rackoff, C.: The knowledge complexity of interactive proof-systems. In: Proceedings of the Seventeenth Annual ACM Symposium on Theory of Computing, STOC 2085, pp. 291–304. Association for Computing Machinery, New York (1985)
37. Kamara, S.: Proofs of storage: theory, constructions and applications. In: Muntean, T., Poulakis, D., Rolland, R. (eds.) CAI 2013. LNCS, vol. 8080, pp. 7–8. Springer, Heidelberg (2013). https://doi.org/10.1007/978-3-642-40663-8_4
38. Liu, J., Jager, T., Kakvi, S.A., Warinschi, B.: How to build time-lock encryption. Des. Codes Crypt. **86**(11), 2549–2586 (2018). https://doi.org/10.1007/s10623-018-0461-x
39. Mahmoody, M., Moran, T., Vadhan, S.: Publicly verifiable proofs of sequential work. In: Proceedings of the 4th Conference on Innovations in Theoretical Computer Science, ITCS 2013, pp. 373–388. Association for Computing Machinery, New York (2013). https://doi.org/10.1145/2422436.2422479
40. Percival, C.: Stronger key derivation via sequential memory-hard functions (2009)
41. Pietrzak, K.: Simple verifiable delay functions. In: Blum, A. (ed.) 10th Innovations in Theoretical Computer Science Conference, ITCS 2019, San Diego, California, USA, 10–12 January 2019. LIPIcs, vol. 124, pp. 60:1–60:15. Schloss Dagstuhl - Leibniz-Zentrum für Informatik (2019). https://doi.org/10.4230/LIPIcs.ITCS.2019.60
42. Poupard, G., Stern, J.: Short proofs of knowledge for factoring. In: Imai, H., Zheng, Y. (eds.) PKC 2000. LNCS, vol. 1751, pp. 147–166. Springer, Heidelberg (2000). https://doi.org/10.1007/978-3-540-46588-1_11
43. Provos, N., Mazières, D.: A future-adaptive password scheme. In: Proceedings of the Annual Conference on USENIX Annual Technical Conference, ATEC 1999, p. 32. USENIX Association (1999)
44. Ren, L., Devadas, S.: Bandwidth hard functions for ASIC resistance. In: Kalai, Y., Reyzin, L. (eds.) TCC 2017. LNCS, vol. 10677, pp. 466–492. Springer, Cham (2017). https://doi.org/10.1007/978-3-319-70500-2_16
45. Rivest, R.L., Shamir, A., Wagner, D.A.: Time-lock puzzles and timed-release crypto. Technical report, Massachusetts Institute of Technology, USA (1996)
46. Rivest, R.L.: Description of the LCS35 time capsule crypto-puzzle (1999). https://people.csail.mit.edu/rivest/lcs35-puzzle-description.txt
47. Schnorr, C.P.: Efficient identification and signatures for smart cards. In: Brassard, G. (ed.) CRYPTO 1989. LNCS, vol. 435, pp. 239–252. Springer, New York (1990). https://doi.org/10.1007/0-387-34805-0_22

48. Thompson, C.D.: Area-time complexity for VLSI. In: Proceedings of the Eleventh Annual ACM Symposium on Theory of Computing, STOC 1979, pp. 81–88. Association for Computing Machinery, New York (1979)
49. Vitto, G.: Factoring primes to factor moduli: backdooring and distributed generation of semiprimes. IACR Cryptology ePrint Archive, p. 1610 (2021). https://eprint.iacr.org/2021/1610
50. Walfish, M., Vutukuru, M., Balakrishnan, H., Karger, D., Shenker, S.: DDoS defense by offense. In: Proceedings of the 2006 Conference on Applications, Technologies, Architectures, and Protocols for Computer Communications, SIGCOMM 2006, pp. 303–314. Association for Computing Machinery, New York (2006)
51. Wesolowski, B.: Efficient verifiable delay functions. J. Cryptol. **33**(4), 2113–2147 (2020). https://doi.org/10.1007/s00145-020-09364-x
52. Youden, W.J.: Index for rating diagnostic tests. Cancer **3**(1), 32–35 (1950)

Post-quantum Cryptography

Implementing Lattice-Based PQC on Resource-Constrained Processors: A Case Study for Kyber/Saber's Polynomial Multiplication on ARM Cortex-M0/M0+

Lu Li[1,2] , Mingqiang Wang[1] , and Weijia Wang[1,2,3(✉)]

[1] School of Cyber Science and Technology, Shandong University, Qingdao, China
lulisdu@mail.sdu.edu.cn, {wangmingqiang,wjwang}@sdu.edu.cn
[2] Quan Cheng Laboratory, Jinan, China
[3] Key Laboratory of Cryptologic Technology and Information Security of Ministry
of Education, Shandong University, Jinan, China

Abstract. This paper studies the implementation of the lattice-based PQC on the 32-bit constrained processors that only have constrained multiplication instructions. A typical example of such constrained processors is the ARM Cortex-M0/M0+, which features a tiny silicon area, low power, and low cost. This paper focuses on implementing polynomial multiplication, the most challenging part in many resource-constrained processors. To achieve an efficient implementation compatible with the target platforms, we first investigate the features of different modular reduction algorithms (e.g., Montgomery reduction, Barrett reduction, and k-reduction) for ARM Cortex-M0/M0+. The investigation suggests a hybrid method combining k-reduction and Montgomery reduction, which is the most efficient with constrained multiplication instructions. Then, we combine two recent techniques, namely Number Theoretic Transform (NTT) multiplication for NTT-unfriendly rings and multi-moduli NTTs, to enable the calculation of NTT for Kyber and Saber in the absence of long multiplication instructions (i.e., the product is larger than 2^{32}). The above combination finally contributes to the fine-grained ARM Cortex-M0/M0+ implementations that significantly outperform state-of-the-art ones. Notably, we increase the running speed of Saber by a factor of ≈ 2.9 on Cortex-M0 and save up to 80% memory requirement for Kyber.

Keywords: Resource-Constrained Processors · Lattice-Based PQC · Kyber KEM · Saber KEM · Number Theoretic Transform

1 Introduction

Though it is still a huge challenge to build a large-scale quantum computer, it is expected that a quantum computer might be realized in the future. When it arrives, Shor's algorithm [32] will break almost all currently deployed public-key cryptography by solving the integer factorization and the discrete logarithms

© The Author(s), under exclusive license to Springer Nature Switzerland AG 2024
A. Chattopadhyay et al. (Eds.): INDOCRYPT 2023, LNCS 14460, pp. 153–176, 2024.
https://doi.org/10.1007/978-3-031-56235-8_8

problems. To cope with the arrival of the post-quantum era, NIST (the U.S. National Institute of Standards and Technology) has launched a PQC (Post Quantum Cryptographic) competition aiming to solicit, evaluate, and standardize quantum-resistant public-key cryptographic algorithms since 2016.

In July 2020, NIST announced 15 cryptosystems as advancing candidates for the third round standardization process [6]. More than half of the seven finalists are lattice-based cryptography. At the moment of writing this paper, Kyber [8], Dilithium [9], and Falcon [18] are selected as candidates to be standardized. In spite of Saber [11] was not selected as the candidate for standardization in the third round, the polynomial multiplication of Saber implemented by NTT technique is instructive. It can be applied to any other polynomial multiplication where the modulus is not NTT-friendly.

Nejatollahi et al. stated in their survey that "Implementation of the modulo arithmetic (multiplication and addition of big numbers) is a bottleneck in lattice-based cryptography" [28]. A very efficient way to perform multiplications in Ring-LWE is using the so-called Number-Theoretic Transform (NTT), with a complexity of $O(n \log(n))$. For example, Kyber deliberately chose its modulus $q = 3329$ with $q - 1 = 2^8 \cdot 13$. NTT-based multiplication is a natural choice for Kyber, and there are many works optimizing it on different platforms. In 2019, [13] presented an improved NTT implementation of Kyber on ARM Cortex-M4. In 2022, [5] applied various optimizations on the polynomial arithmetic to improve the speed of Kyber on the Cortex-M4. Although previous works [7,19, 21] pointed out that other parts of PQC, such as polynomial generation, usually are (even more critical) bottlenecks, the implementation of the NTT is the most challenging part in many resource-constraint processors.

The NTT-unfriendly schemes adopt moduli which are inherently incompatible with straightforward NTT. For example, Saber uses the power-of-two modulus 2^{13}. Most of the previous implementations of Saber use Toom-Cook multiplication [16,33], Karatsuba [22] multiplication and even simple schoolbook multiplication to accelerate the polynomial multiplications. See, e.g., [19,25,30,34] for an incomplete list of works. Meanwhile, Fritzmann et al. applied NTT on Saber for the first time by choosing a sufficiently large modular prime [19]. Then, [15] proposed efficient NTT-based polynomial multiplications with modular a power of two on ARM Cortex-M4 and AVX2. From then on, many works have emerged utilizing NTT to speed up the implementation of lattice-based schemes [4,12,29].

Nevertheless, most previous works consider advanced microcontrollers such as ARM Cortex-M4, ARM Cortex-A series, and even the x86 architecture. Those implementations usually require complex instructions such as multiplication with accumulation and Single Instruction/Multiple Data (SIMD) operations. This paper focuses on the 32-bit constrained microprocessors with the following features/limitations:

- Lack of long multiplication instructions (i.e., the product is $>2^{32}$).
- Lack of one-cycle multiplication instructions.
- Lack of multiplication with accumulation.

- Lack of SIMD instructions.
- A limited number of registers and memory size.

One typical instance of such microprocessors is the ARM Cortex-M0/M0+, released in 2009 [1,2] as the first product that cramps a 32-bit processor with a similar silicon footprint as an 8-bit or 16-bit processor. Cortex-M0/M0+ cores are optimized for low-cost and energy-efficient integrated circuits, which have been embedded in a wide range of products, i.e., microcontrollers, sensors, power management IC, ASSPs, ASICs, etc. They only support 56 instructions, which is quite small compared with Cortex-M3 and Cortex-M4.

1.1   Contributions

In this work, we take the NTT-based implementations of Kyber and Saber on ARM Cortex-M0/M0+ as a case study to investigate the approach to more efficiently implement the lattice-based PQC on constrained processors. Our technique can be applied to other small ideal-lattice-based cryptosystems, such as NTRU [14], Dilithium [9], and the NTT portion of the Falcon [18], which also use NTT-based multiplications.

First, we study the features of different modular reductions (i.e., Montgomery reduction, Barrett reduction, and k-reduction) with respect to the ARM Cortex-M0/M0+ instruction set. Based on the comparison of modular reductions, we combine the Montgomery reduction and k-reduction and come to a hybrid approach that is the most suitable for the constrained processors. We showcase our hybrid approach in Fig. 1, where different reductions are used in multiplications of two layers.

Next, our hybrid reduction approach provides a memory-efficient high-speed implementation of Kyber on Cortex-M0/M0+. It is the first Kyber implementation on Cortex-M0/M0+, with a small footprint and high performance simultaneously.

Then, to reduce the size of intermediate variables to comply with the constrained multiplication instructions, we integrate NTT multiplication on NTT-unfriendly rings and multi-moduli NTT technique and present a fine-grained implementation of Saber targeting ARM Cortex-M0/M0+ cores. These enable the calculation of NTT for Saber without long multiplication instructions. Notably, our new implementations are faster than the state-of-the-art ones by a factor of 2.6, 2.7, and 2.9 on key-generation, encapsulation, and decapsulation of Saber, respectively.

The implementation of this work is publicly available at https://github.com/wjwangcrypto/KyberSaberM0.

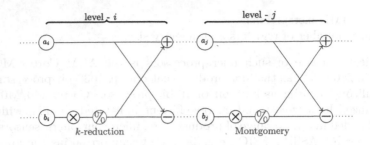

Fig. 1. The hybrid approach on NTT

1.2 Organization

Section 2 describes the backgrounds, i.e., Kyber, Saber, some characteristics of Cortex-M0/M0+ and techniques required to implement Kyber/Saber on Cortex-M0/M0+ using NTT transformations. Section 3 presents different modular reductions and compares these reductions on Cortex-M0/M0+. Section 4 presents NTT/multi-moduli NTT transforms and the rest of Kyber/Saber implementation details on Cortex-M0/M0+. In Sect. 5, we present the performance results of Kyber and Saber on Cortex-M0/M0+. A conclusion is given in Sect. 6.

2 Background

This section begins with an overview of Cortex-M0/M0+. Then, it introduces the cryptographic schemes of Kyber and Saber briefly. After that, NTT techniques are explained, including NTT multiplication on NTT-unfriendly rings and multi-moduli NTTs.

2.1 Cortex-M0/M0+

ARM Cortex-M0/M0+ processors are typical examples of constrained processors. The processors provide a register file of 16 32-bit registers (R0-R12 are generally purposed, R13-R15 have special purposes), and several special registers. Many of the instructions, such as MULS, can only access low registers (R0-R7), while some instructions, like MOV, can be used on all registers. R13 is the Stack Pointer used for accessing the stack memory via PUSH and POP operations. R14 is the Link Register (LR) used for storing the return address of a subroutine or function call. R15 is the PC. The Cortex-M0/M0+ processors do not have integer divide instructions. Furthermore, they only have one integer multiply instruction MULS to complete 32-bit × 32-bit, and the destination register only contains the 32-bit of the product. Cortex-M0/M0+ has neither instruction for long multiplication nor multiplication with optional accumulation.

2.2 Kyber

Kyber is an IND-CCA secure key-encapsulation mechanism (KEM) based on the hardness of solving the learning-with-errors problem in module lattices (LWE problem). The Kyber.KEM is defined as follows with regular font letters representing elements in $\mathcal{R}_q$, bold lower-case letters representing vectors, bold upper-case letters are matrices, and $\circ$ is the multiplication operation in the NTT-field.

- KeyGen() generates (pk, sk) by choosing $\mathbf{s}$ and $\mathbf{e}$ from a binomial sampling, and $\hat{\mathbf{A}}$ from a uniform distribution with a seed σ, then computes $\hat{\mathbf{t}} = \hat{\mathbf{A}} \circ \hat{\mathbf{s}} + \hat{\mathbf{e}}$; pk is $(\sigma, \hat{\mathbf{t}})$ and sk is $\hat{\mathbf{s}}$.
- Enc(pk, m, σ) chooses $\mathbf{r}$, $\mathbf{e}_1$, e_2, and $\hat{\mathbf{A}}^T$ with the seed σ, then computes $\mathbf{u} = \mathrm{NTT}^{-1}(\hat{\mathbf{A}}^T \circ \hat{\mathbf{r}}) + \mathbf{e}_1$ and $v = \mathrm{NTT}^{-1}(\hat{\mathbf{t}}^T \circ \hat{\mathbf{r}}) + e_2 + m$, finally construct the ciphertext as $ct = (\texttt{Encode}(\mathbf{u}), \texttt{Encode}(v))$.
- Dec(ct, sk) computes $m = \texttt{Encode}(v - \mathrm{NTT}^{-1}(\hat{\mathbf{s}}^T \circ \mathrm{NTT}(\mathbf{u})))$, while $\mathbf{u}$ and v are extracted from ct and $\hat{\mathbf{s}}^T$ from sk.

Values of the ring $\mathcal{R}_q = \mathbb{Z}_q[X]/(X^n + 1)$ are fixed to $q = 3329$, $n = 256$. A more detailed description of Kyber is presented in [8].

2.3 SABER

Saber is an CCA-secure KEM based on the Module Learning With Rounding (M-LWR) problem. The ring is $\mathcal{R}_q = \mathbb{Z}_q[X]/(X^n + 1)$ with $q = 2^{13}$ and $n = 256$. The key points of Saber are as follows. For more information about Saber, please refer to [11].

- KeyGen() generates (pk, sk) by sampling $\mathbf{A}$ and $\mathbf{s}$ with a seed σ first, then computing $\mathbf{b} = \texttt{Round}(\mathbf{A}^T \mathbf{s})$; pk is $(\sigma, \mathbf{b})$ and sk is $\mathbf{s}$.
- Enc(pk, m, r) generates $\mathbf{A}$, and $\mathbf{s}'$, then calculates $\mathbf{b}' = \texttt{Round}(\mathbf{A}\mathbf{s}')$, $v' = \mathbf{b}^T \mathbf{s}'$, and $c = \texttt{Round}(v' - 2^{\epsilon_p - 1} m)$. The ciphertext is $ct = (c, \mathbf{b}')$.
- Dec(ct, sk) computes $m = \texttt{Round}(v - 2^{\epsilon_p - \epsilon_T} c)$, while $v = \mathbf{b}'^T \mathbf{s}$.

2.4 Number Theoretic Transform

NTT is a generalization of the classic Discrete Fourier Transform (DFT) to finite fields. A polynomial $\mathbf{a} = \sum_{i=0}^{n-1} a_i X^i \mod q$, $a_i \in \mathbb{Z}_q$, the degree is at most $n-1$, $\gcd(n, q) = 1$ and $n | (q - 1)$. Forward-NTT transforms coefficients into the NTT domain. And, inverse-NTT transforms them from the NTT domain to the regular domain. They are as follows.

$$\hat{\mathbf{a}} = NTT(\mathbf{a}) = \sum_{i=0}^{n-1} \hat{a}_i X^i, \; \hat{a}_i = \sum_{j=0}^{n-1} a_j \zeta^{ij} \mod q,$$

$$\mathbf{a} = NTT^{-1}(\hat{\mathbf{a}}) = \sum_{i=0}^{n-1} a_i X^i, \; a_i = \frac{1}{n} \sum_{j=0}^{n-1} \hat{a}_j \zeta^{-ij} \mod q,$$

where ζ is the n-th primitive root of unity referred to as the twiddle factor. That is, $\zeta^n = 1 \mod q$ and $\zeta^i \neq 1 \mod q$, while $1 \leq i \leq n - 1$.

Similar to the divide-and-conquer technique used in Fast Fourier Transform (FFT), the speed of NTT can be further improved to $O(n \log(n))$. In the rest of this paper, we always use the fast implementation of NTT.

CT-Butterfly and GS-Butterfly. In the forward NTT transformation, the Cooley-Tukey butterfly [17] (CT-butterfly) is used to reduce a polynomial from $\mathbb{Z}_q[X]/(X^n - \zeta^n)$ to $\mathbb{Z}_q[X]/(X^{\frac{n}{2}} - \zeta^{\frac{n}{2}}) \times \mathbb{Z}_q[X]/(X^{\frac{n}{2}} + \zeta^{\frac{n}{2}})$. If a is the i-th coefficient in $\mathbb{Z}_q[X]/(X^n - \zeta^n)$ and b is the $(\frac{n}{2} + i)$-th coefficient, then the i-th coefficients c and d of the two reduced polynomials are given by $c = a + b \cdot \zeta^{\frac{n}{2}}$ and $d = a - b \cdot \zeta^{\frac{n}{2}}$. During inverse NTT transformation, the Gentleman-Sande butterfly [20] (GS-butterfly) is employed to compute $a = \frac{1}{2} \cdot (c + d)$ and $b = \frac{1}{2} \cdot \zeta^{-\frac{n}{2}} \cdot (c - d)$. Moreover, CT-butterfly for the forward NTT and GS-butterfly for the inverse NTT can eliminate the bit-reverse operation.

2.5 NTT Multiplication for NTT-Unfriendly Rings

There are cryptographic algorithms that cannot meet the constraints of NTT directly. For example, in Saber, the finite ring is $\mathcal{R}_q = \mathbb{Z}_q[X]/(X^n + 1)$, $q = 2^{13}$, $n = 256$. q is a power-of-two number that does not inherently support straightforward NTT. To benefit from NTT-based multiplication, we should choose a new modulus q', where $q' > 2 \times 256 \times \frac{q}{2} \times \frac{\mu}{2}$ in signed polynomial multiplication, and μ is shown in [11]. Works of literature [19] and [15] adopt a large new modulus to apply NTT on Saber implementation.

2.6 Multi-moduli NTT

Multi-moduli NTT can be viewed as an application of the Chinese Remainder Theorem (CRT). Compared to NTT, in which computations are all in $\mathcal{R}_q$, computations of multi-moduli NTT are split into smaller rings $\mathcal{R}_{q_i}$, where $\prod_i q_i \geq q$, $\gcd(q_i, q_j) = 1$, when $i \neq j$. For example, in Saber, let $q_0 = 3329$ and $q_1 = 12289$. They are coprime, $q' = 2^{23}$, $q_0 < q'$, $q_1 < q'$, $q_0 q_1 > q'$, and we can compute NTT over $\mathcal{R}_{q_1}$ and $\mathcal{R}_{q_2}$, respectively. In the end, we can compute the result back to the original ring $\mathcal{R}_q$. In [15] and [4], authors adopt multi-moduli NTT on the implementations of Saber and NTRU with SIMD instructions.

3 Modular Reductions

This section compares three reductions, i.e., Montgomery reduction, k-reduction, and Barrett reduction, on Cortex-M0/M0+. Based on the comparison, we come to a hybrid strategy shown to be the most efficient on Cortex-M0/M0+.

3.1 Montgomery Reduction

Montgomery reduction [26] is used to compute modular reduction in most literature. In [31], the author proposed a signed Montgomery reduction. The resource-constrained processors that we consider do not have SIMD instructions but have more limitations on the computational resources. Thus, we adopt the modification of signed Montgomery reduction as shown in Algorithm 1, which is more friendly to ARM Cortex-M0/M0+. It should be noted that NTT transformation contains the modular multiplication $C \mod q = ab \mod q$, where b is the twiddle factor. We can precompute $bR \bmod q$, and the result is exactly $abRR^{-1} \equiv ab \bmod q$ after Montgomery reduction.

Algorithm 1. signed Montgomery reduction

Input: C, q, q^{-1}, R, where $0 < q < \frac{R}{2}$, $-\frac{R}{2}q < C < \frac{R}{2}q$, $qq^{-1} = 1 \bmod R$

Output: $r = CR^{-1} \bmod q$

1: $t_1 \longleftarrow Cq^{-1} \bmod {}^{\pm}R$; $\triangleright t_1 \in (-\frac{R}{2}, \frac{R}{2})$

2: $t_2 \longleftarrow t_1 q$; $\triangleright t_2 \in (-\frac{R}{2}q, \frac{R}{2}q)$

3: $t_3 \longleftarrow C - t_2$; $\triangleright t_3 \in (-qR, qR)$

4: $r \longleftarrow \lfloor t_3/R \rfloor$; $\triangleright r \in (-q, q)$

3.2   Barrett Reduction

Barrett reduction is a reduction algorithm introduced in [10], as Algorithm 2 described. Since $\frac{m}{R}$ is an approximation of $\frac{1}{q}$, the approximation error is $|\frac{1}{q} - \frac{m}{R}|$. As long as $C|\frac{1}{q} - \frac{m}{R}| < 1$, the reduction is valid.

Algorithm 2. Barrett reduction

Input: C, q, m, $R = 2^l$, q is the prime, $m = \lfloor \frac{R}{q} \rfloor$, and $C|\frac{1}{q} - \frac{m}{R}| < 1$

Output: $r = C \bmod q$

1: $t_1 \longleftarrow Cm$; $\triangleright t_1 = Cm$

2: $t_2 \longleftarrow \lfloor t_1/R \rfloor$; $\triangleright t_2 = \lfloor \frac{Cm}{R} \rfloor$

3: $t_3 \longleftarrow t_2 q$; $\triangleright t_3 = \lfloor \frac{Cm}{R} \rfloor q$

4: $r \longleftarrow C - t_3$; $\triangleright r \in [0, 2q]$

3.3 k-Reduction

k-reduction [24] is a modular reduction algorithm widely adopted for the implementations of lattice-based homomorphic encryption [27]. We describe k-reduction in Algorithm 3. The modulus q has a particular form $q = k \cdot 2^m + 1$ in k-reduction, where k is a small positive integer. The output value of k-reduction does not lie in a definite range. It depends on the input value C when q is fixed. In a concrete NTT transform, $C = a \cdot b$, where b is the twiddle factor. We can precompute $bk^{-1} \bmod q$, and the result is $ab \bmod q$, which divides out the factor k.

Algorithm 3. k-reduction

Input: C, q, where q is the prime, and $q = k \cdot 2^m + 1$

Output: $r = Ck \bmod q$

1: $C_1 \longleftarrow \lfloor C/2^m \rfloor$; $\triangleright C_1 \in (-\frac{|C|}{2^m} - 1, \frac{|C|}{2^m})$

2: $C_0 \longleftarrow C \bmod 2^m$; $\triangleright C_0 \in [0, 2^m)$

3: $t_1 \longleftarrow kC_0$; $\triangleright t_1 \in [0, k2^m)$

4: $r \longleftarrow t_1 - C_1$; $\triangleright r \in (-\frac{|C|}{2^m}, \frac{|C|}{2^m} + q - k)$

3.4 Comparison of Reductions

Table 1 compares the signed Montgomery reduction, k-reduction, and Barrett reduction. The constraints of input are the restricted conditions that input value C should obey to maintain validity. There is no input constraint for k-reduction. The output range of signed Montgomery reduction or Barrett reduction is definite, i.e., $(-q,q)$, $(0,2q)$, respectively. The output range of k-reduction is variable. Moreover, signed Montgomery and Barrett reductions both need two multiplication, while k-reduction only needs one multiplication. The detailed description of these reductions is as follows:

- **Signed Montgomery reduction on Cortex-M0/M0+** Algorithm 4 is the signed Montgomery reduction on ARM Cortex-M0/M0+. It needs 8 Cortex-M0/M0+ instructions, including two MULS.
- **Barrett reduction on Cortex-M0/M0+** Barrett reduction costs 7 instructions, as Algorithm 5 shows. Two MULS are employed. There are some restrictions to guarantee the correctness of Barrett reduction. For instance, on $\mathcal{R}_{3329}$, the input value $C < 3329 \times 3329 < 2^{24}$, then $m = \lfloor \frac{2^{24}}{3329} \rfloor = 5039$. The product $Cm < 3329 \times 3329 \times 5039$, and $3329 \times 3329 \times 5039$ is more than 2^{32}. It doesn't meet the restriction that the register is 32-bit on Cortex-M0/M0+. After a thorough analysis of calculations in $\mathcal{R}_{3329}$, it is evident that direct application of Barrett reduction after coefficient multiplication is not suitable. Similarly, this holds true for $\mathcal{R}_{12289}$; Barrett reduction cannot be employed directly.
- **k-reduction on Cortex-M0/M0+** k-reduction needs 6 Cortex-M0/M0+ instructions as presented in Algorithm 6. It only employs one MULS.

Table 1. Comparison of Reductions

	signed Montgomery reduction	Barrett reduction	k-reduction
constraints of input	$C \in (-\frac{R}{2}q, \frac{R}{2}q)$	$C\lvert\frac{1}{q} - \frac{m}{R}\rvert < 1$	none
range of output	$(-q, q)$	$(0, 2q)$	$(-\frac{\lvert C \rvert}{2m}, \frac{\lvert C \rvert}{2m} + q - k)$
range of important intermediate value	$t_2 \in (-\frac{R}{2}q, \frac{R}{2}q)$ $t_3 \in (-qR, qR)$	$t_1 = Cm$	$C_1 \in (-\frac{\lvert C \rvert}{2m} - 1, \frac{\lvert C \rvert}{2m})$ $r \in (-\frac{\lvert C \rvert}{2m}, \frac{\lvert C \rvert}{2m} + q - k)$
Cortex-M0/M0+ instructions	8	7	6
Cortex-M0/M0+ registers[a]	2+(2) or 2+(1)	2+(2) or 2+(1)	3
number of multiplication	2	2	1

[a] It needs 4 Cortex-M0/M0+ registers, such as $2 + (2)$, to keep constants in registers (i.e., q, q^{-1} for signed Montgomery reduction and q, m for Barrett reduction). It needs 3 registers, such as $2+(1)$, to load constants when needed.

Algorithm 4. signed Montgomery reduction on Cortex-M0/M0+	**Algorithm 5.** Barrett reduction on Cortex-M0/M0+	**Algorithm 6.** k-reduction on Cortex-M0/M0+
Input: C, q, q^{-1}, R, where $0 < q < \frac{R}{2}$, $-\frac{R}{2}q < C < \frac{R}{2}q$, $qq^{-1} = 1 \bmod R$, $R = 2^{16}$	**Input:** C, q, m, $R = 2^l$, R is not the world size, q is the prime, and $m = \lfloor \frac{R}{q} \rfloor$, $C\lvert\frac{1}{q} - \frac{m}{R}\rvert < 1$	**Input:** C, q, k, where q is the prime, and $q = k \cdot 2^m + 1$
Output: $r = C \cdot R^{-1}$ $\bmod\ q$	**Output:** $r = C \bmod q$	**Output:** $r = C \cdot k$ $\bmod\ q$
1: MOVS t_1, C	1: LDR $t_1, \#m$	1: LSLS $t_1,$ C, $\#(32 - m)$
2: LDR $t_2, = \#q^{-1}$	2: MOVS t_2, C	2: LSRS $t_1,$ $t_1,$ $\#(32 - m)$
3: MULS t_1, t_2, t_1	3: MULS t_1, t_2, t_1	3: ASRS $C, C, \#m$
4: UXTH t_1, t_1	4: ASRS $t_1, \#l$	4: MOVS $t_2, \#k$
5: LDR $t_2, = \#q$	5: LDR $t_2, \#q$	5: MULS t_1, t_2, t_1
6: MULS t_1, t_2, t_1	6: MULS t_1, t_2, t_1	6: SUBS C, t_1, C
7: SUBS C, C, t_1	7: SUBS C, C, t_1	
8: ASRS $C, C, \#16$		

3.5 Hybrid Approach for Reductions on Cortex-M0/M0+

As described above, k-reduction consumes the least instructions. It conveys that we should use k-reduction as frequently as possible on Cortex-M0/M0+ to achieve a fast implementation. However, the output range in k-reduction will limit its usage. Concretely, if the input of k-reduction is C, the output will be $-\frac{|C|}{2^m} \leq kC_0 - C_1 \leq (\frac{|C|}{2^m} + q - k)$. It will drastically increase if C is too large. Furthermore, the intermediate value must be no more than 32 bits on Cortex-M0/M0+ cores. If we only employ k-reduction, the intermediate values will have a risk of overflow on the target platform. So, we have to combine other techniques to avoid overflow. Here, we have three strategies: pure k-reduction (case-A), k-reduction + Montgomery reduction (case-B), and k-reduction + Barrett reduction (case-C).

- **Pure k-reduction (case-A):** Only use k-reduction as modular reductions in this case. For example, the modulus q is $q = k \cdot 2^m + 1$, $k \geq 1$, and the result of the first coefficient multiplication (before reducing modulo q) is C such that $C \leq (k \cdot 2^m)^2 \leq 2^{32}$. The high part of C is C_1, and $C_1 \leq k^2 \cdot 2^m$. Then, the result of k-reduction, $C \bmod q$, is in the range $[-k^2 2^m, k 2^m]$. If we perform the subsequent multiplication directly, the result will be in the range $[-k^4 2^{2m}, k^4 2^{2m}]$. It eventually comes to an overflow for several rounds of modular multiplications. To avoid the overflow, we can add one more k-reduction to bound the result of the modular multiplication to a smaller range. That is, let $\mathrm{kred}(\cdot)$ be the k-reduction, we have $\mathrm{kred}(\mathrm{kred}(C))$.
- **k-reduction + Montgomery reduction (case-B):** Alternately use k-reduction and Montgomery reduction in this case. We employ k-reduction as far as possible, but there may be an overflow as the instance described

above. Unlike the previous choice, if the previous k-reduction will harm the following operations (mainly the multiplication), Montgomery reduction will be the substitution of the k-reduction in this strategy.

- **k-reduction + Barrett reduction (case-C):** Alternately use k-reduction and Barrett reduction in this case. k-reduction is the first choice indeed. However, the overflow problem remains. Barrett reductions are applies as additional reductions to make the following operations go smoothly.

There are other hybrid cases that can overcome the overflow issue, for example, k-reduction + Montgomery reduction + Barrett reduction. However, Barrett reduction cannot be used directly after the multiplication in $\mathcal{R}_{3329}$ and $\mathcal{R}_{12289}$. If a layer of NTT or NTT^{-1} uses Barrett reduction, other reductions are applied before Barrett reduction. Thus, k-reduction + Montgomery reduction + Barrett reduction involves more instructions than case-B and case-C. Therefore, in Sect. 4, we provide details for case-A, case-B, and case-C, while omitting information about other hybrid approaches.

4   Implementations on Cortex-M0/M0+

We will implement Kyber and Saber by NTT/multi-moduli NTT under the limitation of Cortex-M0/M0+. Kyber is on the ring $\mathcal{R}_{3329} = \mathbb{Z}_{3329}[X]/(X^{256} + 1)$. The rings of the multi-moduli NTT on Saber are $\mathcal{R}_{3329}$ and $\mathcal{R}_{12289} = \mathbb{Z}_{12289}[X]/(X^{256} + 1)$. 7-level NTT is applied on $\mathcal{R}_{3329}$ in Sect. 4.1. 8-level NTT can be applied on $\mathcal{R}_{12289}$, as Sect. 4.2 explored. The rest implementation details are presented in Sect. 4.3.

4.1   NTT over $\mathcal{R}_{3329}$

k-reduction can be further accelerated on $\mathcal{R}_{3329}$, as Algorithm 7 shows. This more efficient algorithm consumes one less instruction than Algorithm 6 because the low 8-bit of product C can be extracted by only one instruction UXTB.

Algorithm 7. k-reduction on Cortex-M0/M0+ over $\mathcal{R}_{3329}$

Input: C, q, k, where q is the prime, and $q = k \cdot 2^m + 1$, $k = 13$, $m = 8$.
Output: $r = C \cdot k \mod q$
1: UXTB t_1, C
2: ASRS $C, C, \#m$
3: MOVS $t_2, \#k$
4: MULS t_1, t_2, t_1
5: SUBS C, t_1, C

Under three different modular reduction strategies, the number of instructions required by forward-NTT over $\mathcal{R}_{3329}$ is analyzed. By pure k-reduction, it needs 8320 instructions. By k-reduction + Montgomery reduction, it needs 6400 instructions. While by k-reduction + Barrett reduction, 8064 instructions are required. Analyzed similarly, the inverse-NTT over $\mathcal{R}_{3329}$ needs 8320 instructions by pure k-reduction. It needs 6784 instructions by k-reduction + Montgomery reduction. And, by k-reduction + Barrett reduction, 9856 instructions are required. Furthermore, the comparison of every level and total instructions of NTT and NTT^{-1} is listed in Table 2 and Table 3, respectively. 'k-red.' represents k-reduction, 'Mont.' is Montgomery reduction, and 'Barr.' is Barrett reduction. Detailed analyses can be found in Appendix A.1.

Table 2. NTT on $\mathcal{R}_{3329}$

	case-A	case-B	case-C
Level-1	128 × k-red.	128 × k-red	128 × k-red.
Level-2	128 × k-red.	128 × k-red	128 × k-red.
Level-3	128 × k-red.+ 256 × k-red.	128×Mont.	128 × k-red.+ 256×Barr.
Level-4	128 × k-red.	128×Mont	128 × k-red.
Level-5	128 × k-red.+ 256 × k-red.	128×Mont.	128 × k-red.+ 256×Barr.
Level-6	128 × k-red.+ 256 × k-red.	128×Mont.	128 × k-red.
Level-7	128 × k-red.	128×Mont	128 × k-red.
Total	8320	6400	8064

Table 3. NTT^{-1} on $\mathcal{R}_{3329}$

	case-A	case-B	case-C
Level-7	128 × k-red.	128 × k-red	128 × k-red.
Level-6	128 × k-red.+ 256 × k-red.	128×Mont.	128 × k-red.+ 256×Barr.
Level-5	128 × k-red.	128×Mont	128 × k-red.
Level-4	128 × k-red.+ 256 × k-red.	128×Mont.	128 × k-red.+ 256×Barr.
Level-3	128 × k-red.+ 256 × k-red.	128 × Mont.	128 × k-red.
Level-2	128 × k-red.	128 × Mont.	128 × k-red.+ 256×Barr.
Level-1	128 × k-red.	128 × Mont.	128 × k-red.
Total	8320	6784	9856

4.2 NTT over $\mathcal{R}_{12289}$

We will also analyze polynomial multiplication over $\mathcal{R}_{12289}$. For 8-level forward-NTT, it needs 9216 instructions by the pure k-reduction. By k-reduction + Montgomery reduction, it requires 7424 instructions. While by k-reduction + Barrett reduction, 9728 instructions are needed. The 8-level inverse-NTT over $\mathcal{R}_{12289}$ is analyzed in the same way. Pure k-reduction consumes 10752. It needs 7936 instructions by k-reduction + Montgomery reduction. Moreover, k-reduction + Barrett reduction is 11520 instructions. Table 4 and Table 5 are the number of instructions of NTT and NTT^{-1}. Refer to Appendix A.2 for more details.

Table 4. NTT on $\mathcal{R}_{12289}$

	case-A	case-B	case-C
Level-1	128 × k-red.	128 × k-red	128 × k-red.
Level-2	128 × k-red.	128 × k-red	128 × k-red.
Level-3	128 × k-red.	128 × k-red	128 × k-red.
Level-4	128 × k-red.+ 256 × k-red.	128×Mont	128 × k-red.+ 256×Barr.
Level-5	128 × k-red.	128×Mont.	128 × k-red.
Level-6	128 × k-red.+ 256 × k-red.	128×Mont.	128 × k-red.+ 256×Barr.
Level-7	128 × k-red.	128×Mont	128 × k-red.
Level-8	128 × k-red.	128×Mont	128 × k-red.
Total	9216	7424	9728

Table 5. NTT^{-1} on $\mathcal{R}_{12289}$

	case-A	case-B	case-C
Level-8	128 × k-red.	128 × k-red	128 × k-red.
Level-7	128 × k-red.+ 256 × k-red.	128×Mont	128 × k-red.+ 256×Barr.
Level-6	128 × k-red.	128×Mont	128 × k-red.
Level-5	128 × k-red.+ 256 × k-red.	128×Mont	128 × k-red.+ 256×Barr.
Level-4	128 × k-red.	128×Mont	128 × k-red.
Level-3	128 × k-red.+ 256 × k-red.	128×Mont	128 × k-red.+ 256×Barr.
Level-2	128 × k-red.	128×Mont	128 × k-red.
Level-1	128 × k-red.	128×Mont	128 × k-red.
Total	10752	7936	11520

4.3   Other Implementation Details

Merging NTT Layers. Due to the limited number of registers, we compute two NTT layers once a time. The 7-level NTT over $\mathcal{R}_{3329}$ can be split into 2-layers + 2-layers + 2-layers + 1-layer. As for the 8-level NTT over $\mathcal{R}_{12289}$, it will be split as 2-layers + 2-layers + 2-layers + 2-layers.

Every level of inverse-NTT will generate a factor of two. To further reduce the load and store operations, we can remove the factor in `pairMul`.

pairMul and baseMul. The product of polynomials **ab** is $\text{NTT}^{-1}(\text{NTT}(\mathbf{a}) \circ \text{NTT}(\mathbf{b}))$. `pairMul` is the pairwise multiplication on the NTT domain. After 8-level forward-NTT, $\text{NTT}(\mathbf{a}) \circ \text{NTT}(\mathbf{b})$ is plain over $\mathcal{R}_{12289}$. Over $\mathcal{R}_{3329}$, after 7-level forward-NTT, **a** and **b** are mapped to 128 components, $\text{NTT}(\mathbf{a}) \circ \text{NTT}(\mathbf{b}) = (a_{2i+1}x + a_{2i}) \times (b_{2i+1}x + b_{2i}) \mod (x^2 - \zeta_i) = ((a_i + a_{2i+1})(b_i + b_{2i+1}) - a_i b_i - a_{2i+1}b_{2i+1})x + (a_i b_i + a_{2i+1}b_{2i+1}\zeta_i)$ with Karatsuba, $i = 0, 1, 2, \cdots, 127$.

The general method for $\sum_{i=0}^{l-1} \mathbf{a}_i \mathbf{b}_i$ is $\sum_{i=0}^{l-1} \text{NTT}^{-1}(\text{NTT}(\mathbf{a}_i)\text{NTT}(\mathbf{b}_i))$, while we compute $\text{NTT}^{-1}(\sum_{i=0}^{l-1}(\text{NTT}(\mathbf{a}_i)\text{NTT}(\mathbf{b}_i)))$ in this paper. The `baseMul` is $\sum_{i=0}^{l-1} \text{NTT}(\mathbf{a}_i)\text{NTT}(\mathbf{b}_i)$. The inverse-NTT is reduced significantly, from l to 1. The efficiency of multi-moduli NTTs will double, from $2l$ inverse-NTT to 2 inverse-NTT.

Memory-Speed Trade-Off. In the reference implementation of Kyber and Saber, the polynomial matrices and vectors are sampled and stored in memory. To trade off memory and efficiency, we merge the sampling and computations. For example, we sample and store polynomial vector **s** in memory and generate one polynomial of matrix $\hat{\mathbf{A}}$ and **e** at a time in Kyber key-generation. So, key-generation of Kyber consumes $k + 1$ polynomial memory footprints. Similarly, encryption requires $k + 1$ polynomial footprints too. And decryption only uses 3 polynomials. Saber adopts a memory-speed trade-off in the same way.

Register Allocation. There are 16 32-bit registers in the cortex-M0/M0+ processors, 14 of which (i.e., R0-R12, R14) can be used in the NTT transforms. Moreover, some instructions, such as MULS, ADDS, SUBS, etc., can only access low registers R0-R7. So, many additional MOV instructions are needed to swap the contents between low and high registers.

In the 2-layers NTT, the detailed registers allocation is described below. We treat all fourteen registers equally. R0 is the address of polynomial **a**; R1 is the address of NTT(**a**). R2, R3, R4, and R5 are the four coefficients needed in the 2-layers NTT. R6 is the address of twiddle factors. R7, R8, and R9 are the twiddle factors employed in the 2-layers NTT. R10 and R11 are used as counters for the loop. R12 and R14 are still available and are used to compute reductions. Table 1 shows that k-reduction needs 3 registers, and Montgomery reduction and Barrett reduction both need 4 or 3 registers. In k-reduction, R12 is the register to maintain intermediate results, and the constant k can be kept in R14 to reduce the load instructions. In Montgomery reduction and Barrett reduction, R12 is used similarly. These reductions both have two constants, but there is only one R14. Therefore, there is one register missing to keep both constants in registers. One solution is loading the constant in R14 once needed. The other solution is to spill values to the stack. Moreover, some 2-layers NTT is composed of k-reduction and Montgomery reduction, where the number of constants increases to three. In our successive hybrid 2-layers NTT, we take the second option keeping constants in registers. Considering additional MOV, PUSH, POP, load, and store instructions, the actual cycle number will be more than the theoretical number in Table 2, 3, 4, and 5.

5 Results

As described above, we adopt case-B, which integrates k-reduction and Montgomery reduction in modular reduction on NTT and multi-moduli NTT, during the implementation of Kyber and Saber on Cortex-M0/M0+. In this section, we will report our results. Our Kyber implementation is the first Cortex-M0/M0+ memory-efficient trade-off implementation. The software baseline measurements are the reference implementations in [8,11]. In this paper, we use Keil uVision 5 to comply the code with the optimization flag 'O3'.

5.1 Polynomial Multiplication

Table 6 presents our speed result for polynomial multiplication over $\mathcal{R}_{3329}$. The speed of $\mathbf{a} \times \mathbf{b}$ has increased by more than 30%/22% on Cortex-M0/M0+ by our optimizations. To be specific, [8] requires a total of $7 \times 128 \times \text{Mont.} + 7 \times 128 \times \text{Barr.} + 256 \times \text{Mont.}$ instructions in NTT^{-1}, while this work only employs $1 \times 128 \times k\text{-red.} + 6 \times 128 \times \text{Mont.}$. Thus, the cycle counts of NTT^{-1} in this work are fewer than those in [8]. As for NTT, [8] requires $7 \times 128 \times \text{Mont.}$, and this work requires $2 \times 128 \times k\text{-red.} + 5 \times 128 \times \text{Mont.}$. They are quite similar.

Table 6. Cycle Counts of Polynomial Multiplication over $\mathcal{R}_{3329}$

	Cortex-M0		Cortex-M0+	
	[8]	this work	[8]	this work
NTT	24 022	20 972	18 870	18 669
NTT^{-1}	45 059	20 303	36 741	18 001
PairMul	13 599	12 101	10 714	10 948
$\mathbf{a} \times \mathbf{b}$	106 702	**74 346** (−30%)	85 195	**66 287** (−22%)

Table 7. Cycle Counts of Polynomial Multiplication over $\mathcal{R}_{2^{13}}$

	Impl.		Cortex-M0	Cortex-M0+		Impl.		Cortex-M0	Cortex-M0+
[11]	Kara.		351 020	238 906	[23]	Kara.		322 369	201 495
this work	$\mathcal{R}_{3329}$	NTT	20 972	18 669	Mont	$\mathcal{R}_{3329}$	NTT	21 805	19 503
		NTT^{-1}	20 303	18 001			NTT^{-1}	20 882	18 580
		PairMul	12 101	10 948			PairMul	14 025	12 872
	$\mathcal{R}_{12289}$	NTT	22 228	19 926		$\mathcal{R}_{12289}$	NTT	23 513	21 211
		NTT^{-1}	25 036	22 734			NTT^{-1}	25 356	22 989
		PairMul	11 075	9 794			PairMul	10 825	9 544
	Garner		10 826	9 801		Garner		11 085	10 060
	Total(a × b)		**165 740**	**148 460**		Total(a × b)		172 808	155 473

The cycle counts of polynomial multiplication on Saber over $\mathcal{R}_{2^{13}}$ are represented in Table 7. [11] and [23] employ Karatsuba techniques. We empoly pure Montgomery reduction ('Mont') and the hybrid strategy case-B ('this work') in this work. The hybrid reduction has a better speed result for the polynomial multiplication on Saber. It requires 53%/38% fewer cycles than [11] and 49%/26% fewer cycles than [23] on Cortex-M0/M0+, respectively. The data also shows that the hybrid reduction is more efficient than the pure Montgomery reduction.

5.2 Kyber Implementation

We employ the reference implementation from [8] as the benchmark on Cortex-M0/M0+. Table 8 displays cycle counts for our Kyber implementations compared to the benchmark. Key generation, encapsulation, and decapsulation of Kyber.KEM are denoted as **K**, **E**, and **D**. Assembly code is utilized in NTT, NTT^{-1}, and `pairmul` to optimize polynomial multiplication within 'opt.NTT'. In addition to using assembly code for polynomial acceleration, 'opt.tradeoff' incorporates the memory-speed trade-off strategy outlined in Sect. 4.3. Furthermore, 'opt.all' encompasses all the techniques mentioned, including the straightforward assembly Keccak permutation from eXtended Keccak Code Package [3].

Table 8. Cycle Counts (in kilo cycles) of Kyber on ARM Cortex-M0/M0+

		Cortex-M0			Cortex-M0+		
		K	E	D	K	E	D
Kyber512	[8]	1 262	1 929	2 033	908	1 430	1 538
	opt.NTT	1 201 (−5%)	1 795 (−7%)	1 847 (−9%)	875 (−4%)	1 345 (−6%)	1 422 (−8%)
	opt.tradeoff	1 173 (−7%)	1 778 (−8%)	1 825 (−10%)	850 (−6%)	1 330 (−7%)	1 403 (−9%)
	opt.all	942 (−25%)	1 489 (−23%)	1 594 (−22%)	755 (−17%)	1 212 (−15%)	1 309 (−15%)
Kyber768	[8]	2 035	2 983	3 074	1 467	2 202	2 308
	opt.NTT	1 908 (−6%)	2 756 (−8%)	2 778 (−10%)	1 394 (−5%)	2 057 (−7%)	2 122 (−8%)
	opt.tradeoff	1 855 (−9%)	2 722 (−9%)	2 739 (−11%)	1 347 (−8%)	2 028 (−8%)	2 089 (−9%)
	opt.all	1 492 (−27%)	2 259 (−24%)	2 359 (−23%)	1 199 (−18%)	1 838 (−17%)	1 933 (−16%)
Kyber1024	[8]	3 153	4 331	4 417	2 268	3 181	3 291
	opt.NTT	2 937 (−7%)	3 987 (−8%)	3 990 (10%)	2 140 (−6%)	2 960 (−7%)	3 018 (−8%)
	opt.tradeoff	2 870 (−9%)	3 948 (−9%)	3 941 (−11%)	2 078 (−8%)	2 927 (−8%)	2 982 (−9%)
	opt.all	2 292 (−27%)	3 246 (−25%)	3 347 (−24%)	1 841 (−19%)	2 638 (−17%)	2 738 (−17%)

'Opt.tradeoff' achieves 7% to 11% (6% to 9%) speed-ups on Cortex-M0 (Cortex-M0+). The speed improvement of Kyber is not so significantly, because the original polynomial multiplication in [8] already utilizes the NTT technique. Furthermore, [8] utilizes special data formats to accelerate performance. For example, key-generation avoids the NTT^{-1} operation and all polynomials of matrix A do not need to do NTT operations either. 'Opt.all' requires 22% to 27% (15% to 19%) fewer cycles on Cortex-M0 (Cortex-M0+). The achievement shows that accelerating the polynomial generation can accelerate Kyber implementation further. In this paper, we only focus on the bottleneck of polynomial multiplication. We will investigate accelerating polynomial generation for further work.

The memory footprint of Kyber is reduced significantly, as seen in Table 9. We apply the same strategy on memory usage to 'opt.tradeoff' and 'opt.all' instances on Cortex-M0/M0+. Moreover, 'opt.NTT' and [8] use the same memory. The memory usage of our optimized Kyber is enhanced by 1.6 to 5.2 times.

Table 9. Memory for Kyber on ARM Cortex-M0/M0+

		K (×B)	E (×B)	D (×B)
Kyber512	opt.all/ opt.tradeoff	3136 (−40%)	2720 (−66%)	2720 (−69%)
	[8]/ opt.NTT	5184	7840	8704
Kyber768	opt.all/ opt.tradeoff	3648 (−61%)	3232 (−74%)	3232 (−77%)
	[8]/ opt.NTT	9280	12448	13664
Kyber1024	opt.all/ opt.tradeoff	4160 (−72%)	3744 (−80%)	3744 (−81%)
	[8]/ opt.NTT	14400	18080	19648

5.3 Saber Implementation

Table 10 and Table 11 present the performance results of Saber. Benchmarks are the C-code and Cortex-M0-based reference implementations [11,23]. [23] only presented the memory-efficient performance of Saber on Cortex-M0, shown as blue integers in Table 10. Table 10 also presents 'opt.NTT', 'opt.tradeoff', and 'opt.all' instances for our optimized implementation.

Compared to the benchmark results, our implementation demonstrates higher efficiency. Optimized NTT with assembly code, 'opt.NTT' achieves 27% to 34% (15% to 18%) fewer cycles on Cortex-M0/M0+ for LightSaber/Saber/FireSaber. Employing a memory-speed trade-off strategy, 'opt.tradeoff' accelerates the implementations by 47%/52%/56% (38%/45%/49%) on Cortex-M0/M0+. 'Opt.all' requires 50% to 63% (39% to 52%) fewer cycles than [11] on Cortex-M0/M0+. Similar to the Kyber implementation, accelerating the polynomial generation of Saber is our further work.

We apply the same memory usage strategy on Cortex-M0 and Cortex-M0+, so we do not distinguish it in Table 11. Our Saber implementation is more efficient, achieving a speedup factor of 2.6, 2.7, and 2.9 on key-generation, encapsulation, and decapsulation of Saber.KEM compared with [23]. Our memory cost is a little more than [23]. Considering [23] is optimized for memory, and our implementation is a memory-efficient trade-off on Cortex-M0/M0+. A small memory sacrifice is reasonable and acceptable with the drastic speed improvement.

Table 10. Cycle Counts (in kilo cycles) of Saber on ARM Cortex-M0/M0+

		Cortex-M0			Cortex-M0+		
		K	E	D	K	E	D
LightSaber	[11]	2 121	3 001	3 532	1 449	2 032	2 381
	opt.NTT	1 544 (−27%)	2 139 (−29%)	2 383 (−33%)	1 234 (−15%)	1 713 (−16%)	1 959 (−18%)
	opt.tradeoff	1 222 (−43%)	1 619 (−46%)	1 762 (−50%)	955 (−34%)	1 259 (−38%)	1 418 (−40%)
	opt.all	1 055 (−50%)	1 401 (−53%)	1 579 (−55%)	893 (−39%)	1 185 (−42%)	1 348 (−43%)
Saber	[23]	4 789	6 328	7 509	N/A	N/A	N/A
	[11]	4 367	5 682	6 484	2 963	3 831	4 363
	opt.NTT	3 071 (−30%)	3 950 (−30%)	4 327 (33%)	2 483 (−16%)	3 193 (−17%)	3 571 (−18%)
	opt.tradeoff	2 178 (−50%)	2 732 (−52%)	2 918 (−55%)	1 708 (−42%)	2 129 (−44%)	2 342 (−46%)
	opt.all[a]	1 877 (−57%) (−61%)	2 359 (−58%) (−63%)	2 598 (−60%) (−65%)	1 595 (−47%)	2 000 (−48%)	2 222 (−49%)
FireSaber	[11]	7 386	9 100	10 205	4 995	6 129	6 863
	opt.NTT	5 077 (−31%)	6 214 (−32%)	6 752 (−34%)	4 144 (−17%)	5 065 (−17%)	5 594 (−18%)
	opt.tradeoff	3 342 (−55%)	4 028 (−56%)	4 284 (−58%)	2 634 (−47%)	3 154 (−49%)	3 441 (−50%)
	opt.all	2 890 (−61%)	3 489 (−62%)	3 809 (−63%)	2 463 (−51%)	2 966 (−52%)	3 263 (−52%)

[a] As blue integers shown, the 'opt.all' variant of Saber requires 61% to 65% fewer cycles than the memory-efficient implementation of Saber on Cortex-M0.

Table 11. Memory for Saber on ARM Cortex-M0/M0+

		K (×B)	E (×B)	D (×B)
LightSaber	opt.tradeoff/ opt.all	5 600 (−28%)	5 664 (−42%)	5 664 (−42%)
	[11]/ opt.NTT	7 766	9 750	9 750
Saber	[23]	5 031	5 119	6 215
	opt.tradeoff/ opt.all	6 624 (−42%)	6 688 (−52%)	6 688 (−52%)
	[11]/ opt.NTT	1 1488	1 3984	1 3984
FireSaber	opt.tradeoff/ opt.all	7 648 (−60%)	7 712 (−62%)	7 712 (−62%)
	[11]/ opt.NTT	1 9008	2 2016	2 2016

6   Conclusions and Future Works

This work investigates the efficient implementation of lattice-based PQC on constrained processors and mainly focuses on NTT-based polynomial multiplication. Cortex-M0/M0+ core is an example of these constrained processors, which lack long multiplication instructions, multiplication with accumulation, SIMD instructions, and so on. We study the features of different modular reductions, i.e., Montgomery reduction, Barrett reduction, and k-reduction. Based on the comparison, we combine the k-reduction and Montgomery reduction as the most suitable hybrid approach on the constrained processor. Then, the hybrid approach is applied to the polynomial multiplication of Kyber and Saber on Cortex-M0/M0+ and gets memory-efficient high-speed trade-off implementations.

Though the 32-bit Cortex-M0/M0+ processor is chosen as the target platform, our hybrid approach can be applied to other more constrained processors, particularly 8-bit processors. To perform 16-bit$\times$16-bit multiplication using 8-bit registers, we can split the two 16-bit operands (a and b) into four 8-bit parts each ($a = a_0 + a_1 \cdot 2^8$, $b = b_0 + b_1 \cdot 2^8$), and then perform a series of multiplications and additions, i.e., $a \times b = (a_0 + a_1 \cdot 2^8) \times (b_0 + b_1 \cdot 2^8) = a_0 b_0 + (a_1 b_0 + a_0 b_1) \cdot 2^8 + a_1 b_1 \cdot 2^{16}$. This involves 4 multiplications on an 8-bit core. Therefore, for 8-bit processors, fewer 16-bit $\times$ 16-bit multiplications will result in better performance for $\mathcal{R}_{3329}$ and $\mathcal{R}_{12289}$.

In Algorithm 1, two long multiplications are needed (line 1 and line 2) on an 8-processor, while Algorithm 2 also demands two long multiplications (line 1 and line 3). In contrast, Algorithm 3 only necessitates one multiplication (line 3). Therefore, k-reduction prefers a better speed performance on 8-bit processors compared with Montgomery reduction and Barrett reduction. Moreover, $k = 13$ for $\mathcal{R}_{3329}$ and $k = 3$ for $\mathcal{R}_{12289}$. The k value less than 8 bits will further reduce the number of multiplication operations on an 8-bit processor. In specific, $k \times C_0 = k \times (C_{00} + C_{01} \cdot 2^8) = kC_{00} + kC_{01} \cdot 2^8$ (Algorithm 3, line 3), requiring only two 8-bit multiplications. On 8-bit processors, k-Reduction is the preferred option, requiring just two 8-bit multiplications, while Montgomery reduction and Barrett reduction demand eight 8-bit multiplications. To address overflow issues during the computation of NTT and NTT^{-1} on 8-bit processors, additional reduction operations are required. Following analysis and comparison, the specific hybrid approach that appropriates for 8-bit processors could be k-reduction+Montgomery reduction or others (such as pure k-reduction).

Additionally, the proposed approach is a favorable choice when multiplication operations are time-consuming. For instance, Cortex-M0 core on LPC43xx and Cortex-M0+ core on LPC5410x implement an iterative multiplier that takes 32 cycles to execute a MULS instruction. In contrast, k-reduction is preferred for requiring only one multiplication, while Montgomery Reduction and Barrett Reduction each demand two. k-Reduction outperforms in speed. Considering the risk of overflow, a hybrid approach that utilizes both k-reduction and other reductions (such as Montgomery reduction) is necessary. As mentioned earlier, the specific approach depends on the particular implementation platform.

At the same time, k-reduction outperforms Montgomery reduction and Barrett reduction in hardware implementation. k-Reduction requires fewer multiplication operations, and the factor k is a relatively small integer. These attributes lead to reduced circuit area overhead and lower power consumption in hardware implementation. Consequently, k-reduction can be a good choice for hardware implementation.

The above discussion illustrates the broader applicability of our approach in resource-constraint environment. Therefore, we believe the utilization of our approach on other scenarios is a promising avenue for future research.

Acknowledgements. We thank all reviewers and editors for their valuable comments. This work was supported by the National Key Research and Development Program of China (Nos 2021YFA1000600), the National Natural Science Foundation of China (Grant Nos 62002202, 62372273), the Program of Taishan Young Scholars of the Shandong Province, the Program of Qilu Young Scholars (Grant No 61580082063088) of Shandong University, Key Research and Development Program of Shandong Province (No. 2022CXGC020101), Department of Science & Technology of Shandong Province (SYS202201) and Quan Cheng Laboratory (QCLZD202306).

A   NTT on Cortex-M0/M0+

A.1   NTT on Cortex-M0/M0+ over $\mathcal{R}_{3329}$

We will analyze NTT over $\mathcal{R}_{3329}$ by pure k-reduction (case-A), k-reduction + Montgomery reduction (case-B), and k-reduction + Barrett reduction (case-C), respectively.

- Level-1: coefficient $0 \le a < 2^{13}$; twiddle factor $\zeta = 133$; CT-butterfly is:
  - Multiply: $0 \le C = a\zeta < 2^{13} \times 133 < 2^{21}$
  - k-reduction: $C = C_0 + C_1 2^8$, $0 \le C_0 < 2^8$, $0 \le C_1 < 2^{13}$, $-2^{13} < 13C_0 - C_1 < 2^{12}$
  - ADDS(SUBS): the range is $(-2^{13}, 2^{14})$.
- Level-2: the largest twiddle factor $\zeta = 1991$; CT-butterfly is:
  - Multiply: $-2^{24} < -2^{13} \times 1991 < C = a\zeta < 2^{14} \times 1991 < 2^{25}$
  - k-reduction: $C = C_0 + C_1 2^8$, $0 \le C_0 < 2^8$, $-2^{16} < C_1 < 2^{17}$, $-2^{17} < 13C_0 - C_1 < 2^{12} + 2^{16}$, $|13C_0 - C_1| < 2^{17}$
  - ADDS(SUBS): the range is $(-2^{18}, 2^{18})$.
- Level-3: the largest twiddle factor $\zeta = 2764$; CT-butterfly is:
  - Multiply: $-2^{30} < -2^{18} \times 2764 < C = a\zeta < 2^{18} \times 2764 < 2^{30}$
  - k-reduction: $C = C_0 + C_1 2^8$, $0 \le C_0 < 2^8$, $-2^{22} < C_1 < 2^{22}$, $-2^{22} < 13C_0 - C_1 < 2^{12} + 2^{22}$, $|13C_0 - C_1| < 2^{12} + 2^{22}$
  - ADDS(SUBS): the range is $(-2^{23}, 2^{23})$.

Product $C = a\zeta$ of the 4-th level will exceed 32 bits leading to an overflow on Cortex-M0/M0+. The remainder forward-NTT will be computed by the above three strategies.

In case-A, we always use k-reduction. The output of the 3-rd level creates a risk of overflow in the subsequent level. Therefore, we should reduce level-3 outputs to a smaller range. So, a successive k-reduction is appended in level-3 to reduce ADDS(SUBS) results. After the additional k-reduction, the output of level-3 is $(-2^{15}, 2^{16})$. k-reduction is applied in level-4 after multiplying the coefficient and twiddle factor. Then we get outputs in $(-2^{20}-2^{15}, 2^{20}+2^{15}+2^{12})$. Level 5 follows the same strategy as level 3. After CT-butterfly, the result is $(-2^{25}, 2^{25})$. Then, an additional k-reduction is applied, and the result is $(-2^{17}, 2^{12} + 2^{17})$. The procedure of level 6 is the same as level 5 too. $(-2^{14}, 2^{12} + 2^{14})$ is its output range. As for level-7, it only employs one k-reduction after multiplication. The result of level-7 is $(-2^{20}, 2^{19})$.

In case-B, we adopt k-reduction and Montgomery reductions. The output of the signed Montgomery modification is in $(-3329, 3329)$ over $\mathcal{R}_{3329}$. If level-1, level-2, and level-3 are all k-reduction as described above, there is an overflow in level-4. In level 3, we adopt Montgomery reduction rather than k-reduction. After Montgomery reduction and ADDS(SUBS), the output of level-3 is $(-2^{18} - 3329, 2^{18} + 3329)$. Level-4, level-5, level-6, and level-7 use Montgomery reduction to reduce multiplications. The forward-NTT output is $(-2^{18} - 3329 \cdot 5, 2^{18} + 3329 \cdot 5) \in (-2^{19}, 2^{19})$.

In case-C, k-reduction and Barrett reductions are used together. After Barrett reductions, the output of level-3 is $(0, 6658)$. Level-4 executes CT-butterfly as usual and gets a result in $(-2^{18}, 2^{18})$. Then in level 5, there is an additional Barrett reduction after CT-butterfly. The result is in $(0, 6658)$. Level-6 and level-7 employ CT-butterfly as level-4. After level 7, the output is $(-2^{23}, 2^{23})$. Since there is no further calculation, the result is sufficient.

Therefore, in forward-NTT over $\mathcal{R}_{3329}$, case-A needs $13 \times 128 \times k$-reduction $= 8320$ instructions. Case-B needs $2 \times 128 \times k$-reduction $+ 5 \times 128 \times$ Montgomery $= 6400$ instructions, and case-C needs $7 \times 128 \times k$-reduction $+ 2 \times 256 \times$ Barrett $= 8064$ instructions. Similarly, analyzing inverse-NTT over $\mathcal{R}_{3329}$, case-A needs $13 \times 128 \times k$-reduction $= 8320$ instructions. While in case-B, there are $1 \times 128 \times k$-reduction $+ 6 \times 128 \times$ Montgomery $= 6784$ instructions. As for case-C, $7 \times 128 \times k$-reduction $+ 3 \times 256 \times$ Barrett $= 9856$ instructions are needed. Upon the comparison, we will employ case-B both in forward-NTT and inverse-NTT over $\mathcal{R}_{3329}$. Table 2 and Table 3 illustrate comparisons over $\mathcal{R}_{3329}$.

A.2   NTT on Cortex-M0/M0+ over $\mathcal{R}_{12289}$

On Saber's multi-moduli NTT-based Cortex-M0/M0+ implementation, $\mathcal{R}_{12289}$ is another smaller ring under the CRT technique. Considering polynomial multiplications over $\mathcal{R}_{12289}$, $12289 = 3 \times 2^{12} + 1$, we will also analyze NTT by the above three strategies.

- Level-1: the coefficient $a < 2^{13}$; the twiddle factor $\zeta = 493$; CT-butterfly is:
  - Multiply: $0 \le C = a\zeta < 2^{13} \times 493 < 2^{22}$
  - k-reduction: $C = C_0 + C_1 2^{12}$, $0 \le C_0 < 2^{12}$, $0 \le C_1 < 2^{10}$, $-2^{10} < 3C_0 - C_1 < 3 \cdot 2^{12}$

- ADDS(SUBS): the range is $(-3 \cdot 2^{12}, 3 \cdot 2^{12} + 12289) \in (-2^{14}, 2^{15})$.
  - Level-2: the largest twiddle factor $\zeta = 5444$; CT-butterfly is:
    - Multiply: $-2^{26} < -3 \cdot 2^{12} \times 5444 < C = a\zeta < (3 \cdot 2^{12} + 12289) \times 5444 < 2^{27}$
    - k-reduction: $C = C_0 + C_1 2^{12}$, $0 \le C_0 < 2^{12}$, $-2^{14} < C_1 < 2^{15}$, $-2^{15} < 3C_0 - C_1 < 2^{15}$
    - ADDS(SUBS): the range is $(-3 \cdot 2^{12} - 2^{15}, 3 \cdot 2^{12} + 12289 + 2^{15}) \in (-2^{16}, 2^{16})$
  - Level-3: the largest twiddle factor $\zeta = 4337$; CT-butterfly is:
    - Multiply: $-2^{28} < (-3 \cdot 2^{12} - 2^{15}) < C = a\zeta < (3 \cdot 2^{12} + 12289 + 2^{15}) \times 4337 < 2^{28}$
    - k-reduction: $C = C_0 + C_1 2^{12}$, $0 \le C_0 < 2^{12}$, $-2^{16} < C_1 < 2^{16}$, $-2^{16} < 3C_0 - C_1 < 2^{14} + 2^{16}$
    - ADDS(SUBS): the range is $(-3 \cdot 2^{12} - 2^{15} - 2^{14} - 2^{16}, 3 \cdot 2^{12} + 12289 + 2^{15} + 2^{14} + 2^{16}) \in (-2^{17}, 2^{18})$
  - Level-4: the largest twiddle factor $\zeta = 11885$; CT-butterfly is:
    - Multiply: $-2^{31} < (-3 \cdot 2^{12} - 2^{15} - 2^{14} - 2^{16}) \times 11885 < C = a\zeta < (3 \cdot 2^{12} + 12289 + 2^{15} + 2^{14} + 2^{16}) \times 11885 < 2^{31}$
    - k-reduction: $C = C_0 + C_1 2^{12}$, $0 \le C_0 < 2^{12}$, $-2^{19} < C_1 < 2^{19}$, $-2^{19} < 3C_0 - C_1 < 2^{14} + 2^{19}$
    - ADDS(SUBS): the range is $(-3 \cdot 2^{12} - 2^{19} - 2^{19}, 3 \cdot 2^{12} + 12289 + 2^{17} + 2^{19}) \in (-2^{20}, 2^{20})$

The 5-th level can not compute as usual because product C will be larger than 2^{32}, and an overflow will occur on Cortex-M0/M0+.

In case-A, we append a k-reduction after CT-butterfly and get the result in $(-2^8, 2^{14} + 2^8)$. On level 5, CT-butterfly executes as usual. The result of level-5 is $(-2^8 - 2^{17}, 2^{14} + 2^8 + 2^{17})$. Level-6 first employs CT-butterfly, then appends a supplemental k-reduction, and finally gets the result in $(-2^8, 2^{14} + 2^8)$. Level-7 is the same as level-5. So, it employs CT-butterfly and gets the result in $(-2^8 - 2^{17}, 2^{14} + 2^8 + 2^{17})$. Moreover, level-8 does not need an additional k-reduction. After CT-butterfly, its output is $(-2^{20}, 2^{20})$. Given that level-8 is the last level of forward-NTT over $\mathcal{R}_{12289}$, the output range is proper.

In case-B, level-4, level-5, level-6, level-7, and level-8 employ Montgomery reductions in modular multiplications. Because the output range of signed Montgomery reduction is $(-12289, 12289)$, the result will be $(-2^{17} - 5 \times 12289, 2^{18} + 5 \times 12289) \in (-2^{18}, 2^{19})$ after forward-NTT over $\mathcal{R}_{12289}$.

In case-C, we employ k-reduction and Barrett reduction in level-4 and level-6, while the rest levels use k-reduction.

Therefore, during the forward-NTT over $\mathcal{R}_{12289}$, case-A needs $8 \times 128 \times k$-reduction $+ 2 \times 256 \times k$-reduction $= 9216$ instructions. Case-B needs $3 \times 128 \times k$-reduction $+ 5 \times 128 \times$ Montgomery $= 7424$ instructions. Case-C needs $8 \times 128 \times k$-reduction $+ 2 \times 256 \times$ Barrett $= 9728$ instructions. We analyze inverse-NTT over $\mathcal{R}_{12289}$ in a similar way. Case-A needs $8 \times 128 \times k$-reduction $+ 3 \times 256 \times k$-reduction $= 10752$ instructions. Case-B needs $1 \times 128 \times k$-reduction $+ 7 \times 128 \times$ Montgomery $= 7936$ instructions. As for case-C, $8 \times 128 \times k$-reduction $+ 3 \times 256 \times$ Barrett $= 11520$ instructions are needed. Table 4 and Table 5 present comparisons over $\mathcal{R}_{12289}$.

References

1. Cortex-M0 technical reference manual. http://infocenter.arm.com/help/topic/com.arm.doc.ddi0432c/DDI0432C_cortex_m0_r0p0_trm.pdf
2. Cortex-M0+ technical reference manual. http://infocenter.arm.com/help/topic/com.arm.doc.ddi0484b/DDI0484B_cortex_m0p_r0p0_trm.pdf
3. The eXtended Keccak code package. https://github.com/XKCP/XKCP/blob/master/lib/low/KeccakP-1600/ARM
4. Abdulrahman, A., Chen, J., Chen, Y., Hwang, V., Kannwischer, M.J., Yang, B.: Multi-moduli NTTs for saber on cortex-M3 and cortex-M4. IACR Trans. Cryptogr. Hardw. Embed. Syst. **2022**(1), 127–151 (2022). https://doi.org/10.46586/tches.v2022.i1.127-151
5. Abdulrahman, A., Hwang, V., Kannwischer, M.J., Sprenkels, D.: Faster Kyber and Dilithium on the cortex-M4. In: Ateniese, G., Venturi, D. (eds.) ACNS 2022. LNCS, vol. 13269, pp. 853–871. Springer, Cham (2022). https://doi.org/10.1007/978-3-031-09234-3_42
6. Alagic, G., et al.: NISTIR8309-status report on the second round of the NIST post-quantum cryptography standardization process (2020). https://doi.org/10.6028/NIST.IR.8309
7. Alkim, E., Evkan, H., Lahr, N., Niederhagen, R., Petri, R.: ISA extensions for finite field arithmetic accelerating Kyber and NewHope on RISC-V. IACR Trans. Cryptogr. Hardw. Embed. Syst. **2020**(3), 219–242 (2020). https://doi.org/10.13154/tches.v2020.i3.219-242
8. Avanzi, R., et al.: CRYSTALS-Kyber (version 3.0). NIST Post-Quantum Cryptography Standardization Project (2020). https://csrc.nist.gov/CSRC/media/Projects/post-quantum-cryptography/documents/round-3/submissions/Kyber-Round3.zip
9. Bai, S., et al.: CRYSTALS-Dilithium. NIST Post-Quantum Cryptography Standardization Project (2020). https://csrc.nist.gov/CSRC/media/Projects/post-quantum-cryptography/documents/round-3/submissions/Dilithium-Round3.zip
10. Barrett, P.: Implementing the Rivest Shamir and Adleman public key encryption algorithm on a standard digital signal processor. In: Odlyzko, A.M. (ed.) CRYPTO 1986. LNCS, vol. 263, pp. 311–323. Springer, Heidelberg (1987). https://doi.org/10.1007/3-540-47721-7_24
11. Basso, A., et al.: SABER: mod-LWR based KEM (round 3 submission). NIST Post-Quantum Cryptography Standardization Project (2020). https://csrc.nist.gov/CSRC/media/Projects/post-quantum-cryptography/documents/round-3/submissions/SABER-Round3.zip
12. Becker, H., Hwang, V., Kannwischer, M.J., Yang, B., Yang, S.: Neon NTT: faster Dilithium, Kyber, and Saber on cortex-A72 and apple M1. IACR Trans. Cryptogr. Hardw. Embed. Syst. **2022**(1), 221–244 (2022). https://doi.org/10.46586/tches.v2022.i1.221-244
13. Botros, L., Kannwischer, M.J., Schwabe, P.: Memory-efficient high-speed implementation of Kyber on cortex-M4. In: Buchmann, J., Nitaj, A., Rachidi, T. (eds.) AFRICACRYPT 2019. LNCS, vol. 11627, pp. 209–228. Springer, Cham (2019). https://doi.org/10.1007/978-3-030-23696-0_11
14. Chen, C., et al.: NTRU. NIST Post-Quantum Cryptography Standardization Project (2020). https://csrc.nist.gov/CSRC/media/Projects/post-quantum-cryptography/documents/round-3/submissions/NTRU-Round3.zip

15. Chung, C.M., Hwang, V., Kannwischer, M.J., Seiler, G., Shih, C., Yang, B.: NTT multiplication for NTT-unfriendly rings new speed records for Saber and NTRU on Cortex-M4 and AVX2. IACR Trans. Cryptogr. Hardw. Embed. Syst. **2021**(2), 159–188 (2021). https://doi.org/10.46586/tches.v2021.i2.159-188
16. Cook, S.A., Aanderaa, S.O.: On the minimum computation time of functions. Trans. Am. Math. Soc. **142**, 291–314 (1969)
17. Cooley, J.W., Tukey, J.W.: An algorithm for the machine calculation of complex Fourier series. Math. Comput. **19**(90), 297–301 (1965)
18. Fouque, P.A., et al.: Falcon: fast-Fourier lattice-based compact signatures over NTRU. NIST Post-Quantum Cryptography Standardization Project (2020). https://csrc.nist.gov/CSRC/media/Projects/post-quantum-cryptography/ documents/round-3/submissions/Falcon-Round3.zip
19. Fritzmann, T., Sigl, G., Sepúlveda, J.: RISQ-V: tightly coupled RISC-V accelerators for post-quantum cryptography. IACR Trans. Cryptogr. Hardw. Embed. Syst. **2020**(4), 239–280 (2020). https://doi.org/10.13154/tches.v2020.i4.239-280
20. Gentleman, W.M., Sande, G.: Fast Fourier transforms: for fun and profit. In: Proceedings of the November 7–10, 1966, fall joint computer conference, pp. 563–578 (1966)
21. Kannwischer, M.J., Rijneveld, J., Schwabe, P., Stoffelen, K.: pqm4: testing and benchmarking NIST PQC on ARM cortex-m4. IACR Cryptol. ePrint Arch. 844 (2019). https://eprint.iacr.org/2019/844
22. Karatsuba, A.: Multiplication of multidigit numbers on automata. Sov. Phys. Dokl. **7**, 595–596 (1963)
23. Karmakar, A., Mera, J.M.B., Roy, S.S., Verbauwhede, I.: Saber on ARM CCA-secure module lattice-based key encapsulation on ARM. IACR Trans. Cryptogr. Hardw. Embed. Syst. **2018**(3), 243–266 (2018). https://doi.org/10.13154/tches. v2018.i3.243-266
24. Longa, P., Naehrig, M.: Speeding up the number theoretic transform for faster ideal lattice-based cryptography. In: Foresti, S., Persiano, G. (eds.) CANS 2016. LNCS, vol. 10052, pp. 124–139. Springer, Cham (2016). https://doi.org/10.1007/ 978-3-319-48965-0_8
25. Mera, J.M.B., Karmakar, A., Verbauwhede, I.: Time-memory trade-off in Toom-Cook multiplication: an application to module-lattice based cryptography. IACR Trans. Cryptogr. Hardw. Embed. Syst. **2020**(2), 222–244 (2020). https://doi.org/ 10.13154/tches.v2020.i2.222-244
26. Montgomery, P.L.: Modular multiplication without trial division. Math. Comput. **44**(170), 519–521 (1985)
27. Natarajan, D., Dai, W.: Seal-embedded: a homomorphic encryption library for the internet of things. IACR Trans. Cryptogr. Hardw. Embed. Syst. **2021**(3), 756–779 (2021). https://doi.org/10.46586/tches.v2021.i3.756-779
28. Nejatollahi, H., Dutt, N.D., Ray, S., Regazzoni, F., Banerjee, I., Cammarota, R.: Post-quantum lattice-based cryptography implementations: a survey. ACM Comput. Surv. **51**(6), 129:1–129:41 (2019). https://doi.org/10.1145/3292548
29. Nguyen, D.T., Gaj, K.: Fast NEON-based multiplication for lattice-based NIST post-quantum cryptography finalists. In: Cheon, J.H., Tillich, J.-P. (eds.) PQCrypto 2021 2021. LNCS, vol. 12841, pp. 234–254. Springer, Cham (2021). https://doi.org/10.1007/978-3-030-81293-5_13
30. Roy, S.S., Basso, A.: High-speed instruction-set coprocessor for lattice-based key encapsulation mechanism: Saber in hardware. IACR Trans. Cryptogr. Hardw. Embed. Syst. **2020**(4), 443–466 (2020). https://doi.org/10.13154/tches.v2020.i4. 443-466

31. Seiler, G.: Faster AVX2 optimized NTT multiplication for ring-LWE lattice cryptography. IACR Cryptol. ePrint Arch. 39 (2018). http://eprint.iacr.org/2018/039
32. Shor, P.W.: Polynomial-time algorithms for prime factorization and discrete Logarithmson a quantum computer. SIAM J. Comput. **26**(5), 1484–1509 (1997). https://doi.org/10.1137/S0097539795293172
33. Toom, A.L.: The complexity of a scheme of functional elements realizing the multiplication of integers. Sov. Math. Dokl. **3**, 714–716 (1963)
34. Zhu, Y., et al.: LWRpro: an energy-efficient configurable crypto-processor for Module-LWR. IEEE Trans. Circuits Syst. I Regul. Pap. **68**(3), 1146–1159 (2021). https://doi.org/10.1109/TCSI.2020.3048395

Algorithmic Views of Vectorized Polynomial Multipliers – NTRU

Han-Ting Chen[1], Yi-Hua Chung[2], Vincent Hwang[2,3]($\boxtimes$), and Bo-Yin Yang[2]($\boxtimes$)

[1] National Taiwan University, Taipei, Taiwan
r10922073@csie.ntu.edu.tw
[2] Academia Sinica, Taipei, Taiwan
yhchiara@gmail.com, vincentvbh7@gmail.com, by@crypto.tw
[3] Max Planck Institute for Security and Privacy, Bochum, Germany

Abstract. The lattice-based post-quantum cryptosystem NTRU is used by Google for protecting Google's internal communication. In NTRU, polynomial multiplication is one of bottleneck. In this paper, we explore the interactions between polynomial multiplications, Toeplitz matrix-vector products, and vectorization with architectural insights. For a unital commutative ring R, a positive integer n, and an element $\zeta \in R$, we reveal the benefit of vector-by-scalar multiplication instructions while multiplying in $R[x]/\langle x^n - \zeta \rangle$.

We aim at designing an algorithm exploiting no algebraic and number-theoretic properties of n and ζ. An obvious way is to multiply in $R[x]$ and reduce modulo $x^n - \zeta$. Since the product in $R[x]$ is a polynomial of degree at most $2n - 2$, one usually chooses a polynomial modulus g such that (i) $\deg(g) \geq 2n - 1$, and (ii) there exists a well-studied fast polynomial multiplication algorithm f for multiplying in $R[x]/\langle g \rangle$.

We deviate from common approaches and point out a novel insight with dual modules and vector-by-scalar multiplications. Conceptually, we relate the module-theoretic duals of $R[x]/\langle x^n - \zeta \rangle$ and $R[x]/\langle g \rangle$ with Toeplitz matrix-vector products, and demonstrate the benefit of Toeplitz matrix-vector products with vector-by-scalar multiplication instructions. It greatly reduces the register pressure, and allows us to multiply with essentially no permutation instructions that are commonly used in vectorized implementation.

We implement the ideas for the NTRU parameter sets ntruhps2048677 and ntruhrss701 on a Cortex-A72 implementing the Armv8.0-A architecture with the single-instruction-multiple-data (SIMD) technology Neon. For polynomial multiplications, our implementation is 2.18× and 2.23× for ntruhps2048677 and ntruhrsss701 than the state-of-the-art optimized implementation. We also vectorize the polynomial inversions and sorting network by employing existing techniques and translating AVX2-optimized implementations into Neon. Compared to the state-of-the-art optimized implementation, our key generation, encapsulation, and decapsulation for ntruhps2048677 are 7.67×, 2.48×, and 1.77× faster, respectively. For ntruhrss701, our key generation, encapsulation, and decapsulation are 7.99×, 1.47×, and 1.56× faster, respectively.

Keywords: Toeplitz matrix · NTRU · Vectorization · Dual Module

© The Author(s), under exclusive license to Springer Nature Switzerland AG 2024
A. Chattopadhyay et al. (Eds.): INDOCRYPT 2023, LNCS 14460, pp. 177–196, 2024.
https://doi.org/10.1007/978-3-031-56235-8_9

1   Introduction

At PQCrypto 2016, the National Institute of Standards and Technology (NIST) announced the Post-Quantum Cryptography (PQC) Standardization Process for replacing existing standards for public-key cryptography with quantum-resistant cryptosystems [15]. For lattice-based cryptosystems, polynomial multiplications had been the most time-consuming operations. In this paper, we investigate the interations between the underlying mathematical structure of polynomial rings and the architectural insights of vector-by-scalar multiplication instructions in instruction set architectures (ISAs).

In the NTRU submission [6] to the NIST PQC Standardization, polynomial rings of the form $\mathbb{Z}_q[x]/\langle x^n - 1 \rangle$ and $\mathbb{Z}_q[x]\big/\big\langle \sum_{i=0}^{n-1} x^i \big\rangle$ are used where $\mathbb{Z}_q$ is an integer ring, and n is a prime. Since $x^n - 1 = (x-1)\sum_{i=0}^{n-1} x^i$, multiplications in $\mathbb{Z}_q[x]\big/\big\langle \sum_{i=0}^{n-1} x^i \big\rangle$ is often implemented as $\mathbb{Z}_q[x]/\langle x^n - 1 \rangle$ followed by reduction modulo $\sum_{i=0}^{n-1} x^i$. In this paper, we focus on the polynomial multiplications in $\mathbb{Z}_q[x]/\langle x^n - 1 \rangle$.

Common approaches for multiplying two size-n polynomials in $\mathbb{Z}_q[x]/\langle x^n - 1 \rangle$ usually multiply in $\mathbb{Z}_q[x]$ and reduce modulo $x^n - 1$. Let a, b be polynomials in $\mathbb{Z}_q[x]/\langle x^n - 1 \rangle$ and f be an algebra monomorphism computing $ab = f^{-1}(f(a)f(b))$ in $\mathbb{Z}_q[x]$. Recent work [13] showed that the module-theoretic dual $f(a)^*$ can be used for multiplying a Toeplitz matrix and a vector. Since polynomial multiplications in $\mathbb{Z}_q[x]/\langle x^n - 1 \rangle$ can be regarded as a Toeplitz matrix-vector multiplication, we don't need the reduction modulo $x^n - 1$ anymore.

In this paper, we point out the architectural benefit of Toeplitz matrix-vector products for ISAs implementing vector-by-scalar multiplication instructions. We show that the outer-product approach multiplying two matrices in cubic time implies efficient Toeplitz matrix-vector products with vector-by-scalar multiplication instructions.

1.1   Contributions

We summarize our contributions as follows.

- We point out the architectural benefit of Toeplitz matrix-vector products for vectorization and implement the ideas on a Cortex-A72 implementing Armv8.0-A where vector-by-scalar multiplication instructions are supported.
- We explain that Toeplitz matrix-vector product is actually a generic approach – it is only tied to the shape of polynomial rings and not the underlying monomorphism. Prior work [13] doesn't seem to observe this and they compared the Toeplitz matrix-vector product with Toom–Cook and the plain polynomial multiplication with number-theoretic transform[1] followed by reduction modulo $x^n - 1$.

---

[1] Number-theoretic transform refers to a broad family of algebra monomorphisms that doesn't contain Toom–Cook.

- For the performance of polynomial multiplications, we outperform the state-of-the-art optimized implementation by 2.18× and 2.23× for the NTRU parameter sets `ntruhps2048677` and `ntruhrss701`, respectively.
- For the overall performance of the scheme, our `ntruhps2048677` key generation, encapsulation, and decapsulation is 7.67×, 2.48×, and 1.77× faster than the state-of-the-art optimized implementation; our `ntruhrss701` key generation, encapsulation, and decapsulation is 7.99×, 1.47×, and 1.56× faster than the state-of-the-art optimized implementation.

1.2 Code

Our source code is publicly available at
https://github.com/vector-polymul-ntru-ntrup/NTRU.

1.3 Structure of This Paper

This paper is structured as follows: Sect. 2 describes our target operations and platforms. Section 3 surveys polynomial transformations used for multiplications. Section 4 goes through the benefit of Toeplizt matrix–vector products. Section 5 describes our implementations. We show the performance numbers in Sect. 6.

2 Preliminaries

Sections 2.1 describe the polynomials rings in NTRU, and Sect. 2.2 describes our target platform Cortex-A72.

2.1 Polynomials in NTRU

The NTRU submission comprises two families NTRU-HPS and NTRU-HRSS. Both operate on polynomial rings $\mathbb{Z}_3[x]/\langle \Phi_n \rangle$, $\mathbb{Z}_q[x]/\langle \Phi_n \rangle$, and $\mathbb{Z}_q[x]/\langle x^n - 1 \rangle$ where q is a power of 2, n is a prime, and Φ_n is the nth cyclotomic polynomial, which for prime n is $\frac{x^n - 1}{x - 1} = \sum_{i<n} x^i$. We target the parameter sets `ntruhps2048677` $((q, n) = (2048, 677))$ and `ntruhrss701` $((q, n) = 8192, 701)$. For more parameter sets and details, we refer to the specification [6]. While NTRU also requires inversions in $\mathbb{Z}_3[x]/\langle \Phi_n \rangle$ and $\mathbb{Z}_q[x]/\langle x^n - 1 \rangle$, we focus on multiplying polynomials in $\mathbb{Z}_{2048}[x]/\langle x^{677} - 1 \rangle$ and $\mathbb{Z}_{8192}[x]/\langle x^{701} - 1 \rangle$.

2.2 Cortex-A72

Our target platform is the ARM Cortex-A72. Cortex-A72 implements the 64-bit Armv8.0-A instruction set architecture. It is a superscalar Central Processing Unit (CPU) with an in-order frontend and an out-of-order backend. We summarize some architectural features relevant to this paper, and refer to [1] for more details about the pipelines.

SIMD Registers. In Armv8.0-A, there are 32 architectural 128-bit SIMD registers each viewable as packed 8-, 16-, 32-, or 64-bit elements. The width of the element is specified the suffices `.16B` `.8H`, `.4S`, and `.2D` respectively on the register name. For referencing a certain lane, we use the annotation `.H[5]` for the 5th (zero-based) halfword of the register and similarly for other lanes and data widths [2, Figure A1-1].

Armv8-A Vector Instructions. A plain `mul` multiplies corresponding vector elements and returns same-sized results. Additionally, `mul` also refers to another instruction encoding — vector-by-scalar multiplication — if the last operand is a lane of a register. In this case `mul` multiplies the vector by a scalar (the lane value). This simple feature plays significant roles on maximizing register utilization and minimizing permutations. There are many variants of multiplications: `mla`/`mls` computes the same product vector and accumulates to or subtracts from the destination. Next, the shifts: `shl` shifts left; `sshr` arithmetically shifts right. For basic arithmetic, the usual `add`/`sub` adds/subtracts the corresponding elements. Then we have permutations — `uzp{1,2}` extracts the even and odd positions respectively from a pair of vectors and concatenates the results into a vector. `zip{1,2}` takes the bottom and top halves of a pair of vectors and riffle-shuffles them into the destination.

3   Polynomial Multiplications

This section surveys the Chinese remainder theorem for polynomial rings and Toom–Cook, and is structured as follows. We assume all the rings are commutative and unital in this paper. Sect. 3.1 reviews the Chinese remainder theorem for polynomial rings. This forms the basis of various fast polynomial ring transformations. Section 3.2 reviews Toom–Cook. Section 3.3 reviews the bit losses of Toom–Cook.

3.1   The Chinese Remainder Theorem for Polynomial Rings

Let $n = \prod_l n_l$ and $\boldsymbol{g}_{i_0,\dots,i_{h-1}} \in R[x]$ be coprime polynomials for $i_l \in [0, n_l)$. The CRT gives us the following the isomorphism

$$\prod_{i_0,\dots,i_{l-1}} \frac{R[x]}{\left\langle \prod_{i_l,\dots,i_{h-1}} \boldsymbol{g}_{i_0,\dots,i_{h-1}} \right\rangle} \cong \prod_{i_0,\dots,i_l} \frac{R[x]}{\left\langle \prod_{i_{l+1},\dots,i_{h-1}} \boldsymbol{g}_{i_0,\dots,i_{h-1}} \right\rangle}$$

for all $l = 1, \dots, h-1$[2]. We call each of the isomorphism "a layer of computation" and "a layer" for short. Usually, multiplications in $\prod_{i_0,\dots,i_{h-1}} R[x] \big/ \left\langle \boldsymbol{g}_{i_0,\dots,i_{h-1}} \right\rangle$

---

[2] For possibly non-commutative unital rings, we only have $R[x]/(\langle \boldsymbol{g}_i \rangle \cap \langle \boldsymbol{g}_j \rangle) \cong R[x]/\langle \boldsymbol{g}_i \rangle \times R[x]/\langle \boldsymbol{g}_j \rangle$ for coprime polynomials $\boldsymbol{g}_i$ and $\boldsymbol{g}_j$. If R is commutative, $R[x]$ is also commutative and we have $\langle \boldsymbol{g}_i \rangle \cap \langle \boldsymbol{g}_j \rangle = \langle \boldsymbol{g}_i \rangle \langle \boldsymbol{g}_j \rangle = \langle \boldsymbol{g}_i \boldsymbol{g}_j \rangle$. This leads to $R[x]/\langle \boldsymbol{g}_i \boldsymbol{g}_j \rangle \cong R[x]/\langle \boldsymbol{g}_i \rangle \times R[x]/\langle \boldsymbol{g}_j \rangle$ in our context.

are cheap. If all the layers are cheap, we have an algorithmic improvement for multiplying polynomials in $R[x]\big/\big\langle\prod_{i_0,\dots,i_{h-1}} g_{i_0,\dots,i_{h-1}}\big\rangle$. If the n_l is a small constant, then it is usually cheap to decompose a polynomial ring into a product of n_l polynomial rings.

3.2   Toom–Cook (TC) and Karatsuba

For a positive integer n, we define $R[x]_{<n}$ as $\{a(x) \in R[x]|\deg(a(x)) < n\}$, the set of polynomials with degree less than n. Toom–Cook [7,17] and Karatsuba [11] are divide-and-conquer approaches for multiplying polynomials in $R[x]$. We can also use them for multiplying polynomials in $R[x]_{<n}$. We introduce $y \sim x^{\frac{n}{k}}$ (zero-pad so that $k|n$) [4], and map $R[x]_{<n} \hookrightarrow R[x]\big/\big\langle x^{\frac{n}{k}} - y\big\rangle [y]_{<k} \hookrightarrow R'[y]_{<k}$ for $R' = R[x]/\langle g\rangle$ with $\deg g \geq \frac{2n}{k} - 1$.

For $a,b \in R'[y]_{<k}$, a k-way Toom–Cook computes $ab \in R'[y]_{<2k-1}$ via evaluating a,b at suitably chosen s_i's in R'. In other words, we apply the map

$$R'[y]_{<k} \hookrightarrow R'[y]\big/\big\langle\prod_{i=0}^{2k-2}(y - s_i)\big\rangle \cong \prod_{i=0}^{2k-2} R'[y]/\langle y - s_i\rangle.$$

If one of the *evaluation points* is $s_i = \infty$, the corresponding map into $R'[y]/\langle y - s_i\rangle$ takes the highest degree coefficient (deg-$(k - 1)$ for a, b, deg-$(2k - 2)$ for ab). [11] chose $k = 2$ at $\{s_i\}_i = \{0, 1, \infty\}$; [17] chose $\{s_i\}_i = \{0, \pm1, \dots, \pm(k - 1)\}$; and [18, Page 31] replaced $-k + 1$ with ∞. We write $\mathbf{TC}_{(2k-1)\times k}$ for the matrix mapping the coefficients of a deg $< k$ polynomial into $\prod_{i=0}^{2k-2} R'[y]/\langle y - s_i\rangle$ and $\mathbf{TC}_{(2k-1)\times(2k-1)}^{-1}$ for the matrix mapping $\prod_{i=0}^{2k-2} R'[y]/\langle y - s_i\rangle$ into $R[y]\big/\big\langle\prod_{i=0}^{2k-2}(y - s_i)\big\rangle$.

A key observation is that while working over $\mathbb{Z}_{2^k}$ for $k = 5$ and $\{s_i\} = \{0, \pm1, \pm2, \pm\frac{1}{2}, 3, \infty\}$, $\mathbf{TC}_{9\times9}^{-1}$ only requires "division by 8". This implies 3-bit losses. The matrix $\mathbf{TC}_{9\times9}^{-1}$ will be stated explicitly in the full version.

3.3   Enlarging Coefficient Rings

We briefly explain how to divide a power of 2 when 2 is not invertible, for example while working over $\mathbb{Z}_{2^k}$. Suppose we want $r \in \mathbb{Z}_{2^k}$. We instead compute $2^\epsilon r \in \mathbb{Z}_{2^{k+\epsilon}}$, and right-shift $2^\epsilon r$ by ϵ bits [4, Section 7, Paragraph "What to do when 2 is not invertible"]. For our Toom–Cook defined over $\mathbb{Z}_{2^k}$, we would compute in $\mathbb{Z}_{2^{16}}$ so $r = \frac{2^{16-k} r}{2^{16-k}} \in \mathbb{Z}_{2^k}$ can be derived by right-shifting $2^{16-k} r \in \mathbb{Z}_{2^{16}}$ by $16 - k$ bits.

4   Toeplitz Matrix–Vector Product

In this section, we go through the benefit of Toeplitz matrix–vector products. The fundamental of using Toeplitz matrix–vector product is best described via R-modules, dual R-modules, and associative R-algebra. When the context is clear, we call an R-module a module and an associative R-algebra an algebra.

Section 4.1 reviews some basics about modules and algebras. Section 4.2 distinguish the inner-product-based and outer-product-based approaches for matrix–vector product. Section 4.3 introduces Toeplitz matrix–vector product. Section 4.4 explains the benefit of vector-by-scalar multiplications. Section 4.5 presents the generic Toeplitz matrix–vector product conversion from ring monomorphisms computing the double-size products.

4.1 Module and Associative Algebra

This section goes through some basics about modules, dual modules, and associative algebras. Readers familiar with these basic algebraic structures can skip this section.

Module and Dual Module. Let $(M, +)$ be an abelian group and R a ring. We turn M into an R-module by introducing a scalar multiplication $\cdot_M : R \times M \to M$ (we write $r \cdot_M \boldsymbol{a}$ for $(\cdot_M)(r, \boldsymbol{a})$) satisfying the following:

- $\forall \boldsymbol{a}, \boldsymbol{b} \in M, \forall r, s \in R, (r + s) \cdot_M (\boldsymbol{a} + \boldsymbol{b}) = r \cdot_M \boldsymbol{a} + r \cdot_M \boldsymbol{b} + s \cdot_M \boldsymbol{a} + s \cdot_M \boldsymbol{b}$.
- $\forall \boldsymbol{a} \in M, 1 \cdot_M \boldsymbol{a} = \boldsymbol{a}$.
- $\forall \boldsymbol{a} \in M, \forall r, s \in R, (rs) \cdot_M \boldsymbol{a} = r \cdot_M (s \cdot_M \boldsymbol{a})$.

We call $(M, +, \cdot_M)$ a left R-module. One can define a right R-module in a similar way by identifying a scalar multiplication from $M \times R$ to M. Since we assume R is commutative, we do not distinguish between left and right R-modules and simply call them R-modules. For elements $\boldsymbol{b}_0, \ldots, \boldsymbol{b}_{n-1} \in M$, if they are linearly independent and every element in M can be expressed as a linear combination of $\boldsymbol{b}_0, \ldots, \boldsymbol{b}_{n-1}$, we call $\{\boldsymbol{b}_0, \ldots, \boldsymbol{b}_{n-1}\}$ a basis of M and n the rank. A free module of rank n is a module with a basis of n elements and is very close to an n-dimensional vector space in our context. We denote by R^n for the free module of rank n. Notice that a ring R and a polynomial ring $R[x]/\langle \boldsymbol{g} \rangle$ are free modules, and the matrix ring $M_{n \times n}(R)$ is an R-module.

An R-module homomorphism is a map $\eta : M \to N$ satisfying:

$$\forall r \in R, \forall \boldsymbol{a}, \boldsymbol{b} \in M, \eta(r \cdot_M \boldsymbol{a} + \boldsymbol{b}) = r \cdot_N \eta(\boldsymbol{a}) + \eta(\boldsymbol{b}).$$

One can verify that the set of R-module homomorphisms $\mathrm{Hom}_R(M, R)$ from M to R is an R-module. We call $\mathrm{Hom}_R(M, R)$ the dual of M, and denote it as M^*. If M is a free R-module of finite rank, it is isomorphic to M^*. For an R-module homomorphism $\eta : M \to N$, we define the transpose of η as the R-module homomorphism $\eta^* : N^* \to M^*$ sending $\boldsymbol{a}^*$ to $\boldsymbol{a}^* \circ \eta$.

Associative Algebra. For rings R and $\mathcal{A}$, we turn $\mathcal{A}$ into an associative R-algebra by introducing a module structure. One identifies the module addition with the ring addition, and provide a scalar multiplication $\cdot_{\mathcal{A}} : R \times \mathcal{A} \to \mathcal{A}$ for the module structure satisfying

$$\forall r \in R, \forall \boldsymbol{a}, \boldsymbol{b} \in \mathcal{A}, r \cdot_{\mathcal{A}} (\boldsymbol{ab}) = (r \cdot_{\mathcal{A}} \boldsymbol{a})\boldsymbol{b} = \boldsymbol{a}(r \cdot_{\mathcal{A}} \boldsymbol{b}).$$

An R-algebra homomorphism is a map that is a ring homomorphism and a module homomorphism at the same time.

Obviously, a polynomial ring is an R-algebra and all the ring monomorphisms in Sect. 3 are also module monomorphisms; therefore, they are algebra monomorphisms.

4.2 Matrix–Vector Products

There are two basic ways to multiply a matrix by a vector. For a matrix M, we denote $M[i_0][i_1]$ for the (i_0, i_1)-th entry, $M[i_0][-]$ for the i_0-th row, and $M[-][i_1]$ for the i_1-th column of M. Let A be an $n_0 \times n_1$ matrix a B be a column vector of n_1 elements. We wish to compute the matrix-vector product $C = AB$. Algorithm 1 computes the result with several inner products of the rows of the matrix and the vector. Algorithm 2 accumulates several products of the columns of the matrix and the corresponding elements of the vector.

Algorithm 1. Inner-product-based matrix–vector multiplication.

1: **for** $i_0 = 0, \ldots, n_0 - 1$ **do**
2:    **for** $i_1 = 0, \ldots, n_1 - 1$ **do**
3:       $C[i_0] = C[i_0] + A[i_0][i_1]B[i_1]$
4:    **end for**
5:                              ▷ Inner product of the vectors $A[i_0][-]$ and $B[-]$.
6: **end for**

Algorithm 2. Outer-product-based matrix–vector multiplication.

1: **for** $i_1 = 0, \ldots, n_1 - 1$ **do**
2:    **for** $i_0 = 0, \ldots, n_0 - 1$ **do**
3:       $C[i_0] = C[i_0] + A[i_0][i_1]B[i_1]$
4:    **end for**
5:                              ▷ Outer product of the vectors $A[-][i_1]$ and $B[i_1]$.
6: **end for**

In the context of a vector instruction set, the former translates into vector-by-vector multiplications with interleaved operands, requiring transposition of the inputs and outputs, and a larger number of registers. The latter can be easily implemented with vector-by-scalar multiplications, requiring much fewer permutation instructions and less rigid instruction scheduling. It is easily seen that in the context of matrix multiplications, Algorithm 1 is a special case of the inner product approach (cf. Algorithm 3), and Algorithm 2 is a special case of the outer product approach (cf. Algorithm 4). We also call them accordingly.

Algorithm 3. Inner-product-based matrix–matrix multiplication.

1: **for** $i_0 = 0, \ldots, n_0 - 1$ **do**
2:     **for** $i_1 = 0, \ldots, n_1 - 1$ **do**
3:         **for** $i_2 = 0, \ldots, n_2 - 1$ **do**
4:             $A[i_0][i_1] = C[i_0][i_1] + A[i_0][i_2]B[i_2][i_1]$
5:         **end for**
6:                                  ▷ Inner product of the vectors $A[i_0][-]$ and $B[-][i_1]$.
7:     **end for**
8: **end for**

Algorithm 4. Outer-product-based matrix–matrix multiplication.

1: **for** $i_2 = 0, \ldots, n_2 - 1$ **do**
2:     **for** $i_0 = 0, \ldots, n_0 - 1$ **do**
3:         **for** $i_1 = 0, \ldots, n_1 - 1$ **do**
4:             $C[i_0][i_1] = C[i_0][i_1] + A[i_0][i_2]B[i_2][i_1]$
5:         **end for**
6:     **end for**
7:                                  ▷ Outer product of the vectors $A[-][i_2]$ and $B[i_2][-]$.
8: **end for**

4.3 Toeplitz Matrices

Let M be an $m \times n$ matrix over the ring R. We call it a Toeplitz matrix if it takes the form

$$M = \begin{pmatrix} a_{n-1} & a_{n-2} & \cdots & a_1 & a_0 \\ a_n & a_{n-1} & \cdots & a_2 & a_1 \\ \vdots & \vdots & \ddots & \vdots & \vdots \\ a_{m+n-3} & a_{m+n-4} & \cdots & a_{m-1} & a_{m-2} \\ a_{m+n-2} & a_{m+n-3} & \cdots & a_m & a_{m-1} \end{pmatrix}, \text{for all possible } i, j, M_{i,j} = M_{i+1,j+1}.$$

We denote M as $\mathbf{Toeplitz}_{m \times n}(a_{m+n-2}, \ldots, a_0)$.

Toeplitz Matrices for Weighted Convolutions. For a weighted convolution $c = ab = \left(\sum_i a_i x^i\right)\left(\sum_i b_i x^i\right) \in R[x]/\langle x^n - \zeta \rangle$, we choose an $n' \geq n$, zero-pad a and c to size-n' polynomials a' and c', respectively, and define $\mathrm{Expand}_{n \to n', \zeta} =$

$$\left(\sum_{i<n} b_i x^i, \zeta\right) \mapsto \left(\underbrace{0, \ldots, 0}_{n'-n}, b_{n-1}, \ldots, b_0, \zeta b_{n-1}, \ldots, \zeta b_1, \underbrace{0, \ldots, 0}_{n'-n}\right). \text{ We have}$$

$$c' = \mathbf{Toeplitz}_{n' \times n'}\left(\mathrm{Expand}_{n \to n', \zeta}(b)\right)a'.$$

$\mathbf{Toeplitz}_{n \times n}\left(\mathrm{Expand}_{n \to n, \zeta}(-)\right)(-)$ is exactly the `asymmetric_mul` by [3, Section 4.2]. See [8, Paragraph "A Toeplitz matrix view of asymmetric multiplication", Sect. 8.3.2] for explanations.

4.4 Small-Dimensional Cases

Toeplitz matrix–vector multiplications are extensively used in our implementations. For a fast polynomial ring transformation resulting weighted convolutions, we apply the outer-product-based Toeplitz matrix–vector multiplication. Existing works [3,14,16] applied the inner product approach with pre-and post-transposes. The Toeplitz structure admits fast construction of the full matrix. For a weighted convolution over $x^4 - \zeta$, we apply $\mathtt{Expand}_{4 \to 4, \zeta}$ with $\mathtt{ext}$ instructions, and accumulate vector-by-scalar products. Algorithm 5 is an illustration.

Algorithm 5. Outer product approach for $R[x]/\langle x^4 - \zeta \rangle$.

Inputs: $a = a_0 + a_1 x + a_2 x^2 + a_3 x^3$, $b = b_0 + b_1 x + b_2 x^2 + b_3 x^3$.
Outputs: $c = ab \bmod (x^4 - \zeta)$.

1: $\mathtt{b} = b_3 || b_2 || b_1 || b_0$
2: $\mathtt{t0} = a_3 || a_2 || a_1 || a_0$
3: Compute $\mathtt{t} = \zeta a_3 || \zeta a_2 || \zeta a_1 || \zeta a_0$ with Barrett multiplication.
4:                    ▷ [3] proposed an interleaved version of this; others [14,16] reduced the interleaved partial results instead.
5:   ▷ The remaining steps are different from [3].
6: $\mathtt{ext\ t1,\ t,\ t0,\ \#3 \cdot 4}$ ▷ $\mathtt{t1} = a_2 || a_1 || a_0 || \zeta a_3$
7: $\mathtt{ext\ t2,\ t,\ t0,\ \#2 \cdot 4}$ ▷ $\mathtt{t2} = a_1 || a_0 || \zeta a_3 || \zeta a_2$
8: $\mathtt{ext\ t3,\ t,\ t0,\ \#1 \cdot 4}$ ▷ $\mathtt{t3} = a_0 || \zeta a_3 || \zeta a_2 || \zeta a_1$
9: $(\mathtt{lo}, \mathtt{hi}) = (\mathtt{smull}, \mathtt{smull2})(\mathtt{t0}, b_0)$
10: $(\mathtt{lo}, \mathtt{hi}) = (\mathtt{lo}, \mathtt{hi})(\mathtt{smlal}, \mathtt{smlal2})(\mathtt{t1}, b_1)$
11: $(\mathtt{lo}, \mathtt{hi}) = (\mathtt{lo}, \mathtt{hi})(\mathtt{smlal}, \mathtt{smlal2})(\mathtt{t2}, b_2)$
12: $(\mathtt{lo}, \mathtt{hi}) = (\mathtt{lo}, \mathtt{hi})(\mathtt{smlal}, \mathtt{smlal2})(\mathtt{t3}, b_3)$
13: $c = \mathtt{Montgomery\_long}(\mathtt{lo}, \mathtt{hi})$

Generally speaking, once the Toeplitz matrix is constructed via $\mathtt{exts}$ or memory loads (recall that we can instead store an $n \times n$ Toeplitz matrix as an array of $2n - 1$ elements), vector-by-scalar multiplications significantly reduce the register pressure and remove the follow up permutation instructions. We illustrate the differences between inner-product-based and outer-product-based Toeplitz matrix–vector multiplication for

$$
\begin{pmatrix} a_0 & a'_1 & a'_2 & a'_3 \\ a_1 & a_0 & a'_1 & a'_2 \\ a_2 & a_1 & a_0 & a'_1 \\ a_3 & a_2 & a_1 & a_0 \end{pmatrix} \begin{pmatrix} b_0 \\ b_1 \\ b_2 \\ b_3 \end{pmatrix}
$$

where $a'_1 = \zeta a_3, a'_2 = \zeta a_2$, and $a'_3 = \zeta a_1$ for the weighted convolutions defined in $R[x]/\langle x^4 - \zeta \rangle$. Figure 2 illustrates the register view of inner-product-based Toeplitz matrix–vector multiplication and Fig. 1 for the outer-product-based one. For Fig. 2, we apply $\log_2 4 \cdot \frac{4}{2} = 4$ pairs of $(\mathtt{trn1}, \mathtt{trn2})$ to each operand to reach the register view. While applying vector-by-vector multiplications, the interleaved operands occupy 11 registers and the interleaved partial results occupy

4 or 8 registers (this depends on the coefficient ring). Finally, we also need to transpose the interleaved results with 4 pairs of (trn1, trn2). On the other hand, Fig. 1 requires no additional permutations and avoids the interleaved operands and results. This implies nearly no permutation instructions and very low register pressure.

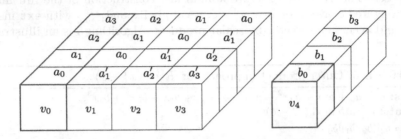

Fig. 1. Outer-product-based Toeplitz matrix–vector multiplication via vector-by-scalar multiplication. No permutations are required once we have data in registers $v_0, \ldots, v_3$. We only need 5 registers $v_0, \ldots, v_4$ for holding the operands and 1 or 2 registers for the partial results. (Color figure online)

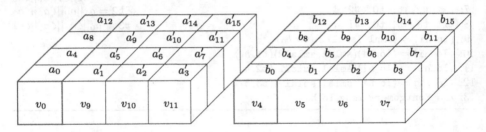

Fig. 2. Inner-product-based Toeplitz matrix–vector multiplication via vector-by-vector multiplication. One load $(a_0, \ldots, a_3), \ldots, (a_{12}, \ldots, a_{15})$ into registers $(v_0, \ldots, v_3)$, and transpose the registers as a 4×4 matrix. Same for $(v_4, \ldots, v_7)$ holding $(b_0, \ldots, b_3), \ldots, (b_{12}, \ldots, b_{15})$ and $(v_8, \ldots, v_{11})$ holding $(c_0, \ldots, c_3), \ldots, (c_{12}, \ldots, c_{15})$. Notice that we need to hold the registers $v_0, v_4, \ldots, v_{11}$ for computing $a_0 b_0 + c_1 b_1 + c_2 b_2 + c_3 b_3$. Therefore, we need 11 registers (we don't need (c_0, c_4, c_8, c_{12})) for the operands. Since we also need registers for holding the partial results (4 registers for $\mathbb{Z}_{2^{16}}$ and 8 registers otherwise), the register pressure is high and forbids us to generalize to size-16 computations. (Color figure online)

4.5   Large-Dimensional Toeplitz Transformation

There are several benefits when working on Toeplitz matrices. Firstly, we only need to store $m + n - 1$ coefficients $M_{m-1,0}, \ldots, M_{0,0}, \ldots, M_{0,n-1}$ of the matrix. Secondly, additions/subtractions of two Toeplitz matrices require only

$m + n - 1$ additions/subtractions in R. Finally, submatrices from adjacent rows and columns are also Toeplitz matrices. These properties enable efficient divide-and-conquer computations when the dimension is large.

For the sake of generality, multiplying two polynomials $a, b \in R[x]_{<k}$ will be considered as $ab \in R[x]_{<n}$ with $n \geq 2k - 1$. Given an $a \in R[x]_{<k}$, we write $(a, -) : R^k \to R^n$ for the module homomorphism $b \mapsto ab$ and $(a, -)^*$ its transpose. Suppose we have an R-algebra S where multiplications are much faster than in $R[x]_{<n}$, the Toeplitz matrix-vector product (TMVP) can be defined for an R-algebra homomorphism $f : R[x]_{<n} \to S$ with $f|_{R[x]_{<k}}$ a monomorphism.

Definition 1. Let S be an R-algebra and $f : R[x]_{<n} \to S$ be an R-algebra homomorphism, with $f_k := f|_{R[x]_{<k}} : R[x]_{<k} \to S$ a monomorphism. Furthermore, let $\mathrm{rev}_{k \to k} : R^k \to R^k$ be the index reversal map and $\mathrm{id}_{m \to n} : R^m \to R^n$ be the inclusion (pad 0's) map for $m \leq n$. The TMVP associated with f refers to the following module homomorphisms:

$$\left(\mathbf{Toeplitz}_{k \times k}(-)\right)(a) = \mathrm{rev}_{k \times k} \circ f_k^* \circ (f_k(a), -)^* \circ (f^{-1})^* \circ \mathrm{id}_{(2k-1) \to n}.$$

We call $(f^{-1})^* \circ \mathrm{id}_{(2k-1) \to n}$ split–matrix, $f_k(a)$ split–vector, $(f_k(a), -)^*$ base multiplication, and f_k^* interpolation. If $n = 2k - 1$, $f = \mathbf{TC}_{(2k-1) \times (2k-1)}$, then this is the k-way Toeplitz-TC matrix–vector product [12,13]. Generally, any R-algebra monomorphism suffices. See Appendices A for a formal proof and B for examples. We go through a higher-level overview of the idea.

Since f is a ring monomorphism, we implement the module homomorphism $(a, -)$ as $\mathrm{id}_{n \to (2k-1)} \circ f^{-1} \circ (f_k(a), -) \circ f_k$, take the transpose of $(a, \ )$, and relate $(a, -)^*$ to the Toeplitzation $\mathbf{Toeplitz}_{k \times k}(-)$ and the right-vector-multiplication $(-)(a)$. This allows us to convert any fast computation for $(a, -)$ into something for $\left(\mathbf{Toeplitz}_{k \times k}(-)\right)(a)$. Since $\left(\mathbf{Toeplitz}(-)\right)(a) = \mathrm{rev}_{k \times k} \circ (a, -)^*$, and $(-, a)_{R[x]/\langle x^k - \zeta \rangle} = \mathrm{id}_{k \to k} \circ \left(\mathbf{Toeplitz}(-)\right)(a) \circ \mathrm{Expand}_{k \to k, \zeta}$ as shown in Fig. 3, we eventually have a fast computation for $(-, a)_{R[x]/\langle x^k - \zeta \rangle}$.

5   Implementations

We propose two implementations for `ntruhps2048677` with 16-bit arithmetic modulo 65536: (i) `Toom-Cook` implements Toom–Cook with the splitting sequence $5 \to 3 \to 3 \to 2$, and (ii) `Toeplitz-TC` computes the Toeplitz matrix-vector product derived from `Toom-Cook`. Our `Toom-Cook` applies a more aggressive divide-and-conquer than prior works [9,14] by carefully choosing the point set for evaluations. Our `Toeplitz-TC` reveals the benefit of vector-by-scalar multiplications, which is more significant than the findings of [13].

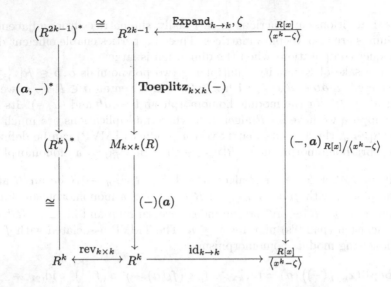

Fig. 3. Relations between $(-, a)_{R[x]/\langle x^k-1\rangle}$, $(\mathbf{Toeplitz}_{k \times k}(-))(a)$, and $(a, -)^*$.

For `ntruhrss701`, we implement the `Toeplitz-TC` approach with the same splitting sequence $5 \rightarrow 3 \rightarrow 3 \rightarrow 2$ with the 3's referring to 3-way Karatsuba instead of Toom-3. We skip the Toom–Cook approach since it is obviously not worth implementing given the experience from `ntruhps2048677`. Since the implementation `Toeplitz-TC` of `ntruhrss701` is very close to the one for `ntruhps2048677`, we skip the description for `ntruhrss701`.

Section 5.1 describes the `Toom-Cook` approach, and Sect. 5.2 describes the `Toeplitz-TC` approach. Additionally, we summarize existing strategies multipliying polynomials in `ntruhps2048677` in Table 1.

5.1   Toom-Cook

We first describe our chosen Toom–Cook splitting sequence and implementation considerations. We then detail our memory optimization for the interpolation of $\mathbf{TC}^{-1}$.

Chosen Splitting Sequence. We choose the splitting sequence Toom–5 $\rightarrow$ two Toom–3's $\rightarrow$ Karatsuba. We first zero-pad the size-677 polynomials to size-720 for ease of vectorization and compute in $\mathbb{Z}_{2^{16}}$. Since the coefficient ring of `ntruhps2048677` is $\mathbb{Z}_{2048}$ and $\frac{2^{16}}{2048} = 2^5$, divisions by 2^e for $e = 0, \ldots, 5$ translate into shifting e bits. We choose the splitting sequence $5 \rightarrow 3 \rightarrow 3 \rightarrow 2$. Our Toom-Cook consists of one layer of $\mathbf{TC}_{9 \times 5}$, two layers of $\mathbf{TC}_{5 \times 3}$'s, one layer of

Table 1. Overview of divide-and-conquer strategies multiplying polynomials in $R[x]_{<720}$ for ntruhps2048677. We first start with $R[x]_{<720}$ and alternatingly list all the number of subproblems of divide-and-conquer and the resulting ring. For example, the sequence $R[x]_{<720}, 4 \to 7, R[x]_{<180}$ means that size-720 polynomials are first sectioned into four size-180 polynomials, and mapped to seven size-180 polynomials, and the resulting polynomial multiplications defined in $R[x]_{180}$.

	[13]	[14]	This work
Ring	$R[x]_{<720}$	$R[x]_{<720}$	$R[x]_{<720}$
Divide-and-conquer	$4 \to 7$	$3 \to 5$	$5 \to 9$
Ring	$R[x]_{<180}$	$R[x]_{<240}$	$R[x]_{<144}$
Divide-and-conquer	$3 \to 5$	$4 \to 7$	$3 \to 5$
Ring	$R[x]_{<60}$	$R[x]_{<60}$	$R[x]_{<48}$
Divide-and-conquer	$3 \to 5$	$2 \to 3$	$3 \to 5$
Ring	$R[x]_{<20}$	$R[x]_{<30}$	$R[x]_{<16}$
Divide-and-conquer	$2 \to 3$	$2 \to 3$	$2 \to 3$
Ring	$R[x]_{<10}$	$R[x]_{<15}$	$R[x]_{<8}$

$\mathbf{TC}_{3\times2}$, 675 size-8 schoolbooks, one layer of $\mathbf{TC}_{3\times3}^{-1}$, two layers of $\mathbf{TC}_{5\times5}^{-1}$'s, and one layer of $\mathbf{TC}_{9\times9}^{-1}$. We choose the point sets $\{0, \pm1, \pm2, \pm\frac{1}{2}, 3, \infty\}$ (cf. Sect. 3.2) for $\mathbf{TC}_{9\times5}$ and $\{0, \pm1, 2, \infty\}$ for $\mathbf{TC}_{5\times3}$. The interpolation matrices $\mathbf{TC}_{9\times9}^{-1}$, $\mathbf{TC}_{5\times5}^{-1}$, and $\mathbf{TC}_{3\times3}^{-1}$ incur 3-, 1-, and 0-bit losses of precision, respectively. These add up to 5 bits, allowing us to invert correctly.

Comparisons to Prior Splitting Sequence [14]. [14] treated each polynomial as a size-720 polynomial, and applied Toom–Cook with the splitting sequence $3 \to 4 \to 2 \to 2$. The polynomial size goes down to 240 after the Toom-3, 60 after the Toom-4, and 15 after two Karatsuba's. Since 60 is not a multiple of 8, [14] basically padded to size-64 polynomials before Karatsuba. In this paper, we instead split via the sequence $5 \to 3 \to 3 \to 2$ down to size-8 schoolbooks. Our evaluation points for Toom-5 has the same precision loss as Toom-4. This is 1 fewer bit than the standard $\{0, \pm1, \pm2, \pm3, 4, \infty\}$. We also avoid zero-padding in vectorization. We merge the two Toom-3 layers (for both $\mathbf{TC}_{5\times3}$ and $\mathbf{TC}_{5\times5}^{-1}$) to reduce memory operations.

Memory Optimizations for Interpolations. Let $k|n$, $\mathbf{g}'$ be a polynomial of degree at least $\frac{2n}{k} - 1$, and $R' = R[x]/\langle \mathbf{g}' \rangle$. Recall that $\mathbf{TC}_{(2k-1)\times k}$ computes $R[x]/\langle x^{\frac{n}{k}} - y \rangle [y] \hookrightarrow R'[y]/\langle \prod_{i=0}^{2k-2} (y - s_i) \rangle \cong \prod_{i=0}^{2k-2} R'[y]/\langle y - s_i \rangle$ and results in computations in $R[x]/\langle \mathbf{g}' \rangle$. After examining the source code, we find that prior works [9,14] inverted the steps $\cong$ and $\hookrightarrow$ separately. Algorithm 6 is an illustration. Inverting $\cong$ means applying the interpolation matrix and inverting $\hookrightarrow$ means accumulating the overlapped coefficients while substituting y with

$x^{\frac{n}{k}}$ in each of the polynomials in $R[x]/\langle g' \rangle$. We instead alternate between the inversions of $\cong$ and $\hookrightarrow$ to reduce memory operations, *in essence merging two layers of computations.*

Algorithm 6. $\mathbf{TC}_{5\times5}^{-1}$ by [9,14].

Input: Size-3 polynomials $p_0, \ldots, p_4$.
Output: c[0-10] =
$\mathbf{TC}_{5\times5}^{-1}(p_0, p_1, p_2, p_3, p_4)$.

1: Declare array mem[5].
2: **for** $i = \{0,1,2\}$ **do**
3:     mem[0-4] =
    $\mathbf{TC}_{5\times5}^{-1}(p_0[i], \ldots, p_4[i])$.
4:                 ▷ Memory read and write.
5:     **for** $j = \{0, \ldots, 4\}$ **do**
6:         c[2j + i] = c[2j + i] +
    mem[j]
7:     **end for**
8:                 ▷ Memory read and write.
9: **end for**

Algorithm 7. Our $\mathbf{TC}_{5\times5}^{-1}$.

Input: Size-3 polynomials $p_0, \ldots, p_4$.
Output: c[0-10] =
$\mathbf{TC}_{5\times5}^{-1}(p_0, p_1, p_2, p_3, p_4)$.

1: Registers r[11].
2: r[0-4] = $\mathbf{TC}_{5\times5}^{-1}(p_0[0], \ldots, p_4[0])$
3: r[6-10] = $\mathbf{TC}_{5\times5}^{-1}(p_0[2], \ldots, p_4[2])$
4:                 ▷ Memory read.
5: r[5] = 0
6: **for** $j = \{0, \ldots, 4\}$ **do**
7:     r[i + 1] = r[i + 1] + r[i + 6]
8: **end for**
9: **for** $j = \{0, \ldots, 5\}$ **do**
10:     c[2j] = r[j]
11: **end for**
12:                 ▷ Memory write.
13: r[0-4] = $\mathbf{TC}_{5\times5}^{-1}(p_0[1], \ldots, p_4[1])$
14:                 ▷ Memory read.
15: **for** $j \leftarrow 0$ to 4 **do**
16:     c[2j + 1] = r[j]
17: **end for**
18:                 ▷ Memory write.

5.2   Toeplitz-TC

We apply the Toeplitz matrix–vector product with **TC**'s as the underlying monomorphisms and choose the same splitting sequence $5 \to 3 \to 3 \to 2$. We call it Toeplitz-TC.

Our Toeplitz-TC with the Splitting Sequence $5 \to 3 \to 3 \to 2$. Algorithm 8 describes our Toeplitz-TC implementation. Essentially, we regard size-677 polynomials a and b as size-720 polynomials. In practice, we zero-pad a and b to length 680 and omit the computations involving the indices $680, \ldots, 719$. Then, we apply one layer of Toeplitz-TC-5, two layers of Toeplitz-TC-3's, and one layer of Toeplitz-TC-2. Algorithm 9 describes our Toeplitz-TC--3--3--2 following Toeplitz-TC--5.

Each steps of Algorithms 8 and 9 is implemented as a subroutine. We merge computations while using all available registers without register spills. Initializations to zeros and the corresponding computations are also omitted for efficiency. While applying **TC**, $\mathbf{TC}'^{-1*}$, and $\mathbf{TC}'^*$, we prefer shifts over multiplications and reuse intermediate values.

Algorithm 8. Toeplitz-TC for ntruhps2048677.

Input: size-720 polynomials a, b.
Output: the size-677 polynomial c = ab mod ($x^{677} - 1$).

1: Declare uint16_t buff_a[9][288], buff_b[9][144], buff_c[9][144].
2: buff_a[0-8][0-287] = $\mathbf{TC}_{9\times9}^{-1*}$ (a)
3:   ▷ See Section 4.5 for definition.
4: buff_b[0-8][0-143] = $\mathbf{TC}_{9\times5}$ (b)
5: for $i = \{0,\dots,8\}$ do
6:    buff_c[i][0-143] = Toeplitz-TC-3-3-2 (buff_a[i][0-287], buff_b[i][0-143])
7: end for
8: c[0-676] = $\mathbf{TC}_{9\times5}^{*}$ (buff_c[0-8][0-143])

Algorithm 9. Toeplitz-TC-3-3-2.

Input: a 144 × 144 Toeplitz matrix M, and a size-144 vector v.
Output: the size-144 vector c = M · v.

1: Declare uint16_t M1[5][96], M2[5][5][3][16].
2: Declare uint16_t v1[5][5][16], c1[5][5][16].
3: M1[0-4][0-95] = $\mathbf{TC}_{5\times5}^{-1*}$ (M[0-143][0-143])
4: for $i = \{0,\dots,4\}$ do
5:    M2[i][0-4][0-2][0-15] = $\left(\mathbf{TC}_{3\times3}^{-1*} \circ \mathbf{TC}_{5\times5}^{-1*}\right)$ (M1[i][0-95])
6: end for
7: v1[0-4][0-4][0-15] = $\left(\mathbf{TC}_{5\times3} \circ \mathbf{TC}_{5\times3}\right)$ (v)
8: c1[i][j][0-15] = $\mathbf{TC}_{3\times2}^{*}$ (M2[i][j][0-2][0-15] · $\mathbf{TC}_{3\times2}$ (b1[i][j][0-15]))
9: c[0-143] = $\left(\mathbf{TC}_{5\times3}^{*} \circ \mathbf{TC}_{5\times3}^{*}\right)$ (c1[0-4][0-4][0-15])

Comparisons to [13]. [13] implemented the Toeplitz matrix–vector product with $\mathbf{TC}_{(2k-1)\times k}$ as the underlying monomorphisms on Cortex-M4, but they chose the splitting sequence $4 \to 3 \to 2 \to 2$. We improve the efficiency by applying a more aggressive splitting sequence. For the first layer, we use Toeplitz–TC–5 instead of Toeplitz–TC–4. Both strategies yield 3-bit losses. Although our $\mathbf{TC}_{9\times9}^{-1*}$, $\mathbf{TC}_{9\times5}$, and $\mathbf{TC}_{9\times5}^{*}$ require more multiplications, we have a smaller number of school-books, which is the bottleneck of the computation. Compared to [13], our Cortex-A72 implementation reaches the best performance with size-8 school-books instead of size-16 ones. Also, [13] used $\mathbf{TC}_{(2k-1)\times(2k-1)}^{-1*}$, $\mathbf{TC}_{(2k-1)\times k}$ and $\mathbf{TC}_{(2k-1)\times k}^{*}$ to compute while we multiply some constants to the precomputed matrices for easier computation. The modified $\mathbf{TC}_{(2k-1)\times(2k-1)}^{-1*}$, $\mathbf{TC}_{(2k-1)\times k}$ and $\mathbf{TC}_{(2k-1)\times k}^{*}$ will be shown in the full version.

6   Results

We present the performance numbers in this section. We focus on polynomial multiplications, leaving the fast constant-time GCD [5] as future work.

6.1   Benchmark Environment

We use the Raspberry Pi 4 Model B featuring the quad-core Broadcom BCM2711 chipset. It comes with a 32 kB L1 data cache, a 48 kB L1 instruction cache, and a 1 MB L2 cache and runs at 1.5 GHz. For hashing, we use the `aes`, `sha2`, and `fips202` from PQClean [10] without any optimizations due to the lack of corresponding cryptographic units. For the randombytes, [3] used the `randombytes` from SUPERCOP which in turn used `chacha20`. We extract the conversion from `chacha20` into `randombytes` from SUPERCOP and replace `chacha20` with our optimized implementations using the pipelines I0/I1, F0/F1. We use the cycle counter of the PMU for benchmarking. Our programs are compilable with GCC 10.3.0, GCC 11.2.0, Clang 13.1.6, and Clang 14.0.0. We report numbers for the binaries compiled with GCC 11.2.0.

6.2   Performance of Vectorized Polynomial Multiplications

Table 2 summarizes the performance of vectorized polynomial multiplications. All of our implementations outperform the state-of-the-art Toom–Cook [14]. For ntruhps2048677, our `Toeplitz-TC` and `Toom-Cook` are 2.18× and 1.56× faster than [14]. Comparing `Toeplitz-TC` and `Toom-Cook` based on the same splitting sequence, the result is consistent to [13]. But the most significant reason is the use of vector-by-scalar multiplications. This finding is new. For `ntruhrss701`, we outperform [14]'s implementation by 2.23×.

Table 2. Overview of `polymuls`.

	ntruhps2048677	ntruhrss701
Implementation	Cycles	
[14]	58 286	70 061
Toeplitz-TC	26 784	31 478
Toom-Cook	37 318	–

6.3   Performance of Schemes

Before comparing the overall performance, we first illustrate the performance numbers of some other critical subroutines. Most of our optimized implementations of these subroutines are not seriously optimized except for parts involving polynomial multiplications. We simply translate existing techniques and AVX2-optimized implementations into Neon. Notice that inversions over $\mathbb{Z}_2$ and $\mathbb{Z}_3$, and sorting networks are implemented in a generic sense. With fairly little effort, they can be used for other parameter sets.

Inversions. For ntruhps2048677, we need one inversion in $\mathbb{Z}_{2048}[x]/\langle x^{677} - 1\rangle$ and one inversion in $\mathbb{Z}_3[x]\big/\big\langle \frac{x^{677}-1}{x-1}\big\rangle$. The inversion in $\mathbb{Z}_{2048}[x]/\langle x^{677} - 1\rangle$ consists of one inversion in $\mathbb{Z}_2[x]/\langle x^{677} - 1\rangle$ and lifting to $\mathbb{Z}_{2048}[x]/\langle x^{677} - 1\rangle$ with eight polynomial multiplications since the coefficient ring is $\mathbb{Z}_{2048}$. We use the 1-bit form of $\mathbb{Z}_2$ for the inversion over $\mathbb{Z}_2$ without any algorithmic improvements and obtain a 20.41× speedup, leading to 10.27× overall speedup for the inversion over $\mathbb{Z}_{2048}$. The rest of the improvement for inversion over $\mathbb{Z}_{2048}$ comes from our improved polynomial multiplications (we use Toeplitz-TC here). For the inversion in $\mathbb{Z}_3[x]\big/\big\langle \frac{x^{677}-1}{x-1}\big\rangle$, we use bitsliced implementation and obtain a 8.6× speedup. For ntruhrss701, we outperform obtain 22.63×, 10.04×, 9.46× performance improvement for inversions over $\mathbb{Z}_2$, $\mathbb{Z}_{8192}$, and $\mathbb{Z}_3$, respectively. Table 3 summarizes the performance of inversions.

Table 3. Performance of inversions in NTRU.

Operation	Ref	Ours	Ref	Ours
	ntruhps2048677		ntruhrss701	
poly_Rq_inv	3 506 621	341 482	3 938 579	392 478
poly_R2_inv	2 791 906	136 776	3 175 330	140 290
poly_S3_inv	4 153 823	482 005	4 765 259	503 590
crypto_sort_int32	104 691	17 819	–	–

Sorting Network. We translate AVX2-optimized sorting network into Neon.

Performance of ntruhps2048677 and ntruhrss701. Table 4 summarizes our ntruhps2048677 and ntruhrss701. We compare our Toeplitz-TC to the existing NTRU implementations on Cortex-A72 [14]. For ntruhps2048677, our key generation is 7.67× faster. The main contribution is our optimized inversions, multiplications lifting the inverse in $\mathbb{Z}_2[x]/\langle x^{677} - 1\rangle$, followed by polynomial multiplications in $\mathbb{Z}_{2048}[x]/\langle x^{677} - 1\rangle$ (for lifting) and sorting network. Our ntruhps2048677 encapsulation is 2.48× faster. The main contribution is the sorting network followed by polynomial multiplications. Our ntruhps2048677 decapsulation is 1.77× faster. The improvement entirely comes from the improved polynomial multiplications. For ntruhrss701, our key generation, encapsulation, and decapsulation are 7.99×, 1.47×, and 1.56× faster than [14], respectively.

Finally, Table 5 details the numbers of ntruhps2048677 and ntruhrss701 with Toeplitz-TC. Notice that only performance-critical subroutines are shown.

Table 4. Overall cycles of `ntruhps2048677` and `ntruhrss701`. **K** stands for key generation, **E** stands for encapsulation, and **D** stands for decapsulation.

Operation	ntruhps2048677			ntruhrss701		
	K	E	D	K	E	D
Ref	8 245 039	227 980	331 274	9 397 305	134 737	365 558
[14]	7 686 272	196 526	212 265	8 599 610	87 380	221 986
Toeplitz-TC	1 002 187	79 213	120 208	1 076 810	59 625	142 174
Toom-Cook	1 127 089	88 037	146 422	–	–	–

Table 5. Detailed performance numbers of `ntruhps2048677` and `ntruhrss701` with `Toeplitz-TC`. Only performance-critical subroutines are shown.

ntruhps2048677		ntruhrss701	
Operation	Cycles	Operation	Cycles
crypto_kem_keypair	1 002 187	crypto_kem_keypair	1 076 810
owcpa_keypair	990 579	owcpa_keypair	1 069 128
poly_S3_inv	482 005	poly_S3_inv	503 590
poly_Rq_mul(×5)	5× 26 784	poly_Rq_mul(×5)	5× 31 478
poly_Rq_inv	341 482	poly_Rq_inv	392 478
poly_R2_inv	136 776	poly_R2_inv	140 290
poly_Rq_mul(×8)	8× 26 784	poly_Rq_mul(×8)	8× 31 478
sort	17 819		
randombytes	12 054	randombytes	6 294
crypto_kem_enc	79 213	crypto_kem_enc	59 625
owcpa_enc	32 501	owcpa_enc	41 559
poly_Rq_mul	26 784	poly_Rq_mul	31 478
randombytes	13 023	randombytes	6 202
sort	18 040		
sha3	5 148	sha3	5 296
crypto_kem_dec	120 208	crypto_kem_dec	142 174
owcpa_dec	100 842	owcpa_dec	120 485
poly_Rq_mul(×2)	2× 26 784	poly_Rq_mul(×2)	2× 31 478
poly_S3_mul	28 341	poly_S3_mul	33 319
sha3	18 867	sha3	21 342

A   Proof for the Toeplitz Transformation

For an algebra homomorphism $f : R[x]_{<n} \to S$ with $f_k := f|_{R[x]_{<k}}$ a monomorphism, and module homomorphism $(\boldsymbol{a}, -) = \begin{cases} R^k \to R^n \\ \boldsymbol{b} \mapsto \boldsymbol{ab} \end{cases}$ where $n \geq 2k - 1$, we have

$$\left(\mathbf{Toeplitz}_{k \times k}(-)\right)(\boldsymbol{a}) = \mathrm{rev}_{k \times k} \circ f_k^* \circ (f_k(\boldsymbol{a}), -)^* \circ (f^{-1})^* \circ \mathrm{id}_{(2k-1) \to n}.$$

Proof. Observe $(a, -)^* = f_k^* \circ (f_k(a), -)^* \circ (f^{-1})^* \circ \mathrm{id}_{(2k-1) \to n}$, it remains to show $(\mathbf{Toeplitz}_{k \times k}(-))(a) = \mathrm{rev}_{k \times k} \circ (a, -)^*$. Let $z = (z_0, \dots, z_{2k-2})$, $[k] = \{0, \dots, k-1\}$, and $\mathbf{0}_{m_0, m_1}$ the $m_0 \times m_1$ matrix of zeros. We have:

$$\left(\mathrm{rev}_{k \times k} \circ \mathbf{Toeplitz}_{k \times k}(z) \right)(a)$$

$$= (z_{i+j})_{(i,j) \in [k]^2} (a_j)_{(j,0) \in [k] \times [1]}$$

$$= \left(\sum_{j \in [k]} z_{i+j} a_j \right)_{(i,0) \in [k] \times [1]}$$

$$= \sum_{j \in [k]} (z_{i+j} a_j)_{(i,0) \in [k] \times [1]}$$

$$= \sum_{j \in [k]} \left(\mathbf{0}_{k,j} \ a_j I_k \ \mathbf{0}_{k,k-j-1} \right) (z_h)_{(h,0) \in [2k-1] \times [1]}$$

$$= \mathbf{Toeplitz}_{k \times (2k-1)} \left(\mathbf{0}_{1,k-1}, a_0, \dots, a_{k-1}, \mathbf{0}_{1,k-1} \right) (z_h)_{(h,0) \in [2k-1] \times [1]}$$

$$= (a, -)^*(z).$$

Applying $\mathrm{rev}_{k \times k}$ from the left finishes the proof (cf. [18, Theorem 6]).

B   Examples of Toeplitz Transformations

We give some examples of f's implementing $\begin{pmatrix} z_1 & z_2 \\ z_0 & z_1 \end{pmatrix} \begin{pmatrix} a_1 \\ a_0 \end{pmatrix}$:

$$\begin{pmatrix} 0 & 1 \\ 1 & 0 \end{pmatrix} \begin{pmatrix} z_1 & z_2 \\ z_0 & z_1 \end{pmatrix} \begin{pmatrix} a_0 \\ a_1 \end{pmatrix}$$

$$= \begin{pmatrix} 1 & 1 & 0 \\ 0 & 1 & 1 \end{pmatrix} \begin{pmatrix} a_0 & 0 & 0 \\ 0 & a_0 + a_1 & 0 \\ 0 & 0 & a_1 \end{pmatrix} \begin{pmatrix} 1 & -1 & 0 \\ 0 & 1 & 0 \\ 0 & -1 & 1 \end{pmatrix} \begin{pmatrix} z_0 \\ z_1 \\ z_2 \end{pmatrix}$$

$$= \begin{pmatrix} 1 & 1 & 1 \\ 1 & \omega_3 & \omega_3^2 \end{pmatrix} \begin{pmatrix} a_0 + a_1 & 0 & 0 \\ 0 & a_0 + \omega_3 a_1 & 0 \\ 0 & 0 & a_0 + \omega_3^2 a_1 \end{pmatrix} \mathbf{F}_3^{-1} \begin{pmatrix} z_0 \\ z_1 \\ z_2 \end{pmatrix}$$

$$= \begin{pmatrix} 1 & 1 & 1 & 1 \\ 1 & \omega_4 & \omega_4^2 & \omega_4^3 \end{pmatrix} \begin{pmatrix} a_0 + a_1 & 0 & 0 & 0 \\ 0 & a_0 + \omega_4 a_1 & 0 & 0 \\ 0 & 0 & a_0 + \omega_4^2 a_1 & 0 \\ 0 & 0 & 0 & a_0 + \omega_4^3 a_1 \end{pmatrix} \mathbf{F}_4^{-1} \begin{pmatrix} z_0 \\ z_1 \\ z_2 \\ 0 \end{pmatrix}$$

where $\mathbf{F}_k^{-1} = (\mathbf{F}_k^{-1})^T$ is the inverse of the cyclic size-k FFT.

References

1. ARM: Cortex-A72 Software Optimization Guide (2015). https://developer.arm. com/documentation/uan0016/a/

2. ARM: Arm Architecture Reference Manual, Armv8, for Armv8-A architecture profile (2021). https://developer.arm.com/documentation/ddi0487/gb/?lang=en

3. Becker, H., Hwang, V., Kannwischer, M.J., Yang, B.Y., Yang, S.Y.: Neon NTT: faster Dilithium, Kyber, and Saber on Cortex-A72 and Apple M1. IACR Trans. Cryptogr. Hardw. Embed. Syst. **2022**(1), 221–244 (2022). https://tches.iacr.org/index.php/TCHES/article/view/9295

4. Bernstein, D.J.: Multidigit multiplication for mathematicians (2001). https://cr.yp.to/papers.html#m3

5. Bernstein, D.J., Yang, B.Y.: Fast constant-time GCD computation and modular inversion. IACR Trans. Cryptogr. Hardw. Embed. Syst. **2019**(3), 340–398 (2019). https://tches.iacr.org/index.php/TCHES/article/view/8298

6. Chen, C., et al.: NTRU. Submission to the NIST Post-Quantum Cryptography Standardization Project [15] (2020). https://ntru.org/

7. Cook, S.A., Aanderaa, S.O.: On the minimum computation time of functions. Trans. Am. Math. Soc. **142**, 291–314 (1969)

8. Hwang, V.B.: Case Studies on Implementing Number-Theoretic Transforms with Armv7-M, Armv7E-M, and Armv8-A. Master's thesis (2022). https://github.com/vincentvbh/NTTs_with_Armv7-M_Armv7E-M_Armv8-A

9. Kannwischer, M.J., Rijneveld, J., Schwabe, P.: Faster multiplication in $\mathbb{Z}_{2^m}[x]$ on Cortex-M4 to speed up NIST PQC candidates. In: Deng, R., Gauthier-Umana, V., Ochoa, M., Yung, M. (eds.) Applied Cryptography and Network Security. ACNS 2019. LNCS, vol. 11464, pp. 281–301. Springer, Cham (2019). https://doi.org/10.1007/978-3-030-21568-2_14

10. Kannwischer, M.J., Schwabe, P., Stebila, D., Wiggers, T.: PQClean. https://github.com/PQClean

11. Karatsuba, A.A., Ofman, Y.P.: Multiplication of many-digital numbers by automatic computers. In: Doklady Akademii Nauk, vol. 145, no. 2, pp. 293–294 (1962)

12. İrem Keskinkurt Paksoy, Cenk, M.: TMVP-based Multiplication for Polynomial Quotient Rings and Application to Saber on ARM Cortex-M4. Cryptology ePrint Archive (2020). https://eprint.iacr.org/2020/1302

13. İrem Keskinkurt Paksoy, Cenk, M.: Faster NTRU on ARM Cortex-M4 with TMVP-based multiplication (2022). https://eprint.iacr.org/2022/300

14. Nguyen, D.T., Gaj, K.: Optimized Software Implementations of CRYSTALS-Kyber, NTRU, and Saber Using NEON-Based Special Instructions of ARMv8 (2021). third PQC Standardization Conference

15. NIST, the US National Institute of Standards and Technology: Post-quantum cryptography standardization project. https://csrc.nist.gov/Projects/post-quantum-cryptography

16. Sanal, P., Karagoz, E., Seo, H., Azarderakhsh, R., Kermani, M.M.: Kyber on ARM64: compact implementations of Kyber on 64-bit ARM Cortex-A processors. Cryptology ePrint Archive, Report 2021/561 (2021). https://eprint.iacr.org/2021/561

17. Toom, A.L.: The complexity of a scheme of functional elements realizing the multiplication of integers. In: Soviet Mathematics Doklady, vol. 3, no. 4, pp. 714–716 (1963)

18. Winograd, S.: Arithmetic Complexity of Computations, vol. 33. Siam, New Delhi (1980)

VDOO: A Short, Fast, Post-quantum Multivariate Digital Signature Scheme

Anindya Ganguly$^{(\boxtimes)}$, Angshuman Karmakar, and Nitin Saxena

Department of Computer Science and Engineering, IIT Kanpur, Kanpur, India
{anindyag,angshuman,nitin}@cse.iitk.ac.in

Abstract. Hard lattice problems are predominant in constructing post-quantum cryptosystems. However, we need to continue developing post-quantum cryptosystems based on other quantum hard problems to prevent a complete collapse of post-quantum cryptography due to a sudden breakthrough in solving hard lattice problems. Solving large multivariate quadratic systems is one such quantum hard problem.

Unbalanced Oil-Vinegar is a signature scheme based on the hardness of solving multivariate equations. In this work, we present a post-quantum digital signature algorithm VDOO (Vinegar-Diagonal-Oil-Oil) based on solving multivariate equations. We introduce a new layer called the diagonal layer over the oil-vinegar-based signature scheme Rainbow. This layer helps to improve the security of our scheme without increasing the parameters considerably. Due to this modification, the complexity of the main computational bottleneck of multivariate quadratic systems i.e. the Gaussian elimination reduces significantly. Thus making our scheme one of the fastest multivariate quadratic signature schemes. Further, we show that our carefully chosen parameters can resist all existing state-of-the-art attacks. The signature sizes of our scheme for the National Institute of Standards and Technology's security level of I, III, and V are 96, 226, and 316 bytes, respectively. This is the smallest signature size among all known post-quantum signature schemes of similar security.

Keywords: Post-quantum · Digital signature · Multivariate Cryptography · Oil-Vinegar · Multivariate root-finding

1 Introduction

Cryptography is the study of different methods to safeguard our sensitive information in the ever-expanding digital world. The security assurances of cryptographic schemes especially public-key cryptographic schemes emanate from the computational intractability of some underlying hard problems. Currently, public-key cryptographic schemes such as Rivest-Shamir-Adleman [51], elliptic-curve discrete logarithm [44] are predominant in our public-key infrastructure. However, in the context of the rapid development of quantum computers, these schemes exhibit a significant drawback. The underlying hard problems of these

A. Chattopadhyay et al. (Eds.): INDOCRYPT 2023, LNCS 14460, pp. 197–222, 2024.
https://doi.org/10.1007/978-3-031-56235-8_10

schemes *i.e.* integer factorization and discrete logarithm problem can be solved *easily* due to the polynomial time quantum algorithms developed by Shor [54] and Proos-Zalka [50] respectively. Therefore, quantum-resistant hard problems have gained popularity among designers for designing public-key cryptosystems for the future. A landmark event in the development of such quantum-resistant or post-quantum cryptography (PQC) is the PQC standardization procedure [19] initiated by the National Institute of Standards and Technology (NIST) to select quantum-safe cryptographic primitives such as key encapsulation mechanisms (KEM), public-key encryption (PKE), and digital signature algorithm. In 2022, NIST standardized [3] one KEM (Crystals-Kyber [15]) and three signature schemes (SPHINCS+ [4], Crystals-Dilithium [26], and Falcon [32]) after rigorous scrutiny spanning multiple years. Among these only SPHINCS+ is based on the hardness of cryptographically secure hash functions, while Crystals-Kyber (KEM), Crystals-Dilithium, and Falcon are based on hard lattice problems. As the majority of these constructions are lattice-based, there is a lingering risk that a breakthrough in the cryptanalysis of lattice-based cryptography can reduce the security of these schemes drastically. Thus putting the whole plan to migrate to post-quantum cryptography in jeopardy. Such incidents are not uncommon. Recently, Decru et al. [18] proposed an attack to completely break the security of supersingular isogeny Diffie-Hellman [31] which was earlier considered quantum-safe and was also a finalist in the NIST's standardization procedure. Therefore, it is prudent to diversify the portfolio of different quantum-safe problems for seamless migration to a post-quantum world. There exist other problems that are considered quantum-safe, such as multivariate quadratic (MQ) [39,46], isogeny-based [22], and code-based [8]. Standardizing cryptographic primitives necessitates a rigorous and comprehensive investigation. NIST reissued a call [20] for quantum-safe signature schemes to standardize some more signature schemes to diversify the portfolio of quantum-resistant schemes. Due to its small signature size, multivariate oil-vinegar construction has gained significant attention during this standardization process.

Multivariate cryptography relies on the intractability of root findings of MQ equations. The goal of the MQ problem is to find a solution to a system of multivariate quadratic polynomials in the finite field $\mathbb{F}_q$. In other words, the hardness classification of this problem is NP-hard [38]. Numerous schemes, such as Matsumoto-Imai encryption scheme [43], Oil-Vinegar [46] signature, Rainbow [24] signature, Triangular [45,53,60] signature, Simple Matrix encryption [56], and Mayo [12], have been developed based on multivariate cryptography. Patarin first proposed the Oil-Vinegar signature [46]. A successful forgery attack was shown by Kipnis and Shamir [40] against this scheme. Further, Kipnis, Patarin, and Goubin upgraded the signature scheme by proposing Unbalanced Oil-Vinegar (UOV) [39].

Rainbow was a third-round NIST candidate [24], which is the first multilayer construction based on unbalanced oil-vinegar. Therefore, the cryptanalysis of Rainbow has been a well-studied area for the last decade. This resulted in many new novel attacks such as direct attack [6,27,28], min-rank attack [5–7,14], band-separation attack [25,55,57], rectangular min-rank and intersection attack [10]. In 2023, Beullens proposed a cryptanalysis and reduced the security of Rainbow significantly. Rainbow team suggested using the old SL-3 (high security) parameter set as new SL-1 (low security) parameters [36] to mitigate the attack. As Beullens' attack only applies to the Rainbow structure, therefore building scheme on the top of the oil-vinegar layer is still believed to be secure.

In 2022, Cartor *et al.* internally perturbed the second layer of Rainbow by mixing oil variables quadratically [17]. However, this mixing significantly increased the signature generation time. Also, parameter sets proposed by designers are not practical in terms of efficiency. Therefore, designing a new signature scheme that can resist the simple attack while being practical, is an interesting open problem.

1.1   Our Contribution and Motivation

In the context of this endeavor, we summarize our contributions below.

- We review related multivariate signature schemes and provide a comprehensive analysis of their design and performance in Sect. 2.
- We present Vinegar-Diagonal-Oil-Oil (VDOO), a novel multivariate signature scheme based on unbalanced oil and vinegar in Sect. 3. Compared to other UOV schemes VDOO boasts three primary benefits: *simplicity, efficiency,* and *security* (see Sects. 4 and 5). To the best of our knowledge, we are the first to introduce a diagonal layer within the UOV framework, demonstrating that it enhances efficiency without compromising security.
- We establish that VDOO effectively withstands all current attacks and outline the EUF-CMA security of our scheme. Through meticulous parameter selection, our findings reveal that it achieves a remarkably compact smallest signature size of 96 bytes (see Sects. 4 and 5), contrasting favorably with NIST-standardized post-quantum signatures (Crystals-dilithium [26], Falcon [32], and SPHINCS+ [4]).

Introduction of a New Simple Design Element. VDOO is a new layer-based construction, which has one diagonal layer and then two UOV layers. We are adding each new variable in the central polynomial one by one diagonally. This offers efficiency. This translates to a reduction of the Gaussian elimination ($\mathsf{GE}_{(q,n)}$[1]) which is the major computational bottleneck in the signature generation process. Suppose $x_1, x_2, \cdots, x_v, x_{v+1}, \cdots, x_{v+d}, \cdots, x_{v+d+o_1}, \cdots, x_{v+d+o_1+o_2=:n}$ are n variables defined over $\mathbb{F}_q$. In our construction, we call first v-variables as *vinegar variables*, next d-variables as *diagonal variables*, then next o_1 variables are *first-layer oil variables*, and last o_2 variables are *second-layer oil variables*. Figure 1 illustrates the distribution of the variables in each layer of the VDOO central polynomial map.

---

[1] $\mathsf{GE}_{(q,n)}$: Gaussian elimination on a linear system with n unknowns and n linear equation over $\mathbb{F}_q$.

Efficiency. To thwart Beullens' simple attack [11], the authors of Rainbow increased the parameter set [36], which results in increasing the Gaussian elimination cost. The complexity of Gaussian elimination becomes approximately $o_1^3 + o_2^3$ where o_1 and o_2 are number of oil variables of Rainbow [24]. In our scheme, we adapt $d \approx (o_1 + o_2)/3$, $o_1' \approx (o_1 + o_2)/3$, and $o_2' \approx (o_1 + o_2)/3$ as the new parameters. This adjustment results in a Gaussian elimination complexity of around $o_1'^3 + o_2'^3$. To illustrate, consider the signature generation process for security level one (SL-1) parameters [19]: UOV requires $\mathsf{GE}_{(256,64)}$, Rainbow requires $\mathsf{GE}_{(256,32)}$ and $\mathsf{GE}_{(256,48)}$, while VDOO needs only $\mathsf{GE}_{(16,34)}$ and $\mathsf{GE}_{(16,36)}$ (for further details, refer to Table 2). Consequently, this modification notably improves our scheme's performance.

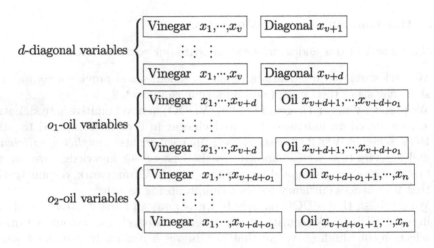

Fig. 1. Variables in each layer of the VDOO central map

Resistance to Existing Attacks. We comprehensively analyze all possible attacks on multivariate cryptographic schemes against our scheme. In an attempt to recover diagonal variables, potential attackers begin by eliminating the uppermost oil layers. Beullens proposed method [11] facilitates the removal of these layers, aiding attackers. For instance, in order to compromise our round-one parameter set, a straightforward attack necessitates 2^{134}-field operations. Furthermore, Beullens combined this simple attack with the rectangular min-rank attack [10,11]. In line with previous efforts, we execute this combined attack against our scheme, determining that it requires 2^{138}-field operations to break SL-1 parameter set. Additionally, we conduct the intersection attack and the direct attack on our scheme, both of which exhibit complexities exceeding 2^{134}-field operations. Consequently, these references collectively imply that VDOO appears to withstand all known attacks securely. We also outline the EUF-CMA security of the VDOO scheme.

Small Signature Size. We present multiple parameters that can withstand the aforementioned attacks. Specifically, our level-one parameters that can provide

128-bit classical and 96-bit post-quantum security has a signature size of 96 bytes and public-key size of 238KB (further elaborated in Table 1). This is the smallest signature size among the majority of all multivariate signature schemes (for additional insights, refer to Tables 2 and 3).

Roadmap. In the upcoming Sect. 2 we present a generic construction of multivariate signatures, and some earlier results. Section 3 proposes a new post-quantum multivariate signature scheme called VDOO. The cryptanalysis of our scheme is presented in Sect. 4. In Sect. 5, we give the parameters for different security levels and we also compare our results with the state-of-the-art. Our Sect. 6 presents conclusions and explores potential future directions for our work.

2   Prior Results

In this section, we introduce some essential mathematical notations and symbols. We then provide a generic construction for multivariate signatures. Following that, we outline the central polynomial for UOV and Rainbow [13,24,39]. Additionally, we describe the subspace representation of Rainbow [10], which is particularly valuable for cryptanalysis purposes. Next, we cover recent multivariate signature schemes [12,23,29,33,34] that were submitted as part of the NIST additional round for post-quantum signature standardization [20]. Finally, we present the required hardness assumptions for these multivariate signatures to understand their cryptanalysis.

Notations. Let, $\mathbb{F}_q$ be the finite field with q elements. We define two invertible affine maps $\mathcal{S} : \mathbb{F}_q^m \to \mathbb{F}_q^m$ and $\mathcal{T} : \mathbb{F}_q^n \to \mathbb{F}_q^n$, and one quadratic map $\mathcal{F} = (f_1, \cdots, f_m) : \mathbb{F}_q^n \to \mathbb{F}_q^m$. We denote $[n]$ for the set $\{1, 2, \cdots, n\}$ and $[i:j]$ denotes $\{i, i+1, \cdots, j\}$. We use lowercase and bold lowercase alphabets to denote field elements and vectors respectively. The notation $a \in_U S$ is used to interpret a *is a random element in the set S*.

2.1   Generic Multivariate Signature Schemes

Here we briefly describe a generic construction for multivariate signature schemes. Due to the NP-hardness of inverting a randomly generated quadratic system [38]. However, signers can leverage a specially structured quadratic system to efficiently perform the inversion. This specialized system is commonly referred to as the *central map* and is typically denoted as $\mathcal{F} = (f_1, \cdots, f_m)$, where each f_i represents a specifically structured multivariate quadratic polynomial. Signers must conceal this unique structure from third parties to prevent forgery attacks. To achieve this objective, signers employ one or two random invertible linear maps: $\mathcal{S}$ and $\mathcal{T}$. Consequently, the public key is constructed by composing these linear maps along with the central map, denoted as $\mathcal{P} = \mathcal{S} \circ \mathcal{F} \circ \mathcal{T} : \mathbb{F}_q^n \longrightarrow \mathbb{F}_q^m$.

The secret key comprises $\mathcal{S}$, $\mathcal{T}$ and $\mathcal{F}$. A hash function, denoted as $\mathcal{H} : \{0,1\}^* \longrightarrow \mathbb{F}_q^m$, is employed to generate a vector $\mathbf{m} \in \mathbb{F}_q^m$ from a message

$msg \in \{0,1\}^*$. The signature generation process unfolds as follows: first, compute $\mathbf{d} \leftarrow \mathcal{S}^{-1}(\mathbf{m})$, then $\mathbf{d}' \leftarrow \mathcal{F}^{-1}(\mathbf{d})$, and finally $\mathbf{s} \leftarrow \mathcal{T}^{-1}(\mathbf{d}')$. The signer sends the signature $\mathbf{s}$ for the message msg to the verifier. The verifier simply evaluates the polynomial map $\mathcal{P}$ on $\mathbf{s}$ and checks whether it matches the hash of the message, i.e., whether $\mathbf{m} = \mathcal{P}(\mathbf{s})$ holds or not.

2.2 Unbalanced Oil-Vinegar (UOV)

The Oil-Vinegar (OV) signature scheme was initially introduced by Patarin [46]. However, due to the Kipnis-Shamir's [40] *invariant subspace* attack, this scheme was modified by increasing the number of vinegar variables. This is known as the Unbalanced Oil-Vinegar (UOV) signature scheme [39].

Consider the OV central map, denoted as $\mathcal{F}$. Split all variables of $\mathbf{x} = (x_1, \cdots, x_v, \cdots, x_n)$ into two buckets: the first bucket has first v variables representing vinegar, and the second bucket contains next o variables representing oil, where $n = v + o$ and $o = m$. To create a multivariate quadratic homogeneous polynomial, combine variables involving vinegar $\times$ vinegar and vinegar $\times$ oil, while excluding all oil $\times$ oil terms.

Definition 1 (OV Central Polynomial Map). *A central map* $\mathcal{F} = (f_1, \cdots, f_m) : \mathbb{F}_q^n \rightarrow \mathbb{F}_q^m$ *is known as OV central polynomial map when each* f_i *is of the form* $f_i(\mathbf{x}) = \sum_{i=1}^{v} \sum_{j=1}^{n} \alpha_{i,j}^{(k)} x_i x_j$ *where* $i \leq j$, $k \in [v+1:n]$, $\mathbf{x} \in \mathbb{F}_q^n$, *and* $\alpha_{i,j}^{(k)} \in_U \mathbb{F}_q$.

Notably, if anyone randomly fixes vinegar variables, then the remaining part would be linear in the oil variables. Therefore, the quadratic system reduces to a linear system of o linear equations with o unknowns.

2.3 Rainbow

Rainbow is a multi-layer variant of UOV [24]. For simplicity consider a two-layer Rainbow. Suppose $n = v + o_1 + o_2$, where the first v variables are vinegar and the next o_1 and o_2 variables are the first and second layer of oil variables respectively. This can be viewed as a UOV map with $v + o_1$ variables and o_1 oil variables and the next layer $v + o_1 + o_2$ variables and o_2 oil variables.

Definition 2 (Rainbow Central Polynomial Map). *The mathematical expression for l-layer Rainbow central polynomial is as follows.*

$$f_k(x_1, x_2, \cdots, x_n) = \sum_{i,j \in [r];\ i \leq j} \alpha_{ij}^{(k)} x_i x_j + \sum_{i \in [r];\ j \in [r+1:r+o_r]} \beta_{ij}^{(k)} x_i x_j$$

where for each $k \in [r+1 : r+o_r]$, *elements* $\alpha_{ij}^{(k)}$, *and* $\beta_{ij}^{(k)}$ *are taken from* $\mathbb{F}_q$; r *denotes the layer, and* $r \leq l$ *where* l *is the total number of layers in Rainbow.*

2.4   Beullens Subspace Description

For a better view of cryptanalysis on Rainbow, Beullens explained the construction of Rainbow via subspaces [10]. Using this description, he derived the simple attack [11]. To elaborate this idea, initially, we define a differential polar form of a polynomial map.

The *differential polar map* of a polynomial map $\mathcal{P}$ is denoted by $\mathcal{DP} : \mathbb{F}_q^n \times \mathbb{F}_q^n \to \mathbb{F}_q^m$ and defined as $\mathcal{DP}(\mathbf{x}, \mathbf{w}) = \mathcal{P}(\mathbf{x}+\mathbf{w}) - \mathcal{P}(\mathbf{x}) - \mathcal{P}(\mathbf{w})$. Note that, we only consider homogeneous quadratic polynomials, so throughout this paper, $\mathcal{P}(0) = 0$.

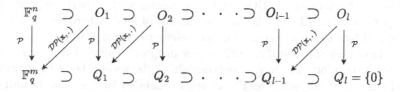

Fig. 2. l layer Rainbow

Trapdoor Information. This part describes the trapdoor information of l-layer Rainbow. At first, signer chooses a secret chain of nested subspaces: input subspaces $O_1 \supset O_2 \supset \cdots \supset O_l$ and output subspaces $Q_1 \supset Q_2 \supset \cdots \supset Q_l = \{0\}$. Using this secret, one can construct a public polynomial map as follows.

– $\mathcal{P}$ maps each O_i to Q_i and
– for any $\mathbf{x} \in_U \mathbb{F}_q^n$, $\mathcal{DP}_\mathbf{x} : O_i \to Q_{i-1}$ is a linear map (see Fig. 2)

Inversion. In this methodology, the goal is to compute $\mathbf{x} \in \mathbb{F}_q^n$ from given $\mathbf{y} \in \mathbb{F}_q^m$ such that $\mathbf{y} = \mathcal{P}(\mathbf{x})$. The knowledge of nested sequences of input and output subspaces is used in this computation. At first glance, for l-layer Rainbow, the value of the unknown $\mathbf{x}$ can be represented as $\mathbf{v}+\mathbf{o_1}+\cdots+\mathbf{o_l}$ where all of the $\mathbf{o_i} \in O_i$. Fix $\mathbf{v} \in_U \mathbb{F}_q^n$. Then $\mathcal{P}$ is used in conjunction with the ith-layer's output subspace Q_i to calculate $\mathbf{o_i}$. For the sake of clarity, let's define the *quotient space* $\overline{O}_i := O_i/O_{i+1}$.

Using the knowledge of sequences of subspaces, the goal is to find $\mathbf{o_i}$ for all i. This will lead to computing the preimage of any element from $\mathbb{F}_q^n$. For computing $\overline{\mathbf{o}}_i \in \overline{O}_i$, use the following relation (note that, from definition, $\mathcal{P}\left(\overline{\mathbf{o}}_i\right) = 0$),

$$\mathcal{P}(\mathbf{v}+\overline{\mathbf{o}}_i) + Q_i = \mathbf{y} + Q_i$$
$$\longrightarrow \mathcal{P}(\mathbf{v}) + \mathcal{P}\left(\overline{\mathbf{o}}_i\right) + \mathcal{DP}(\mathbf{v}, \overline{\mathbf{o}}_i) + Q_i = \mathbf{y} + Q_i.$$

Earlier $\mathbf{v}$ is fixed, so the quadratic system reduces to a linear system. The number of constraints and variables are the same for the linear system. This implies that a unique solution can be obtained with probability $(1-\frac{1}{q})$. Repeatedly running this procedure, one can compute all $\mathbf{o}_i$, which implies that preimage $\mathbf{x}$ will be computed.

In 2022, Beullens [11] reduced the security level of Rainbow. He showed for small $n - m$, recovering all subspaces is significantly efficient. Also, the small finite field size accelerates the attack.

2.5   Concurrent Proposals

The NIST additional signature submission call [20] received a total of eleven multivariate signature schemes e.g. Mayo [12], QR-UOV [34], TUOV [23], etc. Most of them are based on the old *unbalanced Oil-Vinegar* structure. For example, Mayo [12] employed a UOV structure along with a new *whipped-up MQ* (WMQ) approach. QR-UOV is another variant of UOV where the public key is represented by block matrices, with each element corresponding to an element in a quotient ring [34]. Also, in 2022, a new proposal, called IPRainbow [17] was made by perturbing the central polynomials of the second layer by s variables. This change although decreases the attack probability by $1/q^s$, the running time significantly increases due to the usage of Gröbner basis technique for inversion.

2.6   Hardness of Multivariate Cryptography

Here, we describe other approaches used in the cryptanalysis of multivariate signatures apart from the direct solution of MQ equations.

1. **Min-rank.** Let $M_1, M_2, \cdots, M_k \in \mathbb{F}_q^{n \times m}$ be the given matrices and $r \in \mathbb{N}$, find a non-trivial linear combination (with $m_1, m_2, \cdots, m_k \in \mathbb{F}_q$) so that rank $\left(\sum_{i=1}^{k} m_i M_i \right) \leq r$. This problem is called the *min-rank* problem and has been shown to be NP-hard [16]. The min-rank problem appeared as a cryptanalytic tool in multivariate cryptography [6,10,30,41]. This attack helps to find a linear combination of public matrices which sums up to a low-rank matrix.

2. **EIP.** Find an equivalent composition of $\mathcal{P} = \mathcal{S}' \circ \mathcal{F}' \circ \mathcal{T}'$, where $\mathcal{S}'$ and $\mathcal{T}'$ are equivalent affine maps, and $\mathcal{F}'$ is an equivalent central map. The above problem is the *Extended Isomorphism of Polynomials (EIP)* problem. No such hardness classification is known (though it subsumes graph isomorphism problem [1,2]), but for some instances, polynomial time algorithms exist [40].

3   Our Proposal: VDOO Signature Scheme

In our scheme, we introduce a new design element called *diagonals* into the Oil-Vinegar scheme. Let, $\mathbf{x} \in \mathbb{F}_q^n$, we pick the first v variables as vinegar variables. We denote the next d variables as diagonal variables. In this layer, we introduce d quadratic equations. In any i-th ($1 \leq i \leq d$) equation, only $v + i$-th variable is unknown among $v + i$ variables. In the following layers, we apply the Oil-Vinegar technique. This means we can generate o_1 OV polynomials using $v + d$-vinegar variables and newly added o_1-oil variables. Further, we construct o_2 OV polynomials using $v + d + o_1$-vinegar variables and newly added o_2-oil variables. Finally, we have a quadratic system with $n = v + d + o_1 + o_2$ variables and $m = d + o_1 + o_2$ homogeneous quadratic equations.

3.1 VDOOSetUp: Generate Parameters

To construct polynomial maps we need to define parameters associated with this. In this phase algorithm takes input the security parameter λ and output the parameter tuple, that is params $= (q, v, d, o_1, o_2) \leftarrow$ VDOOSetUp(1^λ). Here,

- Finite field $\mathbb{F}_q$ which has q elements.
- Positive integers v, d, o_1, and o_2, where v denotes the number of vinegar variables, d is the number of diagonal variables, o_1 and o_2 stands for the number of first and second layer oil variables respectively. Therefore, total number of variables is $n = v + d + o_1 + o_2$, and number of equations is $m = d + o_1 + o_2$.

3.2 VDOO Central Polynomial Map and Inversion

Construction of central polynomial map $\mathcal{F} : \mathbb{F}_q^n \rightarrow \mathbb{F}_q^m$ plays an important role in the multivariate signature schemes. To the best of our knowledge, we are the first to propose a central polynomial map that involves vinegar, diagonal, and oil variables in a three-layer construction.

- **Diagonal Layer.** Here, we explain the structure of any central polynomial f_k for the diagonal layer $k \in [v + 1 : v + d]$. Each f_k is defined as follows.

$$f_{k-v}(x_1, x_2, \cdots, x_n) = \sum_{i=1}^{k-1} \alpha_{i,k}^{(k)} x_i x_k + \sum_{i,j=1, i \leq j}^{k-1} \beta_{i,j}^{(k)} x_i x_j$$

Each coefficient $\alpha_{ij}^{(k)}$, and $\beta_{ij}^{(k)} \subset_U \mathbb{F}_q$. The subroutine DiagPoly(q, k) is used to generate such central polynomial f_k in the diagonal layer.
- **First Oil Layer.** In this oil layer, we use $v + d$ variables as vinegar variables and next o_1 variables as oil variables. All these variables help us to construct o_1 homogeneous quadratic polynomials of the following form.

$$f_{k-v}(x_1, x_2, \cdots, x_n) = \sum_{i=1}^{v+d} \sum_{j=1}^{v+d} \alpha_{ij}^{(k)} x_i x_j + \sum_{i=1}^{v+d} \sum_{j=v+d+1}^{v+d+o_1} \beta_{ij}^{(k)} x_i x_j$$

where $k \in [v + d + 1 : v + d + o_1]$, $\alpha_{ij}^{(k)}$, and $\beta_{ij}^{(k)} \in_U \mathbb{F}_q$.
- **Second Oil Layer.** The topmost oil layer has $v + d + o_1$ vinegar and o_2 oil variables. That means, it has o_2 quadratic equations. Those equations are of the form

$$f_{k-v}(x_1, x_2, \cdots, x_n) = \sum_{i=1}^{v+d+o_1} \sum_{j=1}^{v+d+o_1} \alpha_{ij}^{(k)} x_i x_j + \sum_{i=1}^{v+d+o_1} \sum_{j=v+d+o_1+1}^{v+d+o_1+o_2} \beta_{ij}^{(k)} x_i x_j,$$

where $k \in [v+d+o_1+1 : v+d+o_1+o_2 = n]$ and $\alpha_{ij}^{(k)}$, and $\beta_{ij}^{(k)} \in_U \mathbb{F}_q$. We denote this as OVPoly(q, v, o) to generate a oil-vinegar central polynomial (according to Sect. 2.2) which has v vinegar variables and o oil variables.

Here, Algorithm 1, uses OVPoly and DiagPoly to generate a VDOO central map $\mathcal{F}$.

Algorithm 1. VDOOCentPoly

Require: Parameter tuple $params = (q, v, d, o_1, o_2)$
Ensure: Central map $\mathcal{F} = (f_1, \cdots, f_m) : \mathbb{F}_q^n \to \mathbb{F}_q^m$
1: Compute $m = d + o_1 + o_2$ and $n = v + m$.
2: **for** $1 \le i \le d$
3:     $f_i \leftarrow$ DiagPoly (q, i)
4: **for** $d+1 \le i \le d+o_1$
5:     $f_i \leftarrow$ OVPoly $(q, v+d, o_1)$
6: **for** $d+o_1+1 \le i \le m$
7:     $f_i \leftarrow$ OVPoly $(q, v+d+o_1, o_2)$
8: **Return** VDOO central polynomial $\mathcal{F}$

Inversion. The main computational bottleneck of UOV-based constructions is the inversion of the central polynomial. It requires Gaussian elimination which runs in $O(N^3)$. However, in our scenario inversion of the diagonal polynomials is straightforward as there is only one unknown variable. Nevertheless, the inversion of OV polynomials in the remaining two layers each needs a Gaussian elimination. Therefore, inverting VDOO central polynomial map needs two Gaussian elimination only. This is shown in Algorithm 2. Following two algorithms help to compute the inverse of the VDOO central polynomial.

ST The subroutine *Substitution* or ST converts a bunch of oil-vinegar polynomials to a bunch of linear polynomials consists of oil variables by fixing the vinegar variables. That means, ST substitutes vinegar variables $x_1, \cdots, x_v$ by random values (in $\mathbb{F}_q$) in the bunch of o oil-vinegar polynomials $(f_i)_{i=1}^o$ and converts it to a bunch of linear polynomials of o oil variables $(\tilde{f}_i)_{i=1}^o$.

GE The GE$_{(q,l)}$ denotes Gaussian elimination for l unknowns over the linear system of equations ($\tilde{f}_i = y_i$)$_{i=1}^l$ over $\mathbb{F}_q$. It returns a failure when the rank of the matrix representing the linear system is less than l.

Algorithm 2. VDOOCentPoly_Inversion

Require: Central map: $\mathcal{F} = (f_1, \cdots, f_m) : \mathbb{F}_q^n \rightarrow \mathbb{F}_q^m$ and $\mathbf{y} \in \mathbb{F}_q^m$, and params.

Ensure: A vector $\mathbf{x} \in \mathbb{F}_q^n$ such that $\mathcal{F}(\mathbf{x}) = \mathbf{y}$.

1: $m \leftarrow d + o_1 + o_2$ and $n \leftarrow v + m$

2: Randomly fix first v-vinegar variables $x_1, \cdots, x_v \leftarrow_\$ \mathbb{F}_q$

3: **for** $1 \leq i \leq d$

4:     compute x_{v+i} using $y_i, x_1, \cdots, x_{v+i-1}$ and f_i.

5: $(\tilde{f}_{d+1}, \cdots, \tilde{f}_{v+d}) \leftarrow \mathsf{ST}(f_{d+1}(x_1, \cdots, x_{v+d}), \cdots, f_{d+o_1}(x_1, \cdots, x_{v+d}))$

6: $(x_{v+d+1}, \cdots, x_{v+d+o_1}) \leftarrow \mathsf{GE}_{(q,o_1)}(\tilde{f}_{d+1} = y_{d+1}, \cdots, \tilde{f}_{d+o_1} = y_{d+o_1})$.

7: $(\tilde{f}_{d+o_1+1}, \cdots, \tilde{f}_m) \leftarrow \mathsf{ST}(f_{d+o_1+1}(x_1, \cdots, x_{n-o_2}), \cdots, f_m(x_1, \cdots, x_{n-o_2}))$

8: $(x_{v+d+o_1+1}, \cdots, x_n) \leftarrow \mathsf{GE}_{(q,o_2)}(\tilde{f}_{d+o_1+1} = y_{d+o_1+1}, \cdots, \tilde{f}_m = y_m)$

9: **Return** $\mathbf{x} \in \mathbb{F}_q^n$

3.3 VDOOKeyGen: **VDOO Key Generation**

The VDOOKeyGen in Algorithm 3 generates two random invertible affine maps $\mathcal{S} : \mathbb{F}_q^m \rightarrow \mathbb{F}_q^m$ and $\mathcal{T} : \mathbb{F}_q^n \rightarrow \mathbb{F}_q^n$ along with the VDOO-central map $\mathcal{F} : \mathbb{F}_q^n \rightarrow \mathbb{F}_q^m$. Here, *secret/signing key* is $\mathcal{S}$, $\mathcal{F}$, and $\mathcal{T}$ and *public/verification key* is the composition map $\mathcal{P}$, where $\mathcal{P} = \mathcal{S} \circ \mathcal{F} \circ \mathcal{T} : \mathbb{F}_q^n \rightarrow \mathbb{F}_q^m$. Note that, the individual information of secret maps allows user to compute the inverse of $\mathcal{P}$ efficiently. We denote $S \leftarrow$ randomMatrix $(q, m, seed)$ to generate a random $m \times m$ matrix over $\mathbb{F}_q$ from a *seed*, invMat (q, m, S) helps to compute the inverse of a $m \times m$ matrix S over $\mathbb{F}_q$, and Affine$(S, \mathbf{a})$ computes $\mathcal{S} \leftarrow S \cdot \mathbf{x} + \mathbf{a}$.

Algorithm 3. VDOOKeyGen

Require: Parameter tuple params.

Ensure: Generate public and private key pair.

    – Public key: pk $= \mathcal{P}$.
    – Secret key: sk $= \mathcal{S}, \mathcal{T}$, and $\mathcal{F}$.

1: $m \leftarrow d + o_1 + o_2$ and $n \leftarrow m + v$

2: $seed \leftarrow PRNG(1^\lambda)$ ▷ λ is the security parameter

3: **while** $(\det(S) \neq 0$ && $\det(T) \neq 0)$ **do**

4:     $S \leftarrow$ randomMatrix $(q, m, seed)$ ▷ $S \in_U \mathbb{F}_q^{m \times m}$

5:     $T \leftarrow$ randomMatrix $(q, n, seed)$ ▷ $T \in_U \mathbb{F}_q^{n \times n}$

6: **end while**

7: $\mathbf{a} \in_U \mathbb{F}_q^m$ and $\mathbf{b} \in_U \mathbb{F}_q^n$ ▷ generate two random vector

8: $invS \leftarrow$ invMat(q, m, S) and $invT \leftarrow$ invMat(q, n, T) ▷ compute inverse of matrices

9: $\mathcal{S} \leftarrow$ Affine$(S, \mathbf{a})$ and $\mathcal{T} \leftarrow$ Affine$(T, \mathbf{b})$ ▷ Constructing invertible affine maps

10: $\mathcal{F} \leftarrow$ VDOOCentPoly(params)   ▷ generate VDOO central map

11: Compute $\mathcal{P} \leftarrow \mathcal{S} \circ \mathcal{F} \circ \mathcal{T}$

12: **Return** pk $= \mathcal{P}$ and sk $= (invS, \mathbf{a}, invT, \mathbf{b})$ (equivalently sending $\mathcal{S}$, and $\mathcal{T}$).

3.4 VDOOSign: **VDOO Signature Generation**

Similar to the other OV based constructions [12,23,24,39], we use the hash-and-sign paradigm for our signature algorithm as shown in Algorithm 4. We use a hash function $\mathcal{H} : \{0,1\}^* \to \mathbb{F}_q^m$. Signer knows each polynomial map, so it can compute the inverse of each map *i.e.* $\mathcal{S}^{-1}$, $\mathcal{F}^{-1}$, and $\mathcal{T}^{-1}$. If GE reports a failure during the computation of $\mathcal{F}^{-1}$, we restart the process by regenerating the salt and repeating the entire procedure. Finally, the signature is computed as $\mathcal{P}^{-1}(\mathcal{H}(\mathcal{H}(msg)\|salt))$.

Algorithm 4. VDOOSign

Require: sk $= (invS, \mathbf{a}, invT, \mathbf{b})$, message msg, and $\mathcal{H} : \{0,1\}^* \to \mathbb{F}_q^m$
Ensure: a signature $\sigma = (\mathbf{s}, salt)$
1: $salt \longleftarrow PRNG$
2: Use hash function $\mathbf{d} \leftarrow \mathcal{H}(\mathcal{H}(msg)\|salt)$
3: Compute $\mathbf{t} = invS \times (\mathbf{d} - \mathbf{a})$ $\triangleright \mathbf{t} = \mathcal{S}^{-1}(\mathbf{d})$
4: Compute $\mathbf{y} = \mathcal{F}^{-1}(\mathbf{t})$ using VDOOCentPoly_Inversion 2.
5: Compute $\mathbf{s} = invT \times (\mathbf{y} - \mathbf{b})$ $\triangleright \mathbf{s} = \mathcal{T}^{-1}(\mathbf{y})$
6: **Return** signature $\sigma = (\mathbf{s}, salt)$

Efficiency Analysis. As mentioned earlier, the major computational overhead of OV-based schemes is the Gaussian elimination procedure. In VDOO, during signing, we have to compute only one Gaussian elimination *i.e.* computation of $\mathcal{F}^{-1}$. The computation of $\mathcal{S}^{-1}$ and $\mathcal{T}^{-1}$ can be done during the key-generation procedure. In VDOO the computation of $\mathcal{F}^{-1}$ is also very efficient compared to other OV-based schemes as the number of unknowns is smaller in VDOO as shown in Table 2.

3.5 VDOOVerif: **VDOO Verification**

Our verification procedure is simple. It needs a polynomial evaluation of $\mathcal{P}$, requiring just $O(N^3)$ field operations. Compute $\mathbf{d}' = \mathcal{P}(\mathbf{s})$ from public key $\mathcal{P}$ and signature $\sigma = (\mathbf{s}, salt)$ The signatures is accepted if $\mathbf{d}' = \mathcal{H}(\mathcal{H}(msg) \| salt)$ holds, else rejected.

Algorithm 5. VDOOVerif

Require: pk $= \mathcal{P}$; message msg; signature $\sigma = (\mathbf{s}, salt)$ and $\mathcal{H} : \{0,1\}^* \to \mathbb{F}_q^m$.
Ensure: *accept* or *reject*
1: Use hash function to compute $\mathbf{d} \leftarrow \mathcal{H}(\mathcal{H}(msg) \| salt)$
2: Compute $\mathbf{d}' = \mathcal{P}(\mathbf{s})$
3: **if** $\mathbf{d} = \mathbf{d}'$ **then** output *accept*
4: **else** *reject*
5: **end if**
6: **Return** *accept* or *reject*

3.6   Key Size Computation

Our VDOO contains one diagonal layer and two UOV layers. The size of the private key is determined first, followed by the size of the public key.

– Size of the central map $\mathcal{F}$ for a diagonal layer having d-diagonal polynomials is
$\sum_{i=1}^{d} \left(\frac{v_i(v_i+1)}{2} + v_i \right)$ field elements. The first diagonal layer has $v_1 = n-m$
vinegar variables. In any diagonal layer, a central polynomial f_i has v_i vinegar variables and f_{i+1}-th polynomial has $v_{i+1} = v_i + 1$ vinegar variables.

– Size of the central map $\mathcal{F}$ for a UOV layer is around $o \times \left(\frac{v(v+1)}{2} + ov \right)$
field elements. Such UOV layer has v vinegar variables and o oil variables.

The sizes of the two affine transformations are as follows: for $\mathcal{S}$ we need $m(m+1)$, while for $\mathcal{T}$ we need $n(n+1)$, field elements. These maps can be generated using a random seed.

Now we are interested in computing the size of the public key of standard VDOO. Each n-variate quadratic polynomial requires $\frac{(n+1)(n+2)}{2}$ field elements. Therefore, the size of the public key is $m\frac{(n+1)(n+2)}{2}$. Further optimization of public key is possible [48,49]. It optimized the public key size from $O(mn^2 \log q)$ to $O(m^3 \log q)$.

3.7   Subspace Description of VDOO Central Polynomial

Our scheme can be explained through Beullens's subspace descriptions [10]. This description is useful to understand the cryptanalysis of VDOO. In this case, we have $d+2$ input and output subspaces. These sequences of nested subspaces are as follows.

– **Input subspaces** $\mathbb{F}_q^n \supset D_1 \supset D_2 \supset \cdots \supset D_d \supset O_1 \supset O_2$.
– **Output subspaces** $\mathbb{F}_q^m \supset Q_{1,1} \supset Q_{1,2} \supset \cdots \supset Q_{1,d} \supset Q_2 \supset Q_3 = \{0\}$.

In the Fig. 3 (single arrow denotes $\mathcal{P}$ and bold arrow denotes $\mathcal{DP}(\mathbf{x}, \cdot)$), these following relations will hold: $\dim(D_i) = \dim(D_{i+1}) + 1$ and $\dim(Q_{1,i}) = \dim(Q_{1,i+1}) + 1$ for $1 \leq i < d$. Also, $\dim(D_1) = m$, $\dim(D_i) = \dim(Q_{1,i-1})$ for $1 < i \leq d$. In addition, $\dim(O_1) = \dim(Q_{1,d}) = o_1 + o_2$, $\dim(O_2) = \dim(Q_2) = o_2$.

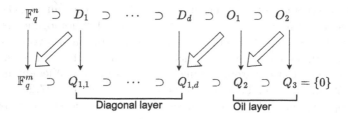

Fig. 3. Central polynomial of VDOO

The signer first fixes $\mathbf{v} \in_U \mathbb{F}_q^n$. Since $\dim(\tilde{D}_i) = \dim(D_i) - \dim(D_{i+1}) = 1$, so for diagonal layer computing $\mathbf{d}_1, \cdots, \mathbf{d}_d$ is very easy. Once these vectors are found, then update $\mathbf{v} \leftarrow \mathbf{v} + \mathbf{d}_1 + \cdots + \mathbf{d}_d$. Now, signer needs to solve for $\tilde{\mathbf{o}}_1 \in \tilde{O}_1 (= O_1/O_2)$, so that the following relation holds. Note that, $\dim(\tilde{O}_1) = o_1$.

$$\mathcal{P}(\mathbf{v}) + \mathcal{DP}(\mathbf{v}, \tilde{\mathbf{o}}_1) = \mathbf{t} \mod Q_2.$$

We know that the above equation is a linear system of o_1 variables and o_1 equations. With the probability $(1 - 1/q)$, the signer will able to compute $\mathbf{o}_1$. Then signer again updates $\mathbf{v} \leftarrow \mathbf{v} + \mathbf{o}_1$ and follow a similar strategy to find $\mathbf{o}_2 \in O_2$. Thus the signer can finally compute the pre-image of $\mathbf{t}$.

4   Security Analysis of VDOO

Cryptanalysis that targets solving the MQ problem directly, is known as the direct attack in multivariate cryptography [6,9,27,28]. Later researchers have used the special structure of the quadratic system and improved the state-of-the-art, like, band-separation attack [25,55,57], intersection attack [10], and simple attack [11].

To determine the complexity of the attacks described below by the number of field multiplications required to perform the attack. One $\mathbb{F}_q$-field multiplication needs $(2(\log_2 q)^2 + \log_2 q)$ gates. Here, each $2(\log_2 q)^2$-bit stands for one $(\log_2 q)^2$-bit multiplication (represented as AND gates) and the same number of additions (represented as XOR gates) during one $\mathbb{F}_q$-multiplication. Additionally, $\log_2 q$ bits are needed for $\log_2 q$-bit additions involved in one $\mathbb{F}_q$-addition, which is required for each field multiplication that occurs during an attack. For example, the cost one $\mathbb{F}_{16}$-multiplication requires 36 gates. Such a strategy to determine the complexity is standard and has been also followed in other MQ-based signature schemes [12,29,34].

Henceforth, in this document, we use the parameter set $(q, v, d, o_1, o_2) = (16, 60, 30, 34, 36)$ as an example to demonstrate the complexity of the following attacks. Incidentally, this is also our SL-1 parameter. Our full parameter set is given in Table 1.

4.1   Direct Attack on VDOO

The direct attack is the fundamental methodology for forging any multivariate signature scheme. To counterfeit a VDOO signature, an attacker aims to solve an underdetermined system with n variables and m homogeneous equations $(n > m)$, to find $\mathbf{s}$ such that $\mathcal{P}(\mathbf{s}) = \mathbf{t}$. The basic approach involves converting this underdetermined system into a determined one by fixing $n - m$ variables. Subsequently, quadratic system-solving techniques like the Wiedemann XL algorithm [21,58] or Gröbner basis methods such as F4 or F5 [27,28] are applied. Another approach named hybrid approach [9] involves guessing k variables prior to solving the system. The time complexity of this attack, using the approach outlined in [9], is expressed in terms of field multiplications as:

$$\min_{0 \le k \le m} q^k \cdot 3 \cdot \binom{m-k+d}{d}^2 \binom{m-k}{2}$$

Here, k denotes the number of variables fixed during the algorithm, and d represents the smallest integer for which the coefficient of t^d in the series $\frac{(1-t^2)^m}{(1-t)^{m-k}}$ is non-positive.

Example for SL-1 Parameters. Our level one parameter set has 160 variables and 100 constraints. According to [9], we fix 60 variables. Now in the algorithm, if we fix twelve variables, then the value of d is 28. The total complexity is around 2^{280}.

4.2   Simple Attack on VDOO

In 2022, Beullens proposed the *simple attack* against Rainbow [24]. For Rainbow, this highly effective attack reduces n-unknown and m-constraints in the quadratic system to $n - m$-unknown and m-constraints. Now an attacker can apply the same methodology on VDOO to recover the secret key. Recall from Fig. 3, $\mathcal{P}$ is the public polynomial map, and sequences of nested input and output subspaces are,

- **Input subspaces** $\mathbb{F}_q^n \supset D_1 \supset D_2 \supset \cdots \supset D_d \supset O_1 \supset O_2$.
- **Output subspaces** $\mathbb{F}_q^m \supset Q_{1,1} \supset Q_{1,2} \supset \cdots \supset Q_{1,d} \supset Q_2 \supset Q_3 = \{0\}$.

The main crux of the simple attack lies in finding a vector within O_2 (as depicted in Fig. 3). To achieve this, the attacker must solve a quadratic system with $n - m$ unknowns and m constraints using the XL algorithm. This computational step constitutes the most significant component of the entire attack. Here is a step-by-step outline detailing the cryptanalysis of our scheme using the simple attack.

Input: Public polynomial map $\mathcal{P}$.
Output: Recover sequences of subspaces.
Find a vector o $\in O_2$: Choose $\mathbf{v} \in_U \mathbb{F}_q^n$. Then from Fig. 3, $\mathcal{DP}_\mathbf{v} : \mathbb{F}_q^n \to \mathbb{F}_q^m$ is a linear map, in particular it maps O_2 to Q_2. The attacker uses this linear relation to reduce the number of unknowns present in the quadratic system. Therefore, to find a vector, an attacker should solve the following system.

$$\mathcal{DP}_\mathbf{v}(\mathbf{o}) = 0$$
$$\mathcal{P}(\mathbf{o}) = 0$$

With probability $\approx 1/q$, the attacker successfully guesses a vector in O_2. Later, the attacker deploys the XL algorithm to solve the quadratic system of $n - m$-unknowns and m-constraints. Thus attacker recovers $\mathbf{o}$.

Recover Q_2: Attacker will retrieve Q_2 using the information $\mathbf{o} \in O_2$. Note that, $\mathcal{DP}_\mathbf{o} : O_2 \to Q_2$ is a linear map. Therefore,

$$\text{Span } \{ \mathcal{DP}_\mathbf{o}(\mathbf{e}_1), \cdots, \mathcal{DP}_\mathbf{o}(\mathbf{e}_n) \} \subseteq Q_1$$

for some linearly independent vectors $\mathbf{e}_i$. For enough such $\mathbf{e}_i$'s equality will hold.

Recover O_2: To recover O_2, solve the following system of linear equations. Because with high probability kernel of $\mathcal{DP}_\mathbf{o}$ matches with O_2.

$$\mathcal{DP}_\mathbf{o}(\mathbf{e}_1) \equiv 0 \mod Q_2$$
$$\mathcal{DP}_\mathbf{o}(\mathbf{e}_2) \equiv 0 \mod Q_2$$
$$\vdots$$
$$\mathcal{DP}_\mathbf{o}(\mathbf{e}_n) \equiv 0 \mod Q_2$$

Recover a vector $\mathbf{o}' \in O_1$: Now the quadratic system $\mathcal{P}$ reduces to $m' = m - o_2$ equations and $n' = n - o_2$ variables. To recover O_1, the goal of the attacker is to find a vector in $\mathbf{o} \in O_1$. Again attacker will guess a vector $\mathbf{v}' \in \mathbb{F}_q^{n'}$. Like above, a similar argument shows that $\mathcal{DP}_{\mathbf{v}'} : O_1 \to Q_{1,1}$ is a linear map and the attacker tries to solve the following systems mod Q_2.

$$\mathcal{DP}_{\mathbf{v}'}(\mathbf{o}') = 0 \mod Q_2$$
$$\mathcal{P}(\mathbf{o}') = 0 \mod Q_2$$

The attacker runs the XL algorithm to solve the quadratic system of $n' - m'$-unknowns and m'-constraints.

Recover O_1: Attacks follows same approach as recovering O_2 to recover O_1. Here, an attacker solves a system $\mathcal{DP}_{\mathbf{o}'}(\mathbf{e}'_i) \equiv 0 \mod Q_1$ for $i \leq n'$.

Recovering vectors from diagonal layer: The only task that remains is to find all the diagonal vectors. The attacker can apply Wolf et al.'s [59] trick to find all the diagonal vectors in the layer. Here observe that the computation of finding a vector in O_2, dominates the computation of finding a vector in O_1.

Attack Complexity. The complexity of the first steps dominates the complexity of other steps involved in this algorithm. Basically, a system of n variables and m non-linear equations reduces to a system of m homogeneous equations with $n - m$ variables. This computation can be performed via XL algorithm and it requires

$$3 \cdot q \binom{n - m - 1 + d}{d}^2 \binom{n - m - 1}{2}$$

field operations, where d is the operating degree of the algorithm. It means, d-is the smallest positive integer so that the coefficient of t^d in the power series $(1 - t^2)^m / (1 - t)^{n-m}$ is non-positive.

Example for SL-1 Parameters. Apply Beullens' trick to guess a vector in O_2, which happens with probability $1/q$. Finding one vector on O_2 asks to solve a quadratic system of 100-variables 60-unknowns. This computation is the most costly in the entire algorithm. Solving this quadratic system needs 2^{134} field operations. The guessing needs $1/q$ search and cost of one $\mathbb{F}_{16}$- multiplication needs 36 gates. Therefore, this parameter set provides approximately at-least 128-bit security.

4.3 Rectangular Min-Rank Attack on VDOO

Rectangular min-rank attack is proposed by Beullens [10]. We first describe the attack against VDOO and then compute the required attack complexity to perform this attack against VDOO. Attacker starts with $n \times m$-rectangular matrices $M_1, M_2, \cdots, M_n$ over $\mathbb{F}_q$ where each M_i is defined as

$$M_i = \begin{bmatrix} \mathcal{DP}(\mathbf{s}_1, \mathbf{s}_i) \\ \mathcal{DP}(\mathbf{s}_2, \mathbf{s}_i) \\ \vdots \\ \mathcal{DP}(\mathbf{s}_n, \mathbf{s}_i) \end{bmatrix}$$

where $(\mathbf{s}_i)_{i=1}^n$ is a basis of $\mathbb{F}_q^n$.

Let $\mathbf{o}_2 \in \mathbb{F}_q^n$. The bi-linearity of $\mathcal{DP}$ implies

$$M := \sum_{i=1}^n o_{2i} M_i := \begin{bmatrix} \mathcal{DP}(\mathbf{s}_1, \mathbf{o}_2) \\ \mathcal{DP}(\mathbf{s}_2, \mathbf{o}_2) \\ \vdots \\ \mathcal{DP}(\mathbf{s}_n, \mathbf{o}_2) \end{bmatrix}.$$

Hence, the maximum rank of M is o_2, since $\mathbf{o}_2 \in O_2$. This observation provides attacker a min-rank instance to find o_{2i}'s in $\mathbb{F}_q$.

To enhance the performance of the simple attack, Beullens combined the rectangular min-rank attack with the simple attack [11]. Like earlier, the attacker fixes $\mathbf{v}$ to get a linear map $\mathcal{DP}_\mathbf{v}$. This helps to find $\mathbf{o}_2 \in O_2$ using $\mathcal{DP}_\mathbf{v}(\mathbf{o}_2) = 0$.

This system of linear equations reduces the number of matrices by m in the rectangular min-rank instance. Thus, the basis of $\text{Ker}(\mathcal{DP}_\mathbf{v})$ is $\mathbf{b}_1, \cdots, \mathbf{b}_{n-m}$. Hence, the new min-rank instance has $n - m$ matrices $\widetilde{M_i}$, where

$$\widetilde{M_i} := \sum_{j=1}^n b_{ij} M_j := \begin{bmatrix} \mathcal{DP}(\mathbf{s}_1, \mathbf{b}_i) \\ \mathcal{DP}(\mathbf{s}_2, \mathbf{b}_i) \\ \vdots \\ \mathcal{DP}(\mathbf{s}_n, \mathbf{b}_i) \end{bmatrix}, \quad \text{for } i = 1 \text{ to } n - m.$$

If $\mathbf{y}$ is a solution of the new min-rank problem having $n - m$ matrices then $\mathbf{o}_2 = \sum_{i=1}^{n-m} y_i \mathbf{b}_i$ is a solution of the old min-rank problem. Hence, the attack needs to be repeated approximately q times, until it finds $\mathbf{o}_2 \in \text{Ker}(\mathcal{DP}_\mathbf{x}) \cap O_2 \neq \{0\}$.

Attack Complexity. The number of field multiplications required to perform this attack is

$$3 \cdot q \cdot (n - m - 1)(o_2 + 1) \binom{n}{r}^2 \cdot \binom{n - m + b - 3}{b}^3$$

where b is the operating degree for the algorithm [7].

Example for SL-1 Parameters. The attacker needs to guess a good $\mathcal{DP}_x$. After then the attacker gets a min-rank instance of 60 matrices which has 159 rows and 100 columns and the span of these matrices has a matrix of rank 36. Bardet et al.'s [7] algorithm provides an efficient way to solve this min-rank instance. This computation needs 2^{133}-field operations.

4.4 Kipnis-Shamir Attack on VDOO

The attacker targeting VDOO can employ a technique similar to the one devised by Kipnis and Shamir [40] to retrieve the subspace O_2. This approach effectively aids in the separation of oil and vinegar variables, ultimately leading to the recovery of the private key. The complexity of this attack can be roughly estimated as $O(o_2^4 \cdot q^{n-o_2-1})$ field multiplications. To expedite this assault, the attacker leverages Grover's algorithm, which serves to reduce the complexity to $O(o_2^4 \cdot q^{(n-o_2-1)/2})$.

Example for SL-1 Parameters. Attacker needs to perform approximately 2^{348}-field operations in classical settings and 2^{174}-field operations in quantum computer.

4.5 Intersection Attack on VDOO

Beullens introduced the intersection attack [10], which effectively reduced the claimed security level of the Rainbow signature scheme by approximately 20 bits compared to the original design. In this attack, Beullens improved upon the Rainbow band separation attack [25] using the analysis proposed by Perlner [47]. The intersection attack helps to identify k-vectors simultaneously within the oil-space O_2 by solving a system of quadratic equations for a vector within the intersection $\cap_{i=1}^{k} L_i O_2$, where L_i's are invertible matrices. This attack performs well when the intersection is non-empty, which occurs when $n < \frac{2k-1}{k-1} o_2$. The computational cost of this attack involves solving a quadratic system with $\binom{k+1}{2}^2 o_2 - 2\binom{k}{2}$ equations in $k(no_2) - (2k-1)o_2$ variables.

However, in the case of VDOO where $n \geq 3o_2$, there is no guarantee that the subspace (for more details, see [10]) namely $L_i O_2 \cap L_j O_2$ will exist. Consequently, the attack becomes probabilistic for VDOO and will succeed with a probability of $\frac{1}{q^{(n-3o_2+1)}}$.

Example for SL-1 Parameters. The complexity to break SL-1 parameters, attacker needs 2^{131}-field multiplications.

4.6   Quantum Attacks

The attacker can accelerate certain aspects of the classical attacks using a quantum computer. For MQ- or OV-based schemes the only quantum algorithm that can help in cryptanalysis is Grover's search [37]. This algorithm reduces the search space, thereby reducing the number of field multiplications by a factor of $q^{k/2}$. This specifically does not threaten the post-quantum security of our scheme [19].

4.7   Provable Security: EUF-CMA Security

Our VDOO scheme, similar to UOV, Rainbow, and other UOV-based signature schemes, offers universal unforgeability [24]. Like these other schemes, we incorporate a salt in the signature generation process to demonstrate the EUF-CMA security of our scheme. We have followed the established methodology for this purpose, as seen in prior work such as [12,52]. Here, we have only provided an outline of the proof. The full proof can be done using similar strategies as Mayo [12], QR-UOV [34], PROV [29], etc. Our security proof relies on the well-understood hardness of the UOV problem. We begin by defining the UOV problem and then introduce the VDOO problem.

For security reasons, we recommend that each salt value should be used for no more than one signature. Consequently, we fix the salt length at 16 bytes, assuming up to 2^{64} signature generations within the system [19].

Definition 3 (UOV Problem). *Suppose* $\mathsf{UOV}_{(n,v,o,q)}$ *denotes a family of UOV public polynomial maps where* n *is the number variables,* $v + o$ *is number of equations and* q *is the size of the finite field, and* $\mathsf{MQ}_{(q,n,m)}$ *denotes a family of random quadratic systems with* n *unknowns and* m *constraints over* $\mathbb{F}_q$. *The UOV problem asks to distinguish* $\mathcal{P}$ *from* $\mathsf{UOV}_{(q,n,v,o,)}$ *and* $\mathsf{MQ}_{(q,n,m)}$. *Suppose* $\mathcal{A}_{\mathsf{UOV}}$ *be the adversary solves the distinguishing problem and it has a distinguishing advantage as:*

$$\mathsf{Adv}_{\mathsf{UOV}}(\mathcal{A}_{\mathsf{UOV}}) = \left| Pr[\mathcal{A}_{\mathsf{UOV}}(\mathcal{P}) = 1 \mid \mathcal{P} \in \mathsf{MQ}] - Pr[\mathcal{A}_{\mathsf{UOV}}(\mathcal{P}) - 1 \mid \mathcal{P} \in \mathsf{UOV}] \right|$$

It is widely believed that there is no probabilistic polynomial-time adversary, including quantum adversaries, denoted as $\mathcal{A}$, that can efficiently solve the UOV problem.

Definition 4 (VDOO Problem). *Suppose* VDOO *be a family of VDOO public polynomial map. Now given a random* $\mathcal{P} \in \mathsf{VDOO}$ *and* $\mathbf{t} \in \mathbb{F}_q^m$ *VDOO problem asks to find* $\mathbf{s}$ *such that* $\mathcal{P}(\mathbf{s}) = \mathbf{t}$. *If* $\mathcal{A}$ *is such an adversary to compute the inverse of the VDOO public map then the advantage of this computation is*

$$\mathsf{Adv}_{\mathsf{VDOO}}(\mathcal{A}_{\mathsf{VDOO}}) = Pr\left[\mathcal{P}(\mathbf{s}) = \mathbf{t} \mid \mathcal{P} \in \mathsf{VDOO}, \mathcal{A}_{\mathsf{VDOO}}(\mathcal{P}, \mathbf{t}) = \mathbf{s} \right]$$

Now we are going to state our main theorem which establishes the EUF-CMA security of the VDOO. To understand the security notion, we refer to [12,42,52].

Theorem 1. *Suppose the adversary $\mathcal{A}$ runs in time T to solve the EUF-CMA game of VDOO in the random oracle model. This adversary makes q_s signing queries and q_h random oracle queries. Then there exists $\mathcal{A}_{UOV}$ and $\mathcal{A}_{VDOO}$ running in time $T + O((q_s + q_t) \cdot poly(q, v, d, o_1, o_2))$ with*

$$\mathsf{Adv}_{\mathsf{VDOO}}^{\mathrm{EUF\text{-}CMA}}(\mathcal{A}) \leq \mathsf{Adv}_{\mathsf{UOV}_{(q,v',o')}}(\mathcal{A}_{UOV}) + q_h \cdot \mathsf{Adv}_{\mathsf{VDOO}_{(q,v,d,o_1,o_2)}}(\mathcal{A}_{VDOO})$$
$$+ (q_s + q_h)q_s \cdot 2^{-|salt|} + q^{-m}.$$

Proof Idea. Here, we informally sketch the proof. We can adopt the proof methodology used in Mayo (see theorem 6 from [12]). In the first step, we can establish a reduction from the EUF-CMA security of the VDOO signature scheme to EUF-KOA (Existential unforgeability against key-only attack) security by simulating the signing oracle. Note that, the adversary does not have access to the signing oracle in the EUF-KOA game. Once this reduction is established, we can easily show a reduction from the UOV problem and VDOO problem to the EUF-KOA security game in the second step. Like the security proof of Mayo [12], we can use the hybrid proof system to establish both reductions. This proof style has also been adopted by many state-of-the-art OV-based constructions [23,29,34,35]. Finally, we can combine both of these two steps to establish the above theorem.

5   Parameters and Performance

This section describes our chosen parameters based on the security analysis described in Sect. 4. We assess the *practicality* of the VDOO signature scheme, which involves a finely tuned trade-off among computation time, security, and communication costs. For most multivariate schemes, computation time is dominated by either the Gaussian elimination (solving linear system[2]) or the Gröbner basis method (solving quadratic system[3]). Communication cost is proportional to signature size + public key size.

5.1   Parameter Selection

Table 1, shows the signature, public-key, and private-key sizes of VDOO for different security levels as determined by the parameter tuple (q, v, d, o_1, o_2). We follow the NIST classification [19] to categorize the parameters. We consider the complexity of two primary attacks: the simple attack [11] (SA) and the rectangular min-rank attack [10] (RA). From the attacker's point of view, these two attacks exhibit the most optimistic complexity among all other known attacks. Here, the complexity represents the number of field multiplications required for their execution.

---

[2] $\mathsf{GE}_{(q,n)}$: Gaussian elimination on a linear system with n unknowns and n linear equation over $\mathbb{F}_q$. This computation needs $O(n^3)$-field operations.

[3] $\mathsf{XL}_{(q,n)}$: eXtended Linearization or Gröbner basis method to solve a quadratic system of n variables and n constraints over $\mathbb{F}_q$. This computation needs 2^{2^n}-field operations.

Table 1. VDOO parameter set for different NIST prescribed security level

Security level (B)	params $(q,\ v,\ d,\ o_1,\ o_2)$	Signature size (B)	Private key size (KB)	Public key size (KB)	Attacks (SA, RA)
SL-I	(16, 60, 30, 34, 36)	96	243	236	(134, 138)
SL-III	(256, 100, 30, 40, 40)	226	1056	2437	(207, 191)
SL-V	(256, 120, 50, 60, 70)	316	3524	8127	(270, 264)

5.2 Comparison with Other Post-quantum Schemes

In response to the NIST's last [19] and the latest [20] standardization call multiple post-quantum signatures schemes have been proposed based on MQ problem or its derivatives. For our comparative analysis, we focus on schemes with small signature sizes and well-established hardness assumptions only in Table 2. For fairness, we compare with the parameters which provide at least 128-bit of classical security [19]. For details about the parameters of a scheme and their role in security and key sizes we kindly request interested readers to the original publications.

Table 2. Compare with other multivariate signature for security level one (at least 128-bit) [19]

Signature schemes	Computational bottleneck	Signature size (B)	Public key size (KB)
VDOO $(16, 40, 30, 34, 36)$	$GE_{(16,34)} + GE_{(16,36)}$	96	238
Rainbow [24,36] $(256, 148, 80, 48)$	$GE_{(256,32)} + GE_{(256,48)}$	164	258
IPRainbow [17] $(257, 32, 32, 38, 7)$	$GE_{(257,32)} + GE_{(257,38)} + XL_{(257,7)}$	120	342.784
Mayo [12] $(16, 66, 65, 7, 11)$	$GE_{(16,65)}$	387	1
QR-UOV [34,35] $(7, 740, 100, 10)$	$GE_{(7,100)}$	331	20.657
PROV [29] $(256, 136, 46, 8)$	$GE_{(256,46)}$	160	68.326
TUOV [23] $(16, 160, 64, 32)$	$GE_{(16,64)} + GE_{(16,32)}$	80	65.552
VOX [33] $(251, 8, 9, 6, 6)$	$XL_{(251,6)}$	102	9.1
UOV [13] $(256, 160, 64, 16)$	$GE_{(256,64)}$	96	66.576

In Table 3, we compare VDOO with recently standardized Crystals Dilithium [26], Falcon [32], SPHINCS+ [4] and recently submitted some signature schemes (see [20]) which are not based on MQ problem.

Table 3. Comparisons with other signatures for NIST security level 1

Comparisons/ Algorithms	VDOO	Crystals Dilithium	Falcon	Sphincs+	FuLeeca	LESS
Signature size (B)	96	2420	666	7856	1100	8400
Public key size (B)	23813	1312	897	32	1318	13700
Comparisons/ Algorithms	SQISign	Hawk	ASCON-Sign	MIRA	MiRitH	RYDE
Signature size (B)	177	555	7856	7376	7661	7446
Public key size (B)	64	1024	32	84	129	86

From the above tables, it is evident that VDOO outperforms the majority of existing multivariate signature schemes. This superiority stems from the smaller number of variables involved in Gaussian eliminations in VDOO. Furthermore, the signature generation process in VDOO does not rely on the Gröbner basis technique, which further confirms its practicality. Further Table 3 illustrates that VDOO has one of the smallest signature sizes with respect to other quantum-safe signature schemes.

6   Conclusion

We have introduced a post-quantum signature algorithm, leveraging well established cryptanalysis techniques to devise a parameter set for VDOO. In order to ensure a minimum of 128-bit security, our scheme achieves a compact 96-byte signature size, which outperforms numerous existing signature schemes. Nonetheless, it does grapple with a sizable public key size, a challenge that is prevalent in a significant number of multivariate signature schemes.

Our immediate future endeavors will be centered around further compressing the public key size within the VDOO scheme. Additionally, we intend to delve into the exploration of VDOO's security within the quantum random oracle model (QROM). Subsequently, our focus will shift towards realizing hardware implementations and assessing potential physical attacks against our scheme.

Acknowledgements. Authors thanks to anonymous reviewers for their valuable feedback. A.G. wish thanks the Tata Consultancy Service for funding. N.S. thanks the funding support from DST-SERB (CRG/2020/000045) and N.Rama Rao Chair (CSE-IITK).

References

1. Agrawal, M., Saxena, N.: Automorphisms of finite rings and applications to complexity of problems. In: Diekert, V., Durand, B. (eds.) STACS 2005. LNCS, vol. 3404, pp. 1–17. Springer, Heidelberg (2005). https://doi.org/10.1007/978-3-540-31856-9_1

2. Agrawal, M., Saxena, N.: Equivalence of F-algebras and cubic forms. In: Durand, B., Thomas, W. (eds.) STACS 2006. LNCS, vol. 3884, pp. 115–126. Springer, Heidelberg (2006). https://doi.org/10.1007/11672142_8

3. Alagic, G., et al.: Status report on the third round of the NIST post-quantum cryptography standardization process (2022). https://nvlpubs.nist.gov/nistpubs/ir/2022/NIST.IR.8413-upd1.pdf. Accessed 26 June 2023

4. Aumasson, J.P., et al.: SPHINCS+ submission to the NIST post-quantum project, v.3.1 (2018). https://sphincs.org/data/sphincs+-r3.1-specification.pdf. Accessed 10 June 2023

5. Baena, J., Briaud, P., Cabarcas, D., Perlner, R., Smith-Tone, D., Verbel, J.: Improving support-minors rank attacks: applications to GeMSS and Rainbow. In: Dodis, Y., Shrimpton, T. (eds.) CRYPTO 2022. LNCS, vol. 13509, pp. 376–405. Springer, Cham (2022). https://doi.org/10.1007/978-3-031-15982-4_13

6. Bardet, M., et al.: Algebraic attacks for solving the rank decoding and min-rank problems without Gröbner basis (2020). https://arxiv.org/pdf/2002.08322.pdf. 3, 22–30 (2002)

7. Bardet, M., et al.: Improvements of algebraic attacks for solving the rank decoding and MinRank problems. In: Moriai, S., Wang, H. (eds.) ASIACRYPT 2020. LNCS, vol. 12491, pp. 507–536. Springer, Cham (2020). https://doi.org/10.1007/978-3-030-64837-4_17

8. Bernstein, D.J., et al.: Classic McEliece: conservative code-based cryptography. NIST submissions (2017)

9. Bettale, L., Faugere, J.C., Perret, L.: Hybrid approach for solving multivariate systems over finite fields. J. Math. Cryptol. 3(3), 177–197 (2009)

10. Beullens, W.: Improved cryptanalysis of UOV and rainbow. In: Canteaut, A., Standaert, F.-X. (eds.) EUROCRYPT 2021. LNCS, vol. 12696, pp. 348–373. Springer, Cham (2021). https://doi.org/10.1007/978-3-030-77870-5_13

11. Beullens, W.: Breaking rainbow takes a weekend on a laptop. Cryptology ePrint Archive (2022)

12. Beullens, W.: MAYO: practical post-quantum signatures from oil-and-vinegar maps. In: AlTawy, R., Hülsing, A. (eds.) SAC 2021. LNCS, vol. 13203, pp. 355–376. Springer, Cham (2022). https://doi.org/10.1007/978-3-030-99277-4_17

13. Beullens, W., et al.: UOV: unbalanced oil and vinegar algorithm specifications and supporting documentation version 1.0 (2018). https://csrc.nist.gov/csrc/media/Projects/pqc-dig-sig/documents/round-1/spec-files/UOV-spec-web.pdf. Accessed 5 Sept 2023

14. Billet, O., Gilbert, H.: Cryptanalysis of rainbow. In: De Prisco, R., Yung, M. (eds.) SCN 2006. LNCS, vol. 4116, pp. 336–347. Springer, Heidelberg (2006). https://doi.org/10.1007/11832072_23

15. Bos, J., Ducas, L., et al.: CRYSTALS – Kyber: a CCA-secure module-lattice-based KEM. Cryptology ePrint Archive, Report 2017/634 (2017). https://ia.cr/2017/634

16. Buss, J.F., Frandsen, G.S., Shallit, J.O.: The computational complexity of some problems of linear algebra. J. Comput. Syst. Sci. 58(3), 572–596 (1999)

17. Cartor, R., Cartor, M., Lewis, M., Smith-Tone, D.: IPRainbow. In: Cheon, J.H., Johansson, T. (eds.) PQCrypto 2022. LNCS, vol. 13512, pp. 170–184. Springer, Cham (2022). https://doi.org/10.1007/978-3-031-17234-2_9

18. Castryck, W., Decru, T.: An efficient key recovery attack on SIDH. In: Hazay, C., Stam, M. (eds.) EUROCRYPT 2023, Part V. LNCS, vol. 14008, pp. 423–447. Springer, Cham (2023). https://doi.org/10.1007/978-3-031-30589-4_15

19. Chen, L., Moody, D., Liu, Y.: NIST post-quantum cryptography standardization. Transition **800**, 131A (2017)

20. Chen, L., Moody, D., Liu, Y.K.: Post-quantum cryptography: digital signature schemes. round 1 additional signatures. https://csrc.nist.gov/Projects/pqc-dig-sig/round-1-additional-signatures

21. Courtois, N., Klimov, A., Patarin, J., Shamir, A.: Efficient algorithms for solving overdefined systems of multivariate polynomial equations. In: Preneel, B. (ed.) EUROCRYPT 2000. LNCS, vol. 1807, pp. 392–407. Springer, Heidelberg (2000). https://doi.org/10.1007/3-540-45539-6_27

22. De Feo, L., Kohel, D., Leroux, A., Petit, C., Wesolowski, B.: SQISign: compact post-quantum signatures from quaternions and isogenies. In: Moriai, S., Wang, H. (eds.) ASIACRYPT 2020, Part I. LNCS, vol. 12491, pp. 64–93. Springer, Cham (2020). https://doi.org/10.1007/978-3-030-64837-4_3

23. Ding, J.: TUOV: triangular unbalanced oil and vinegar (2023)

24. Ding, J., Schmidt, D.: Rainbow, a new multivariable polynomial signature scheme. In: Ioannidis, J., Keromytis, A., Yung, M. (eds.) ACNS 2005. LNCS, vol. 3531, pp. 164–175. Springer, Heidelberg (2005). https://doi.org/10.1007/11496137_12

25. Ding, J., Yang, B.-Y., Chen, C.-H.O., Chen, M.-S., Cheng, C.-M.: New differential-algebraic attacks and reparametrization of rainbow. In: Bellovin, S.M., Gennaro, R., Keromytis, A., Yung, M. (eds.) ACNS 2008. LNCS, vol. 5037, pp. 242–257. Springer, Heidelberg (2008). https://doi.org/10.1007/978-3-540-68914-0_15

26. Ducas, L., et al.: Crystals-dilithium: a lattice-based digital signature scheme. IACR Trans. Cryptogr. Hardw. Embedd. Syst. **2018**(1), 238–268 (2018). https://doi.org/10.13154/tches.v2018.i1.238-268, https://tches.iacr.org/index.php/TCHES/article/view/839

27. Faugere, J.C.: A new efficient algorithm for computing Gröbner bases (F4). J. Pure Appl. Algebra **139**(1–3), 61–88 (1999)

28. Faugere, J.C.: A new efficient algorithm for computing Gröbner bases without reduction to zero (F5). In: Proceedings of the 2002 International Symposium on Symbolic and Algebraic Computation, pp. 75–83 (2002)

29. Faugere, J.C., Fouque, P.A., Macario-Rat, G., Minaud, B., Patarin, J.: PROV: PRovable unbalanced Oil and Vinegar specification v1. 0–06/01/2023 (2023)

30. Faugère, J.-C., Levy-dit-Vehel, F., Perret, L.: Cryptanalysis of MinRank. In: Wagner, D. (ed.) CRYPTO 2008. LNCS, vol. 5157, pp. 280–296. Springer, Heidelberg (2008). https://doi.org/10.1007/978-3-540-85174-5_16

31. Feo, L.D., Jao, D., Plût, J.: Towards quantum-resistant cryptosystems from super-singular elliptic curve isogenies. J. Math. Cryptol. **8**(3), 209–247 (2014). https://doi.org/10.1515/jmc-2012-0015

32. Fouque, P.A., et al.: Falcon: fast-Fourier lattice-based compact signatures over NTRU (2018), https://falcon-sign.info/. Accessed 10 June 2023

33. France, T.D., et al.: Principal submitter: Jacques patarin (2023)

34. Furue, H., Ikematsu, Y., Hoshino, F., Kiyomura, Y., Saito, T., Takagi, T.: QR-UOV (2023)

35. Furue, H., Ikematsu, Y., Kiyomura, Y., Takagi, T.: A new variant of unbalanced oil and vinegar using quotient ring: QR-UOV. In: Tibouchi, M., Wang, H. (eds.) ASIACRYPT 2021, Part IV. LNCS, vol. 13093, pp. 187–217. Springer, Cham (2021). https://doi.org/10.1007/978-3-030-92068-5_7
36. Groups, G.: Rainbow round3 official comment (2022)
37. Grover, L.K.: A fast quantum mechanical algorithm for database search. In: Proceedings of the Twenty-Eighth Annual ACM Symposium on Theory of Computing, pp. 212–219 (1996)
38. Johnson, D.S., Garey, M.R.: Computers and Intractability: A Guide to the Theory of NP-completeness. WH Freeman (1979)
39. Kipnis, A., Patarin, J., Goubin, L.: Unbalanced oil and vinegar signature schemes. In: Stern, J. (ed.) EUROCRYPT 1999. LNCS, vol. 1592, pp. 206–222. Springer, Heidelberg (1999). https://doi.org/10.1007/3-540-48910-X_15
40. Kipnis, A., Shamir, A.: Cryptanalysis of the oil and vinegar signature scheme. In: Krawczyk, H. (ed.) CRYPTO 1998. LNCS, vol. 1462, pp. 257–266. Springer, Heidelberg (1998). https://doi.org/10.1007/BFb0055733
41. Kipnis, A., Shamir, A.: Cryptanalysis of the HFE public key cryptosystem by relinearization. In: Wiener, M. (ed.) CRYPTO 1999. LNCS, vol. 1666, pp. 19–30. Springer, Heidelberg (1999). https://doi.org/10.1007/3-540-48405-1_2
42. Kosuge, H., Xagawa, K.: Probabilistic hash-and-sign with retry in the quantum random oracle model. Cryptology ePrint Archive (2022)
43. Matsumoto, T., Imai, H.: Public quadratic polynomial-tuples for efficient signature-verification and message-encryption. In: Barstow, D., et al. (eds.) EUROCRYPT 1988. LNCS, vol. 330, pp. 419–453. Springer, Heidelberg (1988). https://doi.org/10.1007/3-540-45961-8_39
44. Miller, V.S.: Use of elliptic curves in cryptography. In: Williams, H.C. (ed.) CRYPTO 1985. LNCS, vol. 218, pp. 417–426. Springer, Heidelberg (1986). https://doi.org/10.1007/3-540-39799-X_31
45. Moh, T.: A public key system with signature and master key functions. Comm. Algebra **27**(5), 2207–2222 (1999)
46. Patarin, J.: The oil and vinegar signature scheme. In: Dagstuhl Workshop on Cryptography September 1997 (1997)
47. Perlner, R., Smith-Tone, D.: Rainbow band separation is better than we thought. Cryptology ePrint Archive (2020)
48. Petzoldt, A., Bulygin, S., Buchmann, J.: CyclicRainbow – a multivariate signature scheme with a partially cyclic public key. In: Gong, G., Gupta, K.C. (eds.) INDOCRYPT 2010. LNCS, vol. 6498, pp. 33–48. Springer, Heidelberg (2010). https://doi.org/10.1007/978-3-642-17401-8_4
49. Petzoldt, A., Bulygin, S., Buchmann, J.: Selecting parameters for the rainbow signature scheme. In: Sendrier, N. (ed.) PQCrypto 2010. LNCS, vol. 6061, pp. 218–240. Springer, Heidelberg (2010). https://doi.org/10.1007/978-3-642-12929-2_16
50. Proos, J., Zalka, C.: Shor's discrete logarithm quantum algorithm for elliptic curves. Quantum Inf. Comput. **3**(4), 317–344 (2003). https://doi.org/10.26421/QIC3.4-3
51. Rivest, R.L., Shamir, A., Adleman, L.: A method for obtaining digital signatures and public-key cryptosystems. Commun. ACM **21**(2), 120–126 (1978)
52. Sakumoto, K., Shirai, T., Hiwatari, H.: On provable security of UOV and HFE signature schemes against chosen-message attack. In: Yang, B.-Y. (ed.) PQCrypto 2011. LNCS, vol. 7071, pp. 68–82. Springer, Heidelberg (2011). https://doi.org/10.1007/978-3-642-25405-5_5

53. Shamir, A.: Efficient signature schemes based on birational permutations. In: Stinson, D.R. (ed.) CRYPTO 1993. LNCS, vol. 773, pp. 1–12. Springer, Heidelberg (1994). https://doi.org/10.1007/3-540-48329-2_1

54. Shor, P.W.: Algorithms for quantum computation: discrete logarithms and factoring. In: Proceedings 35th Annual Symposium on Foundations of Computer Science, pp. 124–134. IEEE (1994)

55. Smith-Tone, D., Perlner, R., et al.: Rainbow band separation is better than we thought (2020)

56. Tao, C., Diene, A., Tang, S., Ding, J.: Simple matrix scheme for encryption. In: Gaborit, P. (ed.) PQCrypto 2013. LNCS, vol. 7932, pp. 231–242. Springer, Heidelberg (2013). https://doi.org/10.1007/978-3-642-38616-9_16

57. Thomae, E.: A generalization of the rainbow band separation attack and its applications to multivariate schemes. Cryptology ePrint Archive (2012)

58. Wiedemann, D.: Solving sparse linear equations over finite fields. IEEE Trans. Inf. Theory **32**(1), 54–62 (1986)

59. Wolf, C., Braeken, A., Preneel, B.: On the security of stepwise triangular systems. Des. Codes Crypt. **40**(3), 285–302 (2006)

60. Yang, B.-Y., Chen, J.-M.: Building secure tame-like multivariate public-key cryptosystems: the new TTS. In: Boyd, C., González Nieto, J.M. (eds.) ACISP 2005. LNCS, vol. 3574, pp. 518–531. Springer, Heidelberg (2005). https://doi.org/10.1007/11506157_43

Secure Boot in Post-Quantum Era
(Invited Paper)

Megha Agrawal[(✉)] , Kumar Duraisamy , Karthikeyan Sabari Ganesan ,
Shivam Gupta , Suyash Kandele , Sai Sandilya Konduru ,
Harika Chowdary Maddipati , K. Raghavendra , Rajeev Anand Sahu ,
and Vishal Saraswat

Bosch Global Software Technologies, Bangalore, India
{megha.agrawal,kumar.d,karthikeyansabari.ganesan,shivam.gupta,
suyash.kandele,fixed-term.konduru.saisandilya,harikachowdary.maddipati,
k.raghavendra3,rajeevanand.sahu,vishal.saraswat}@in.bosch.com

Abstract. Secure boot is a standard feature for ensuring the authentication and integrity of software. For this purpose, secure boot leverages the advantage of Public Key Cryptography (PKC). However, the fast-developing quantum computers have posed serious threats to the existing PKC. The cryptography community is already preparing to thwart the expected quantum attacks. Moreover, the standardization of post-quantum cryptographic algorithms by NIST have advanced to 4th round, after selecting and announcing the post-quantum encryption and signature schemes for standardization. Hence, considering the recent developments, it is high time to realize a smooth transition from conventional PKC to post-quantum PKC. In this paper, we have implemented the PQ algorithms recently selected by NIST for standardization- CRYSTALS-Dilithium, FALCON and SPHINCS$^+$ as candidate schemes in the secure boot process. Furthermore, we have also proposed an idea of double signing the boot stages, for enhanced security, with signing a classical signature by a post-quantum signature. We have also provided efficiency analysis for various combinations of these double signatures.

Keywords: Secure Boot · Post-Quantum Cryptography · Dilithium ·
FALCON · SPHINCS$^+$

1 Introduction

In this digital age, we are surrounded by several electronic devices which help make our lives easier, smarter and more accessible. As in the process, these devices mainly leverage the potential of software associated with them possibly containing sensitive personal data. Hence, the aspect of their security naturally becomes a prime concern. For example, modern vehicles nowadays come with several Electronic Control Units (ECUs) installed in them. For security, integrity and authenticity of software in such units must be ensured. Secure boot is a standard security feature and method for such a purpose.

© The Author(s), under exclusive license to Springer Nature Switzerland AG 2024
A. Chattopadhyay et al. (Eds.): INDOCRYPT 2023, LNCS 14460, pp. 223–239, 2024.
https://doi.org/10.1007/978-3-031-56235-8_11

Secure boot is a foundational first step in modern multi-layered embedded system security. In the process of secure boot, software is verified against integrity and authenticity before execution. In other words, secure boot facilitates the detection of unauthentic or modified software when booting an embedded device. Secure boot reduces an attacker's ability to gain persistence in a device. If the intended software does not conform to what is expected, a defined set of instructions is executed, which may include denying access to a cryptographic key or device, resetting the processor, or running a device rescue or recovery program. The authenticity and data integrity can be verified by using either a Message Authentication Code (MAC) or a public key cryptographic algorithm viz. RSA or ECDSA. The choice of method depends on the design requirements or factors such as boot time. However, both methods have their challenges. The main bottleneck of MAC is key management, which can be solved in the PKC-based approach, however, they are susceptible to upcoming quantum threats. A large-scale practical quantum computer will be able to solve the underlying mathematical problems, e.g. the integer factorization problem, the discrete logarithm problem, or the elliptic-curve discrete logarithm problem of the public key cryptosystems.

Therefore, there is an increasing need to focus on the development of quantum-safe algorithms to ensure a smooth transition from the current state of the art PKC to the quantum-safe ones. The cryptographic community has already started developing quantum-safe algorithms to be prepared in advance for this transition. In fact, in 2016 NIST announced a Post-Quantum Cryptography (PQC) competition to standardize the quantum-secure algorithms. The submissions to the competition fall into several categories including Lattice-based, Code-based, Hash-based, Multivariate and Isogeny-based. For standardization, NIST has currently selected CRYSTALS-Kyber [5] for encryption and 3 candidates– CRYSTALS-Dilithium [10], FALCON [16] and SPHINCS$^+$ [8] for digital signature. The former two signature schemes are lattice-based and the latter one is hash-based (*stateless*). In addition to this, NIST has already standardized and released a document on *stateful* hash-based signatures which are quantum-secure [7]. In this work, we have given a software-based demonstration of the secure boot process using the digital signature schemes recently selected by NIST for PQC standardization– CRYSTALS-Dilithium [10], FALCON [16] and SPHINCS$^+$ [8].

1.1   Our Contribution

We summarize our contributions as follows:

1. We present a detailed efficiency analysis of post-quantum signatures as candidate signature schemes in a quantum-secure boot implementation.
2. We analyze and provide the execution time for certificate verification and signature verification along with the corresponding heap usage for secure boot process comprising post-quantum algorithms.
3. We consider both post-quantum stand-alone certificates as well as mixed certificate schemes for our implementations. In the mixed certificate approach, we use different post-quantum signature schemes at different certificate levels.

4. We also propose a double signature approach in which all the signed images at each stage of the secure boot process, using a classical signature, are signed again using a post-quantum signature scheme.
5. We analyze and present the performance of our double signature approach for certificate and signature verification.

1.2   Organization of the Paper

In Sect. 2, we present the related work on post-quantum secure boot implementations. Section 3 briefly describes the working principle of the signature schemes selected for standardization in the third round of the NIST PQC competition. The process of secure boot is explained in Sect. 4. In Sect. 5, we analyze and present the results of our post-quantum secure boot implementations. Finally, in Sect. 6 we present a brief conclusion of our contributions.

2   Related Work

In this section, we discuss the existing literature on post-quantum certificate generation, and implementation of post-quantum signature schemes in secure boot applications.

In 2022, Paul et al. [15] introduced a mixed certificate chain protocol which involves more than one signature scheme within a single certificate chain. The proposed work comprised two groups: control certificate chains– consisting of single signature scheme certificate chains based on a single signature scheme, and mixed certificate chains– consisting of certificate chains based on multiple signature schemes; both of these chains taken together were called the two-step migration strategy. The control group contains six regular certificate chains with only one (i.e., same) signature scheme at all three levels– root CA, intermediate CA, and end entity alongside the post-quantum KEM Kyber. The signature schemes used here are ECDSA, XMSS [3], SPHINCS+ fast, SPHINCS+ small, CRYSTALS-Dilithium and FALCON. The mixed certificate chains include combinations of hash-based schemes XMSS, SPHINCS+ fast, SPHINCS+ small at root CA, with ECDSA at both intermediate CA and end entity in the first step, followed by combinations of lattice-based PQC schemes– Dilithium and FALCON– at intermediate CA and end entity in the second step, resulting in a total of 15 combinations.

The proposed migration strategy based on mixed certificates in TLS 1.3 is feasible, with promising results. The combination of hash-based signature schemes, particularly XMSS, with conventional ECDSA (XMSS+EDS-Kyb) in the first step, shows fast connection establishment consuming 12.9 milliseconds (ms), where the embedded device acts as a client, with the smallest handshake size (10.1 KB), lowest overhead code size of 209 KB, and minimal memory usage. For the final step, the combination of XMSS+Dil-Kyb is feasible for both client devices even outperforming the control combination EDS-EDH on the embedded device [15].

Further, Marzougui et al. [18] proposed a solution to improve the TLS handshake performance and to minimize the size of the chain of trust by combining XMSS and Dilithium signature schemes in the chain of certificates as a substitute for the pure classical chain. They observed a substantial reduction in signature verification time (two ICA) to 1.9 ms [18] while it took 2.10 ms for pure ECDSA384, 0.12 ms for pure RSA3072, and 4.3 ms for FALCON and SPHINCS+ [17]. The size of the certificate for the server is 4.096 KB for the XMSS chain of trust, which is relatively small compared to those for FALCON and SPHINCS+ which are almost 9.89 KB [17]. One of the advantages of this work is that the hardware accelerators can be used to speed up the signing and verification processes. By using Dilithium in the endpoints, Marzougui et al. [18] minimized the signing time and chain size compared to previous approaches. However, a limitation of their work is a potential overload on the root CA, which needs addressing in future work to optimize the XMSS-based chain of trust and enable a complete handshake protocol.

Kumar et al. [12] presented a hardware-based post-quantum boot solution that uses XMSS for enhanced security and performance. The solution is integrated into a secure System-on-Chip (SoC) platform with RISC-V cores and evaluated on a Field Programmable Gate Array (FPGA). The proposed XMSS verification hardware unit is managed by the Signature Verification Unit (SVU), which includes a four-step process for secure boot and manages the inputs and outputs of this component over AXI interfaces. The process of XMSS verification involves various hash calculations, and is integral to XMSS verification for secure boot. The optimization proposed in the paper aims at parallel computation of WOTS chains, and the multiple instances of WOTS chains can be re-used, if the key and signature generation functions in XMSS are implemented.

In [6] Joppe et al. have shown how to safeguard the boot process in a vehicle network processor from quantum attacks. They integrated the Dilithium signature scheme into the secure boot process of the S32G platform, developed a low-memory Dilithium signature verification algorithm resistant to fault attacks (single targeted faults), and integrated it into the S32G *Hardware Security Engine (HSE)* secure boot flow. They observed that their fault-attack-resistant Dilithium verification code needs an extra stack space of less than 3 KB for all parameter sets. The benchmarking results show that Dilithium's verification for small images is just 5–10 times slower than RSA-4K and just 2–5 times slower than ECDSA-p256, depending on the chosen post-quantum security level. This innovative approach not only bolsters the security of vehicle network processors but also sets a promising precedent for the integration of post-quantum cryptographic techniques in critical systems across various industries.

Kampanakis et al. [11] implemented a secure boot using hash-based signatures. They considered two different types of hash-based signatures– *stateful* and *stateless*. They considered the LMS signature scheme [13] as a stateful scheme and SPHINCS+ as a stateless scheme. Both of the results are compared against the classical scheme RSA. They implemented these schemes on an FPGA board and showed that implementing these schemes on hardware-based applications is not as efficient as the RSA scheme.

In another work [14] Kampanakis et al. studied the viability of hash-based signatures in UEFI secure boot. They evaluated various parameter sets and proved that the post-quantum signature schemes that take less than 1 s for signing and less than 10 ms for verification would not have much impact on secure boot. They also analyzed the impact of LMS and SPHINCS$^+$ algorithms in various use-cases viz. hardware, virtual secure boot and FPGA, and compared them against the classical RSA. It was observed that switching to such signatures is possible with immaterial impact on the verifier, and an acceptable impact on the signer.

A similar study on the impact of variants of hash-based signatures– stateful and stateless– on secure boot is done by Wagner et al. [19]. In their work, they used LMS [13] and XMSS signatures as stateful signature schemes, and SPHINCS$^+$ as stateless signature scheme. These were compared against the classical schemes like RSA and ECDSA. They also provided a flexible hardware-software design which can support both stateful and stateless signature schemes. The selection of the scheme depends on the specification of the hardware.

3   Post-quantum Signature Schemes

In this section, we provide a high-level overview of the post-quantum signature schemes recently selected by NIST for standardization. In particular, we briefly describe CRYSTALS-Dilithium, FALCON and SPHINCS$^+$, and provide their comparisons for different NIST security levels.

3.1   CRYSTALS-Dilithium

CRYSTALS-Dilithium [10] is a lattice-based digital signature scheme and is a member of the CRYSTALS (Cryptographic Suite for Algebraic Lattices) suite of algorithms. This signature scheme is constructed on the "Fiat-Shamir with aborts" framework [9], and has been proven to be secure based on *Module-Learning with Errors (MLWE)* and *Module-Small Integer Solution (MSIS)* problems in the random oracle model. The strength of the CRYSTALS-Dilithium key is represented by the size of its matrix of polynomials. For example, CRYSTALS-Dilithium (6, 5) has a matrix size of 6×5. The larger the matrix size, the stronger the key. The polynomial ring used by Dilithium is $Z_q[x]/(X^n + 1)$, where q is $(2^{23} - 2^{13} + 1)$ and $n = 256$, and the parameters are maintained by simply changing the dimension of the public matrix $\mathbf{A}$, according to the security level. Therefore, the core process of CRYSTALS-Dilithium is the operation to generate the open matrix $\mathbf{A}$ and polynomial multiplication to generate the LWE-based problem.

The key generation process generates a $k \times \ell$ matrix $\mathbf{A}$, each of whose entries is a polynomial in the ring $R_q = Z_q[x]/(X^n + 1)$. Afterwards, it samples random vectors s_1 and s_2, and generates the secret key as $sk = (s_1, s_2)$. Each coefficient of these vectors is an element of R_q with small coefficients of size at most η. Finally, the second component of the public key is computed as $\mathbf{t} := \mathbf{A}s_1 + s_2$. All algebraic operations in this scheme are assumed to be over the polynomial ring R_q.

During the the signing process, a masking vector for polynomial $\mathbf{y}$ is generated and $\mathbf{A}\mathbf{y}$ is calculated. Further a challenge is generated by hashing the message with $\mathbf{w}_1$, which is the high-order bit of $\mathbf{A}\mathbf{y}$, and then using the challenge and secret key, signature vector $\mathbf{z}$ is computed.

For verification, the verifier computes $\mathbf{w}_1'$ and checks if it gives a valid challenge.

3.2 FALCON

FALCON [16] is a lattice-based signature scheme, which is based on solving the Short Integer Solution (SIS) problem [1] over the NTRU lattices. Given $n = 2^k$, $q \in \mathbb{N}^*$ and ring R defined using $\phi(x) = x^n + 1$, the problem consists of determining $f, g, G, F \in R$ such that f is invertible modulus q (this condition is equivalent to requiring that $NTT(f)$ does not contain 0 as a coefficient) and such that the following equation (NTRU equation) is satisfied:

$$fG - gF = q \mod \phi$$

If $h = g \cdot f^{-1} \mod q$, it is possible to verify that the matrices $P = \begin{bmatrix} 1 & h \\ 0 & q \end{bmatrix}$ and $Q = \begin{bmatrix} f & g \\ F & G \end{bmatrix}$ generate the same lattice:

$$\Lambda(P) = zP|z \in R_q = zQ|z \in R_q = \Lambda(Q),$$

But, if f and g are sufficiently small, then h should seem random, so, given h, the hardness of this problem consists of finding f and g. Each coefficient of the polynomials $f = \sum_{n=0}^{n-1} f_i x^i$ and $g = \sum_{n=0}^{n-1} g_i x^i$ is generated from a distribution close to a Gaussian of center 0 and standard deviation $\sigma \in [\sigma_{min}, \sigma_{max}]$ where $\sigma, \sigma_{min}, \sigma_{max}$ are signature parameters. The following general property is fundamental to solving the NTRU equation defined above, in particular, if $f = \sum_{n=0}^{n-1} a_i x^i \in Q[x]$, then f can be decomposed in a unique way as:

$$f(x) = f_0(x^2) + x f_1(x^2)$$

where $f_0 = \sum_{n=0}^{n/2-1} a_{2i} x^i$ and $f_0 = \sum_{n=0}^{n/2-1} a_{2i+1} x^i$. Given f and g, it is easy to obtain (F, G) solution of NTRU equation. Indeed, there is a recursive procedure that uses the previous property and allows to solve an NTRU equation in the ring $\mathbb{Z} = \mathbb{Z}[x]/(x+1)$ and then transforms this solution $(F, G) \in \mathbb{Z} X \mathbb{Z}$ into two polynomials of $\mathbb{Z}[x]/(\phi)$. Thanks to the FFT (Fast Fourier Transform), it is possible to define the matrix $\overline{B} = \begin{bmatrix} FFT(g) & FFT(-f) \\ FFT(G) & FFT(-F) \end{bmatrix}$. Moreover, we also need to consider the LDL decomposition of $\mathcal{G} = B \cdot \overline{B}^T = LDL^T$, where $L =$

$\begin{bmatrix} 1 & 0 \\ \overline{L} & 1 \end{bmatrix}$ and $D = \begin{bmatrix} D_{11} & 0 \\ 0 & D_{22} \end{bmatrix}$. Starting from $\mathcal{G} \in M_{2,2}(\mathbb{Q}[x]/(\phi))$, it is possible to construct the so-called FALCON tree T: the root of T is $\overline{L}$ and its two child nodes $G_0, G_1 \in M_{2,2}(\mathbb{Q}[x]/(x^{n/2} + 1))$ are obtained considering the decomposition of D_{11} and D_{22}. Iterating this procedure on G_0 and G_1, it is possible to obtain the whole tree T, where each leaf $l \in \mathbb{Q}$ is normalized, i.e. $l' = \sigma/l$.

3.3 SPHINCS+

SPHINCS+ [4,8] is a stateless hash-based signature scheme in which already used private keys are not monitored. It is a combination of XMSS^{MT} [3] (the multi-tree variant of XMSS) and Forest of Random Subsets (FORS). The SPHINCS+ hypertree consists of d layers of XMSS trees, followed by a layer of FORS.

It is important to note that SPHINCS+ is a *many-time* signature (MTS) scheme built using FORS, a *few-time* signature (FTS) scheme, where one private key can used to sign multiple messages without compromising the security. The SPHINCS+ signature consists of FORS signature and there is no requirement to update the index (of the next key to be used for the signature generation) in comparison to XMSS. Figure 1 shows an overview of the SPHINCS+ hypertree.

In SPHINCS+, the private key contains a secret seed to generate the WOTS+ secret keys and the FORS secret keys. It also contains another seed value to generate a randomization value for randomized message hashing. The SPHINCS+ public key is the root node of the hypertree.

In the SPHINCS+ signature generation, first of all the signer splits the message into partial messages, and then signs the partial messages using the FORS key pair of an XMSS tree, which is at the lowest height in the hypertree. After that, the public key of the FORS key pair is signed using the hypertree. For signature generation in FORS signature scheme, first the signer converts the message into k a-bit strings, where k is the total number of FORS binary hash trees of height a in the last layer. If the integer-equivalent of an a-bit message is i, then i^{th} private key is chosen as the signature for the message. After all the signature values are chosen, they are concatenated to form a single signature. In other words, the FORS signature contains selected private key elements and associated authentication paths.

In the SPHINCS+ signature verification, the verifier uses the signature and authentication path to compute the FORS public key, and then verifies the signature of hypertree on that public key. Verification will succeed if and only if the computed public key is the same as the known public key.

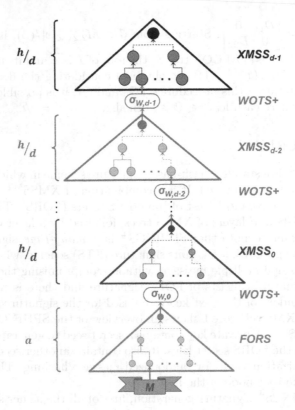

Fig. 1. The SPHINCS$^+$ hypertree consisting of d $XMSS$ trees ((root nodes are •, •, $\cdots$, • and intermediate nodes are •, •, $\cdots$, •), $WOTS+$ signatures ($\bigcirc$), $FORS$ (root node is • and intermediate nodes are •), and message M.

3.4   Comparison of Post-Quantum Signature Schemes

Table 1 shows the comparison of different post-quantum signature schemes selected for standardization by NIST. The comparison is based on the size of the public key, private key and signature for the NIST security levels 1, 3 and 5. In particular, we have considered the variants– FALCON512 (Fal512), FALCON1024 (Fal1024)[1], Dilithium3 (Dil3), Dilithium5 (Dil5)[2], and SPHINCS$^+$-fast (SPf) and SPHINCS$^+$-small (SPs)[3] with SHAKE128, SHAKE192 and SHAKE256. The experiments are performed with the best possible parameters available.

---

[1] https://falcon-sign.info/. Last accessed on October 20, 2023.

[2] https://github.com/pq-crystals/dilithium. Last accessed on October 20, 2023.

[3] https://github.com/Sphincs/Sphincsplus. Last accessed on October 20, 2023.

From Table 1, it can be observed that FALCON512 has the largest public and private key sizes with the smallest signature compared to the other schemes in security level 1. Similarly for security level 3, the public and private key sizes of Dilithium3 are significantly larger than other schemes, but its signature size is much smaller when compared with other hash-based schemes. For the highest security level, i.e., for level 5, Dilithum5 has the largest public and private key sizes when compared with other schemes, and FALCON1024 has the smallest signature size.

Overall, lattice-based schemes have larger public and private keys, but smaller signatures, and hash-based signatures have smaller public and private keys, but larger signatures.

Table 1. Comparison of different post-quantum signature schemes for the NIST security levels 1, 3 and 5.

Security Level	Scheme	Public key size (in bytes)	Private key size (in bytes)	Signature size (in bytes)
Level 1	Fal512	897	1281	690
	SPfshake128	32	64	17088
	SPsshake128	32	64	7856
Level 3	3	1952	4000	3293
	SPfshake192	48	96	35664
	SPsshake192	48	96	16224
Level 5	Dil5	2592	4864	4595
	Fal1024	1793	2305	1330
	SPfshake256	64	128	49856
	SPsshake256	64	128	29792

Table 2 presents the benchmark results, comparing post-quantum signature schemes across keypair generation, signature generation, and signature verification for NIST security levels 1, 3, and 5.

All the experiments were performed on a 64-bit notebook equipped with Intel i7 11th generation processor with 32 GB RAM and x-86 architecture. The benchmarking results of these experiments are obtained by executing the scripts available on open-quantum-safe[4].

4   Secure Boot

Secure boot is defined as a boot sequence in which each software image that is loaded and executed on a device is authorized using software previously authorized by the system [2]. A secure boot process checks the integrity of the installed software, and detects if any segment of the code is altered, by verifying the chain

---

[4] https://github.com/open-quantum-safe/liboqs.

Table 2. Number of keypairs, signature and verification generated per second by different post-quantum signature schemes for the NIST security levels 1, 3 and 5.

Security Level	Scheme	Keypair/s	Sign/s	Verify/s
Level 1	Fal512	203	5654	32299
	SPfshake128	1938	83	1144
	SPsshake128	30	4	3131
Level 3	Dil3	28273	11334	28996
	SPfshake192	1317	51	778
	SPsshake192	20	2	2123
Level 5	Dil5	17712	9176	18214
	Fal1024	67	2821	16740
	SPfshake256	509	25	774
	SPsshake256	30	2	1510

of digital certificates and the corresponding signatures. It is to be noted that the public key is store in the hardware during fabrication, usually in One-Time Programmable (OTP) memory, so the root certificate is immutable.

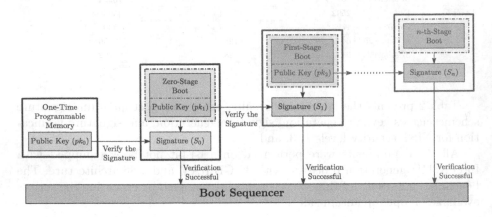

Fig. 2. Secure boot process with $(n + 1)$ Stages.

The functioning of a secure boot process is depicted in Fig. 2. A secure boot process has several stages starting from zero-stage boot, followed by first-stage boot, then second-stage boot, and so on, up to the nth stage. Each stage of the boot has an unprecedented image file of discrete size. Also, it can be observed that before loading the boot image for any stage, first of all, the system verifies the corresponding signature by extracting the public key from the previous stage. For zero-stage boot, the public key is stored in OTP memory, as a part of the certificate. If the signature verification is successful, the boot sequencer continues with the next stage, otherwise, it aborts the boot process.

5 Performance Results

In this section, we present the results for both– secure boot with single signing and with double signing at each of the boot stages.

5.1 Secure Boot with Single Signature

In this section, we analyze and discuss the results of secure boot process considering the post-quantum signature schemes recently selected by NIST for standardization– FALCON512 (Fal512), FALCON1024 (Fal1024), Dilithium3 (Dil3), Dilithium5 (Dil5), and SPHINCS$^+$-fast (SPf) and SPHINCS$^+$-small (SPs) with SHAKE128, SHAKE192 and SHAKE256 variants.

Our results comprise two types of certificate chains:

1. Stand-alone certificate chain– all the certificates i.e. root certificate, intermediate certificate authority (ICA) and leaf certificates, are generated using same (single) signature algorithm.
2. Mixed certificate chain– the certificate chain comprising more than one signature scheme. For example, Falcon for root certification and SPHINCS$^+$ for ICA and leaf certification.

The length of the certificate chain for all the experiments is 3 – root, intermediate and leaf. Our experimental setup and specifications are as follows:

Experimental Setup: All the certificates are generated using OQS-OpenSSL. We considered 3 stages for our secure boot process– zero-stage boot, first-stage boot and second-stage boot. We have used bootloader[5] image for Raspberry Pi of size 52 MB as zero-stage boot, Raspberry Pi OS image[6] of size 364 MB as first-stage boot and application files which are compatible with Raspberry Pi of size 137 MB as second-stage boot.

Hardware Specification: All the experiments are performed on a 64-bit notebook equipped with Intel i7 11th generation processor with 32 GB RAM and x-86 architecture. The operating system is Ubuntu 20.04.

Memory usage is calculated using Valgrind open source tool[7]

---

[5] https://github.com/raspberrypi/firmware/tree/master/boot.
[6] https://www.raspberrypi.com/software/operating-systems/.
[7] https://valgrind.org/.

Table 3 shows the timing of certificate and signature verification for different post-quantum signature algorithms for NIST security levels 1, 3 and 5. Figures 3 and 4 show the certificate and signature verification time, respectively, for the top three schemes for security levels 1 and 3 and the top five schemes for security level 5.

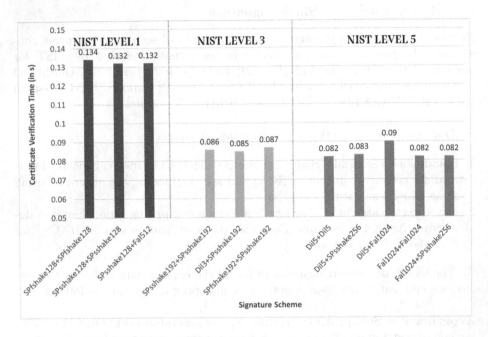

Fig. 3. Certificate Verification Time for single signature.

5.2 Secure Boot with Double Signing

In this section, we discuss the results achieved while performing a double signature– a signature of signature on boot images. In double signing, first the original boot images are signed using a classical signature scheme (RSA or ECDSA), further, this signature of the boot image is concatenated with hash of the boot image and given as an input message to one of the post-quantum signature schemes - Fal512, Fal1024, Dil3, Dil5, SPfshake128, SPfshake192, SPfshake256, SPsshake128, SPsshake192 and SPsshake256. Table 4 and Fig. 5 show the results of double signing boot images. The hardware specification and experimental setup for our analysis were as follows:

Experimental Setup: All the certificates are generated using OQS-OpenSSL. We considered 3 stages for our secure boot process– zero-stage boot, first-stage boot and second-stage boot. We have used the bootloader[8] image for Raspberry

---
[8] https://github.com/raspberrypi/firmware/tree/master/boot.

Table 3. Certificate verification time, signature verification time and heap usage for different stand-alone post-quantum signature schemes and mixed schemes.

NIST Security Level	Scheme		Certificate Verification time	Signature Verification time	Heap usage
	(root)	(ICA, leaf)	(in seconds)	(in seconds)	(in KB)
Level 1	Fal512	Fal512	0.139	1.599	202
	Fal512	SPfshake128	0.150	1.644	202
	Fal512	SPsshake128	0.137	**1.563**	202
	SPfshake128	Fal512	0.144	1.644	202
	SPfshake128	SPfshake128	0.134	1.662	202
	SPfshake128	SPsshake128	0.150	1.707	202
	SPsshake128	Fal512	**0.132**	1.647	202
	SPsshake128	SPfshake128	0.152	1.708	205
	SPsshake128	SPsshake128	**0.132**	1.666	203
Level 3	Dil3	Dil3	0.129	1.566	202
	Dil3	SPfshake192	0.133	1.559	204
	Dil3	SPsshake192	0.085	1.067	202
	SPfshake192	Dil3	0.099	1.083	202
	SPfshake192	SPfshake192	0.109	1.057	203
	SPfshake192	SPsshake192	0.087	1.053	203
	SPsshake192	Dil3	0.099	1.057	202
	SPsshake192	SPfshake192	0.089	**1.037**	203
	SPsshake192	SPsshake192	**0.086**	1.056	203
Level 5	Dil5	Dil5	**0.082**	1.039	202
	Dil5	Fal1024	0.090	1.024	202
	Dil5	SPfshake256	0.093	1.031	202
	Dil5	SPsshake256	0.083	1.031	202
	Fal1024	Dil5	0.133	1.621	202
	Fal1024	Fal1024	**0.082**	1.037	202
	Fal1024	SPfshake256	0.097	1.054	202
	Fal1024	SPsshake256	**0.082**	1.014	202
	SPfshake256	Dil5	0.135	1.597	202
	SPfshake256	Fal1024	0.162	1.610	202
	SPfshake256	SPfshake256	0.174	1.620	203
	SPfshake256	SPsshake256	0.177	1.687	203
	SPsshake256	Dil5	0.141	1.650	202
	SPsshake256	Fal1024	0.149	1.689	202
	SPsshake256	SPfshake256	0.148	1.618	203
	SPsshake256	SPsshake256	0.131	1.610	203

Pi of size 52 MB as zero-stage boot, Raspberry Pi OS image[9] of size 364 MB as first-stage boot and application files which are compatible with Raspberry Pi of size 137 MB as second-stage boot.

Hardware Specification: All the experiments are performed on a 64-bit notebook equipped with Intel i7 11th generation processor with 32 GB RAM and x-86 architecture. The operating system is Ubuntu 20.04.

Memory usage is calculated using Valgrind open source tool[10].

---

[9] https://www.raspberrypi.com/software/operating-systems/.
[10] https://valgrind.org/.

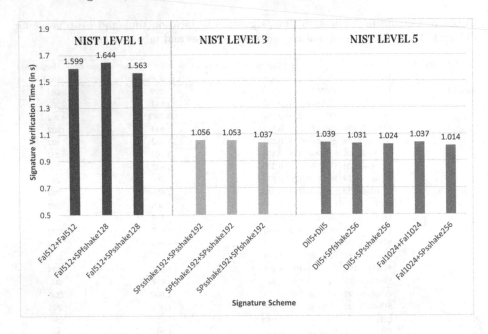

Fig. 4. Signature Verification Time for single signature.

As mentioned above, the double signing process contains two digital signatures– a classical signature scheme which signs the original boot file and a post-quantum signature scheme which is used to sign the previously obtained *classical* signature. These dual signatures provide a robust and multi-layered security mechanism for boot images and reduce the possibility of vulnerabilities due to a stand-alone signature.

Table 4 shows the verification time and heap usage for the double signatures. The results incorporate two types of signature schemes: classical schemes i.e. RSA and ECDSA, and post-quantum schemes Fal512, Fal1024, Dil3, Dil5, SPfshake128, SPfshake192, SPfshake256, SPsshake128, SPsshake192 and SPsshake256. Although in our double signing process we intend to consider combination of classical and post-quantum signatures, however, our results also include combination of two classical signatures for the sake of comparison with the post-quantum variants.

Table 4. Double signature verification times and their heap usage with different stand-alone post-quantum signature schemes, classical schemes.

Security Level	Scheme $Sign_2(Sign_1(bootloader))$		Verification time (in seconds)	Heap usage (in KB)
	$Sign_1(\cdot)$	$Sign_2(\cdot)$		
-	ECDSA	ECDSA	**2.00**	222
	RSA	RSA	2.13	221
Level 1	ECDSA	Fal512	2.10	222
	ECDSA	SPfshake128	2.02	223
	ECDSA	SPsshake128	2.04	223
	RSA	Fal512	**2.00**	222
	RSA	SPfshake128	2.16	223
	RSA	SPsshake128	2.08	223
Level 3	ECDSA	Dil3	**2.03**	221
	ECDSA	SPfshake192	2.05	223
	ECDSA	SPsshake192	2.05	223
	RSA	Dil3	2.08	221
	RSA	SPfshake192	2.12	223
	RSA	SPsshake192	2.05	223
Level 5	ECDSA	Dil5	2.05	222
	ECDSA	Fal1024	2.06	223
	ECDSA	SPfshake256	2.13	223
	ECDSA	SPsshake256	2.06	223
	RSA	Dil5	**2.03**	221
	RSA	Fal1024	2.08	221
	RSA	SPfshake256	2.05	223
	RSA	SPsshake256	2.08	223

From the table it can be observed that verification time for all the hybrid schemes, i.e. combination of classical and post-quantum schemes, are closely align with those of the classical schemes combination. Also, in our single signing and as well as double signing experiments, memory usage remained below 256 KB, well within the acceptable range for standard embedded systems. From this observations we conclude that the post-quantum signature schemes do not compromise efficiency and are compatible with the embedded systems deploying classical signature schemes.

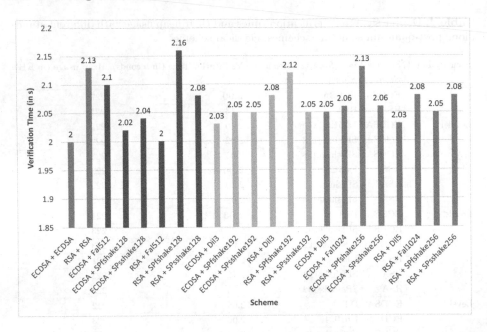

Fig. 5. Double Signature Verification Time.

6   Conclusion

In this paper, we have presented efficiency analysis of the signature schemes recently selected by NIST for post-quantum standardization, as the candidate schemes in secure boot implementation. We have explored several possibilities of considering post-quantum signatures for secure boot. We have examined the efficiency by considering stand-alone (i.e., same) signature for certification at each level of booting, as well as two different post-quantum signature schemes at different levels. Moreover, we have also proposed the idea of double signature to realize quantum-resistant secure boot, comprising a classical and a post-quantum signature for enhanced security. We believe that the results presented in this paper will be helpful in recognizing suitable candidate signature(s) to implement a quantum-resistant secure boot.

References

1. Ajtai, M.: Generating hard instances of lattice problems (Extended Abstract). In: Proceedings of the Twenty-Eighth Annual ACM Symposium on Theory of Computing, STOC '96, pp. 99–108. Association for Computing Machinery, New York, NY, USA (1996)
2. Alexander W. Dent. Secure Boot and Image Authentication (2019). https://www.qualcomm.com/content/dam/qcomm-martech/dm-assets/documents/secure-boot-and-image-authentication-version_final.pdf. Accessed 07 Dec 2023

3. Hülsing, A., Butin, D., Gazdag, S.-L., Rijneveld, J., Mohaisen, A.: XMSS: eXtended merkle signature scheme. RFC 8391 (2018)
4. Bernstein, D.J., Hülsing, A., Kölbl, S., Niederhagen, R., Rijneveld, J., Schwabe, P.: The SPHINCS+ signature framework. In: Proceedings of the 2019 ACM SIGSAC Conference on Computer and Communications Security, pp. 2129–2146 (2019)
5. Bos, J., et al.: CRYSTALS - Kyber: a CCA-secure module-lattice-based KEM. In: 2018 IEEE European Symposium on Security and Privacy (EuroS&P), pp. 353–367 (2018)
6. Bos, J.W., Carlson, B., Renes, J., Rotaru, M., Sprenkels, D., Waters, G.P.: Post-quantum secure boot on vehicle network processors. Cryptology ePrint Archive (2022)
7. Cooper, D.A., Apon, D.C., Dang, Q.H., Davidson, M.S., Dworkin, M.J., Miller, C.A.: Recommendation for stateful hash-based signature schemes. NIST Spec. Publ. **800**, 208 (2020)
8. Bernstein, D.J., et al.: SPHINCS+ Submission to the NIST post-quantum project (2017)
9. Das, D.: Fiat-Shamir with aborts: from identification schemes to linkable ring signatures. In: Batina, L., Picek, S., Mondal, M. (eds.) SPACE 2020. LNCS, vol. 12586, pp. 167–187. Springer, Cham (2020). https://doi.org/10.1007/978-3-030-66626-2_9
10. Ducas, L., et al.: CRYSTALS-Dilithium: a lattice-based digital signature scheme. IACR Trans. Cryptographic Hardw. Embed. Syst. **2018**, 238–268 (2018)
11. Kampanakis, P., Panburana, P., Curcio, M., Shroff, C.: Post-quantum hash-based signatures for secure boot. In: Park, Y., Jadav, D., Austin, T. (eds.) SVCC 2020. CCIS, vol. 1383, pp. 71–86. Springer, Cham (2021). https://doi.org/10.1007/978-3-030-72725-3_5
12. Kumar, V.B., Gupta, N., Chattopadhyay, A., Kasper, M., Krauß, C., Niederhagen, R.: Post-quantum secure boot. In: 2020 Design, Automation & Test in Europe Conference & Exhibition (DATE), pp. 1582–1585 (2020)
13. Leighton, F.T., Micali, S.: Large provably fast and secure digital signature schemes based on secure hash functions. Patent number US5432852 A (1995)
14. Kampanakis, P., Panburana, P., Curcio, M., Shroff, C., Alam, M.M.: Post-quantum LMS and SPHINCS+ Hash-based signatures for UEFI secure boot. IACR Cryptology ePrint Archieve, p. 41 (2021)
15. Paul, S., Kuzovkova, Y., Lahr, N., Niederhagen, R.: Mixed certificate chains for the transition to post-quantum authentication in TLS 1.3. In: Proceedings of the 2022 ACM on Asia Conference on Computer and Communications Security, pp. 727–740 (2022)
16. Fouque, P.A. et al.: Falcon: fast-Fourier lattice-based compact signatures over NTRU (2020)
17. Sikeridis, D., Kampanakis, P., Devetsikiotis, M.: Post-quantum authentication in TLS 1.3: a performance study. Cryptology ePrint Archive (2020)
18. Marzougui, S., Seifert, J.P.: XMSS-based chain of trust. In: Kühne, U., Zhang, F., (ed.) Proceedings of 10th International Workshop on Security Proofs for Embedded Systems, EPiC Series in Computing, vol. 87, pp. 66–82. EasyChair (2022)
19. Wagner, A., Oberhansl, F., Schink, M.: To be, or not to be stateful: post-quantum secure boot using hash-based signatures. In: Proceedings of the 2022 Workshop on Attacks and Solutions in Hardware Security, pp. 85–94 (2022)

Patent Landscape in the field of Hash-Based Post-Quantum Signatures
(Invited Paper)

Megha Agrawal⬥, Kumar Duraisamy⬥, Karthikeyan Sabari Ganesan⬥,
Shivam Gupta⬥, Suyash Kandele(✉)⬥, Sai Sandilya Konduru⬥,
Harika Chowdary Maddipati⬥, K. Raghavendra⬥, Rajeev Anand Sahu⬥,
and Vishal Saraswat⬥

Bosch Global Software Technologies, Bangalore, India
{megha.agrawal,kumar.d,karthikeyansabari.ganesan,shivam.gupta,
suyash.kandele,fixed-term.konduru.saisandilya,harikachowdary.maddipati,
k.raghavendra3,rajeevanand.sahu,vishal.saraswat}@in.bosch.com

Abstract. Post-Quantum Cryptography (PQC) is one of the most fascinating topics of recent developments in cryptography. Following the ongoing standardization process of PQC by NIST, industry and academia both have been engaged in PQC research with great interest. One of the candidate algorithms finalized by NIST for the standardization of post-quantum digital signatures belongs to the family of Hash-based Signatures (HBS). In this paper, we thoroughly explore and analyze the state-of-the-art patents filed in the domain of post-quantum cryptography, with special attention to HBS. We present country-wise statistics of the patents filed on the topics of PQC. Further, we categorize and discuss the patents on HBS based on the special features of their construction and different objectives. This paper will provide scrutinized information and a ready reference in the area of patents on hash-based post-quantum signatures.

Keywords: Post-Quantum Cryptography · Patents · Hash-based signatures · XMSS · LMS · SPHINCS$^+$

1 Introduction

The current *Public Key Infrastructure (PKI)* is based on *Public Key Cryptography (PKC)*. However, the realization of large-scale practical quantum computers will render the conventional *PKC* unsafe, as the underlying mathematical problems like integer factorization and discrete logarithm, on which the security of classical *PKC* relies, can be then solved by those computers using Shor's algorithm [28,29]. Thus, with the expected advent of capable quantum computers, in approximately a decade, the current *PKI* will be devastated, leading to the collapse of the current secure communication.

Though current quantum computers lack the processing power to break a real cryptographic algorithm, continuous advances in realizing real-world quantum

© The Author(s), under exclusive license to Springer Nature Switzerland AG 2024
A. Chattopadhyay et al. (Eds.): INDOCRYPT 2023, LNCS 14460, pp. 240–261, 2024.
https://doi.org/10.1007/978-3-031-56235-8_12

computers have motivated practitioners to get involved in implementing post-quantum algorithms for practical purposes. Altogether the integrated efforts from academia and industries will lead to the construction of a practical large-scale quantum computer very soon. Therefore, there is an urgent need to focus on the development of quantum-safe algorithms to ensure secure communications despite the upcoming quantum threats.

Post-Quantum Cryptography (PQC), also known as *quantum-resistant cryptography*, is a domain of cryptography that deals with the design and analysis of cryptographic algorithms that can be used in classical computers and are assumed to be secure against both quantum and classical adversaries. The current efforts in *PQC* are all about preparing for the era of quantum computing, by updating existing mathematical-based algorithms and standards.

1.1   Current Progress in PQC

Due to several distinctive advantages and applications of quantum computers (such as the study of molecular behavior, drug discovery for treatment of incurable diseases, and forecasting), a huge amount of capital and efforts are being invested to develop these industries by the major players. Moreover, numerous start-up companies are taking shape to build the required hardware and software for quantum computers and post-quantum implementations. In this way, quantum and post-quantum research is extending its reach from academia to industry. With the development of quantum computers and their quick scaling up in the last decade, the *sufficiently* large quantum computers –capable of solving several 'hard' problems– are anticipated to be up and running in another decade, or so, for public use. This calls for immediate actions aligned towards the development and migration to *PQC*.

Foreseeing the requirement of *PQC* algorithms to safeguard sensitive data with larger shelf-life, the cryptography community has already started developing quantum-safe algorithms to be prepared in advance for this transition.

In 2016, the *National Institute of Standards and Technology (NIST)* started a *PQC* competition to standardize the quantum secure algorithms. In this competition, 69 algorithms were submitted and the best shall be chosen until 2024. These submissions fall into five categories: Lattice-based, Code-based, Hash-based, Multivariate-based, and Isogeny-based. At present, the *PQC* competition has advanced to the fourth round with 3 finalists for digital signature schemes – *CRYSTALS-Dilithium*, *FALCON* and *SPHINCS*$^{+}$. Out of these, the *CRYSTALS-Dilithium* and *FALCON* are Lattice-based schemes, whereas *SPHINCS*$^{+}$ is a stateless *HBS* scheme.

Out of these five categories of submission, *HBS* schemes have received special attention due to their simple construction. As a result, significant progress has been made towards standardizing *HBS*. *NIST*, in its special publication, has released a document on stateful *HBS* that are post-quantum secure [6].

1.2   Hash-Based Signatures

The hash-based digital signature schemes use cryptographic hash function(s), and their security completely relies on the *collision resistance* of the underlying hash function(s). In other words, the existence of *collision-resistant* hash function(s), is sufficient to design a quantum-safe digital signature scheme. They generally feature small private and public keys, as well as fast signature generation and verification. However, they are also accompanied by large signatures and relatively slow (and computationally intensive) key generation algorithms. They are suitable for compact implementations (which is beneficial for several applications) and are naturally resistant to most kinds of side-channel attacks.

In 1979, Merkle initiated the study of the *OTS* schemes (that allows the use of a key pair to produce a signature for exactly one message), and the *MTS* schemes (which can be used to produce signature for multiple messages) designed from them [23]. After lying dormant for about a decade, these schemes were further explored in the 1990s and then in the mid-2000s, due to their resistance against quantum-computer-aided attacks. In parallel, Buchmann, Dahmen and Hülsing, have proposed the *eXtended Merkle Signature Scheme (XMSS)* [4], which not only surpassed Merkle's scheme in performance but was also accompanied by modern security proofs in the standard model. We describe this in more detail in Sect. 3.1.

Independently working, Frank Thomson Leighton and Silvio Micali proposed *LMS* in 1995 [20], another hash-based *hierarchical signature scheme (HSS)*, that combines *OTS* schemes with Merkle tree, to design a *MTS*. The security of *LMS* is completely dependent on the security of the underlying cryptographic hash functions.

McGrew, Curcio, and Fluhrer authored an Internet draft describing the *Lamport-Diffie-Winternitz-Merkle (LDWM)* scheme in 2017 [21], and specified the *Leighton-Micali Signature (LMS)* scheme in an RFC, in 2018 [22], that is inspired by the remarkable works by Lamport and Diffie [19], Winternitz [17] and Merkle [23]. *LMS*, discussed further in Sect. 3.2, was based on a fairly different approach (in comparison to *XMSS*) and relies entirely on security arguments in the random oracle model. In 2015, Bernstein *et al.* proposed *SPHINCS*, a stateless *HBS* scheme that is easy to deploy in the current applications [2]. Its successor, *SPHINCS*$^+$ is one of the finalists of digital signature schemes in the current round of *NIST* standardization [1,3] and is elaborated in Sect. 3.3.

The trend of patents being filed in the area of *HBS* schemes is an indicator of the pace of innovation and the interest – of both industry and academia – right at the stage of invention. Empirical studies on real-world patent filing activity can provide valuable evidence to help assess and guide policy proposals related to intellectual property rights (IPRs), innovation and governance of post-quantum proposals.

1.3   Organization of the Paper

Section 2 describes the patent landscape and presents various insights. In Sect. 3, we discuss the three most popular *HBS* schemes – *XMSS*, *LMS* and *SPHINCS*⁺. Section 4 describes the innovations in several patents in the area of the *HBS* scheme. In Sect. 5, we discuss the direction of research, and conclude our paper in Sect. 6.

2   Trend of Filing Patents

In this section, we discuss the trend of patents being filed in the area of *PQC* along with the major contributing countries, which acts as an indicator of the pace of innovation and interest exhibited at the invention stage. We have used various search terms for *PQC*-related areas to explore the existing patent families.

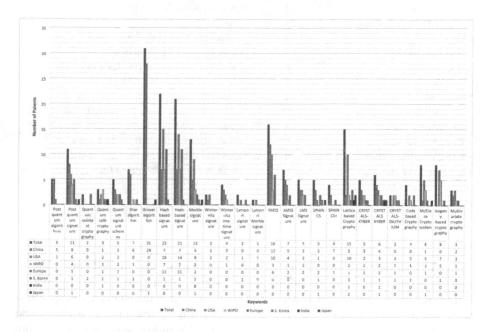

Fig. 1. Patent Landscape: A comparison of number of patents filed by each Country.

In Table 1, we present the search results of the *Quick Search* on *PatBase Express* (https://www.patbase.com), conducted on September 11, 2023. The table enlists the total number of patent families found, along with the top three (or four) contributing countries.

The search terms were placed in double quotes (for instance, "Post-quantum", "Shor algorithm", "Hash-based signature", etc.) to filter out the irrelevant results that consisted of one or more keywords instead of the entire search

Table 1. Statistics on patents filed in various areas of *PQC*.

Search term	Number of results	Countries with highest number of patents
Post-quantum	240	China (121), USA (99), WIPO (47)
Post-quantum algorithms	5	China (5), USA (1), Germany (1)
Post-quantum cryptography	94	USA (38), China (36), WIPO (16)
Post-quantum signature	11	China (8), USA (6), Europe (5)
Quantum-resistant cryptography	2	South Korea (2)
Quantum-safe cryptography	3	WIPO (3), USA (2), China (1), Europe (1)
Quantum-signature schemes	5	China (3), USA (2), Europe (2), WIPO (2)
Shor algorithm	7	China (6), WIPO (1), South Korea (1)
Grover algorithm	31	China (28), Japan (1), Singapore (1), S. Africa (1)
Hash-based signature	21	USA (14), Europe (11), WIPO (7), China (7)
Hash based signature	22	USA (15), Europe (11), WIPO (7), China (7)
Merkle signature	13	USA (9), China (5), WIPO (3)
Winternitz signature	2	USA (2), China (1), Germany(1)
Winternitz one-time signature	4	China (3), USA (2), WIPO (1)
Lamport signature	2	USA (1), South Korea (1)
Lamport-Merkle signature	1	USA (1)
XMSS	16	China (12), USA (10), Europe (6)
XMSS signature	7	China (5), USA (4), Europe (2)
LMS signature	5	China (3), USA (3), WIPO (2), Europe (2)
SPHINCS	5	China (3), Europe (2), USA (1)
SPHINCS$^+$	4	China (2), Europe (1), Spain (1)
Lattice-based cryptography	15	USA (10), South Korea (3), Germany (3)
CRYSTALS-KYBER	5	China (3), USA (2), WIPO (2)
CRYSTALS KYBER	6	China (4), USA (3), WIPO (2), Europe (2)
CRYSTALS-DILITHIUM	2	USA (2), WIPO (2), Europe (1), South Korea (1)
Code-based cryptography	4	South Korea (2), USA (2), WIPO (1), Germany (1)
McEliece cryptosystem	8	USA (5), WIPO (3), China (1)
Isogeny-based cryptography	8	USA (7), WIPO (5), South Korea (1)
Multivariate cryptography	3	USA (3), China (2), WIPO (1), Europe (1)

term. Also, for some search terms, a manual refinement was carried out to obtain precise results. Furthermore, additional care was taken to check for the search terms that may or may not contain the hyphen '-' in various occurrences (such as "Post-quantum algorithms" and "Post quantum algorithms", "Lattice-based cryptography" and "Lattice based cryptography"); unless specified, the results were identical for the search with and without hyphens.

From Table 1, we observe that the research on *PQC* and related areas has already gained substantial momentum, and a significant number of patents have already been filed worldwide. We also conclude that, out of all the search terms, the search result for "Hash-based signature" is notably high. This brings us to the conclusion that the *HBS* schemes have gained a lot of attention as a potential replacement for the state-of-the-art signature schemes in the post-quantum

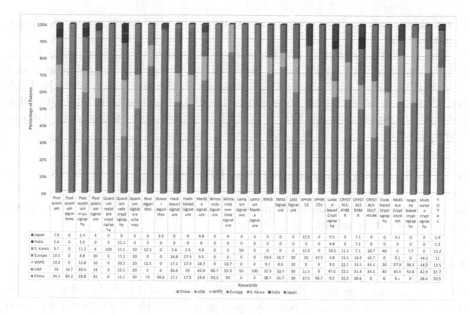

	Post quant um algori thms	Post quant um cryt ograp hy	Post quant um signat ure	Post quant um resist ant signat ure	Quant um resist ant crypt ograp hy	Quant um safe crypt ograp hy	Quant um signat ure sche me	Shor algori thm	Grove r algori thm	Hash based signat ure	Hash-based signat ure	Merkl e signat ure	Winte rnitz one-time signat ure	Winte rnitz signat ure	Lamp ort signat ure	Lamp ort Merkl e signat ure	XMSS Signat ure	XMSS Signat ure	LMS CS	SPHIN CS+	SPHIN CS+	Lattic e based Crypt ograp hy	CRYST ALS-KYBE R	CRYST ALS KYBE R	CRYST ALS-DILIT HIUM	Code based Crypt ograp hy	McEli ece Crypt osyst em	Isoge ny based crypt ograp hy	Multi variat e crypt ograp hy	T O T A L
Japan	2.8	0	2.4	4	0	0	0	0	3.4	0	0	4.8	0	0	0	0	0	0	12.5	0	9.5	0	7.1	0	0	9.1	0	0	1.9	
India	5.4	0	5.6	0	0	11.1	0	0	0	0	0	0	0	0	0	0	0	0	0	0	4.8	0	7.1	0	0	0	0	1.2		
S. Korea	5.7	0	11.2	4	100	11.1	10	12.5	0	2.4	2.5	4.8	0	0	50	0	0	0	12.5	0	14.3	11.1	7.1	16.7	40	0	7.7	0	11.2	
Europe	10.5	0	8.8	20	0	11.1	20	0	26.8	27.5	9.5	0	0	0	19.4	16.7	20	25	33.3	4.8	11.1	14.3	16.7	0	9.1	0	14.3	11		
WIPO	13.3	0	12.8	16	0	33.3	20	12.5	0	17.1	17.5	14.3	0	16.7	0	0	9.7	8.3	20	0	0	9.5	22.7	14.3	33.3	20	27.3	38.5	14.3	13.5
USA	28	16.7	30.4	24	0	22.2	20	0	0	36.6	35	42.9	66.7	33.3	50	100	32.3	33.3	30	12.5	0	47.6	22.2	21.4	33.3	40	45.5	53.8	42.9	31.7
China	34.3	83.3	28.8	32	0	11.1	30	75	96.6	17.1	17.5	23.8	33.3	50	0	0	38.7	41.7	30	37.5	66.7	9.5	33.3	28.6	0	0	9.1	0	28.6	29.5

Keywords

■ China ■ USA ■ WIPO ■ Europe ■ S. Korea ■ India ■ Japan

Fig. 2. Patent Landscape: Share of each Country in the patents filed.

era. Although the advantages are numerous, we believe that the main reason is the reliable security estimates, especially for the security against quantum-computer-aided attacks. This distinguishes *HBS* from other post-quantum signature schemes. Additionally, *HBS* schemes do not require any computationally expensive mathematical operation, such as large integer arithmetic; the only requirement is a secure cryptographic hash function.

Figure 1 presents the visual representation of the number of patents filed in each country/area against the search terms. For the sake of clarity and readability, we have removed the search results for the terms "Post-quantum" and "Post-quantum cryptography" from the comparison. The search term "Post-quantum" has a total of 240 patent families, including those filed in China (121), USA (99), WIPO (47), Europe (37), South Korea (20), India (19) and Japan (10), and the term "Post-quantum cryptography" has a total of 94 patent families including those filed in China (36), USA (38), WIPO(16), Europe (11), South Korea (14), India (7) and Japan (3).

In Fig 2, we have presented the share of contribution of the major countries/areas in *PQC*-related innovation. We observe that China and the USA contribute to the maximum in each of the search terms, except the "Quantum-resistant cryptography". China dominates and shares a majority of the innovation in the areas of "Post-quantum algorithms" (83.3%), "Shor algorithm" (75%), "Grover algorithm" (96.6%), "Winternitz one-time signature" (50%) and "SPHINCS+" (66.7%). The USA dominates and shares a majority in the areas of "Winternitz signature" (66.7%), "Lamport signature" (50%), "Lamport-Merkle signature" (100%) and "Isogeny-based cryptography" (53.8%). South

Korea dominates in "Quantum-resistant cryptography" (100%) and "Lamport signature" (50%).

The overall contribution of each country, as shown in the last bar in Fig. 2, is USA leading with 31.7% share, followed by China with 29.5% share, and then WIPO with 13.5% share. This brings us to the conclusion that the USA and China are investing in huge efforts in all the technologies of *PQC*, with paramount importance on *HBS – XMSS* in particular – followed by lattice-based signatures.

3 Hash-Based Signature Candidates

In this section, we elaborately discuss the three popular hash-based post-quantum signature schemes.

3.1 XMSS

Extended Merkle Signature Scheme (XMSS) is a stateful hash-based digital signature scheme; a stateful scheme keeps track of all the keys that have been already used to generate various signatures. *XMSS* is a magnificient combination of *Winternitz One-Time Signature Plus (WOTS+)* scheme [17] and Merkle tree [24,25]. *XMSS* offers several advantages such as: compact implementations, small key and signature sizes, resists side-channel attacks, and resists Quantum-computer-aided attacks.

The *XMSS* consists of three basic parameters: the Winternitz parameter w, the length n (in bytes) of message digest produced by the hash function, and the height of the Merkle tree h. The Winternitz parameter w is of the order of power of 2, i.e., $w := 2^{2^i}$, where, $i \geq 0$. According to the RFC 8391 documentation [16], the recommended value for w is 16. The height of Merkle tree h is recommended to be 10, 16 or 20, depending on the number of *WOTS+* key required as 2^{10}, 2^{16} and 2^{20}, respectively. The length of the message digest n can be 32, 48 or 64 bytes, depending on the level of security required.

The *XMSS* can provide a maximum of 2^{20} signatures, which, in several applications, can be exhausted much before the lifetime of the product. Therefore, the multi-tree variant of *XMSS*, denoted by $XMSS^{MT}$, based on hypertree, was invented to cater to such a huge number of signatures. In a hypertree, the Merkle tree of height h is divided into d layers; each layer consists of several *XMSS* sub-trees of height h/d, All the sub-trees (except those at the lowest layer) are used to sign the root nodes of the sub-trees on the corresponding layer below them, and the sub-trees on the lowest layer are used to generate the signatures on messages, thereby increasing the total number of *WOTS+* keys.

3.2 LMS

Leighton-Micali Signature (LMS) scheme is a stateful hash-based digital signature scheme designed to provide strong security guarantees while maintaining efficient signature generation and verification processes. Introduced by Frank Thomson Leighton and Silvio Micali in 1995 [20], *LMS* is a tactful combination of *Leighton-Micali One-Time Signature (LM-OTS)* scheme [20] and Merkle tree [24,25], and its security relies on the security of cryptographic hash functions – rather than number-theoretic problems like *RSA* or *Elliptic Curve Cryptography (ECC)*. *LMS* offers several advantages that make it suitable for various use cases in modern cryptographic protocols and systems, such as: compact and efficient implementations, small key and signature sizes, resists side-channel attacks, and resists Quantum-computer-aided attacks.

The *LMS* consists of three basic parameters: the width of the Winternitz coefficients w, the length of message digest produced by hash function n (in bytes), and the height of the Merkle tree h. The width of the Winternitz coefficients w is of the order of 2, i.e., $w := 2^i$, where, $i \geq 0$. According to the RFC 8554 documentation [22], the recommended values for w are 1, 2, 4 and 8. The height of Merkle tree h is recommended to be 5, 10, 15, 20 or 25, depending on the number of *LM-OTS* keys required. The length of message digest n can be 32 bytes.

3.3 SPHINCS+

SPHINCS+ is a stateless hash-based digital signature scheme [1,3]; a stateless scheme does not keep track of the private keys that have already been used, and may choose the index of the next key randomly (sometimes, using the same key multiple times). *SPHINCS+* is an elegant combination of $XMSS^{MT}$ (the multi-tree variant of *XMSS*) and *Forest of Random Subsets (FORS)*. The hypertree of *SPHINCS+* is composed of d layers of *XMSS* trees, followed by a layer of *FORS*.

Here, it is worth noting that the *SPHINCS+* is an *MTS* scheme, built using a few-time signature scheme, which means that one private key can used to sign multiple messages without compromising security (unlike *XMSS* and *LMS* which are stateful schemes based on *OTS* schemes). In *SPHINCS+*, all the messages are signed using the *FORS* signature scheme, similar to the use of the *WOTS+* scheme in *XMSS* [17]. Also, the *SPHINCS+* signature *Sig* consists of *FORS* signature *sig*, identical to the *XMSS* signature *Sig* consisting of *WOTS+* signature *siy*. However, the additional cost of updating the index (of the next key to be used) in the private key SK, is completely avoided in the *SPHINCS+*.

4 Overview of Patents

In this section, we take a closer look at the innovation in the *HBS* schemes. Here, we have considered the stateful schemes – *XMSS* and *LMS* – and the stateless schemes – *SPHINCS* and *SPHINCS+*. We classify the existing patents into the following six categories based on their features, objectives and functionalities:

1. Hardware accelerator
2. GPU-based optimization

3. Platform-dependent optimization
4. Hash function-based optimization
5. Application-based optimization
6. Substitution attack detection

In the following subsections, we briefly discuss the patents with their high-lights.

4.1   Hardware Accelerator

A hardware accelerator is specialized hardware, designed to perform a specific function (such as calculation, wired/wireless communication, rendering images/video, etc.) with a much higher efficiency and accuracy in comparison to when used on a general-purpose machine. In some cases, the hardware accelerators may additionally aim to reduce the power consumption and the turn-around time.

Optimizing the performance of a scheme by designing a hardware accelerator is the most preferred technique, among others, to boost the computation power. Below we discuss the novelty of some patents that aim to design hardware accelerators to increase productivity.

1. **CN113225185 A: Key generation hardware acceleration architecture and method based on hash post-quantum signature** [33].
   In this work, the inventors propose a unique way of designing a hardware accelerator to speed up the process of key generation–the heaviest operation in any *HBS* scheme–for the *LMS* scheme. The hardware accelerator consists of two components: the *Leaf node generation unit* and the *Root node generation unit*.

   The *Leaf node generation unit* is responsible for the computation of *OTS* public key identifier K, which is the hash of the *OTS* public key $y := y[0] \parallel y[1] \parallel \cdots \parallel y[p-1]$, and consists of three modules: the *OTS* public key module, the HASH_LS module and the first HASH_X module. The value of y is computed using the *OTS* public key module, by using p parallel sequences of *Multiplexer-Hash-Demultiplexer* circuits with loops, each computing the value of private key $sk[i]$, and then computing $y[i]$ iteratively (using the loop) and synchronously, to achieve maximum hardware utilization. The HASH_LS module computes K from $y[0], y[1], \cdots, y[p-1]$, and HASH_X finalizes the computation of the leaf node of the Merkle tree.

   The *Root node generation unit* is responsible for the computation of the *LMS* public key PK in the *Depth-first* method. PK is the root node *root* of the Merkle tree formed by the *OTS* public key identifiers–computed by the *Leaf node generation unit*. The unit consists of: the second demultiplexer, the registers, the multiplexer, the second HASH_X module and the third demultiplexer. The second demultiplexer selects whether to store the calculated leaf node in the registers or to send it to the multiplexer. The registers are used to temporarily store the computed leaf nodes and the intermediate values. The multiplexer selects the second input to the HASH_X module as either the generated leaf node or the computed intermediate node. The second HASH_X module computes the nodes of the Merkle tree, and the third demultiplexer chooses to output the generated node as the *root* or to send it back to the multiplexer for further calculations.

2. **CN113922955: A full hardware implementation architecture of an *XMSS* algorithm and system thereof** [37].
   In this work, the inventors provide a full hardware implementation of the *XMSS* algorithm with improved efficiency. The system may consist of: a *service device*, for key and signature generation; or a *terminal device*, for signature verification; or both.
   A *service device* consists of all the sub-modules required for key and signature generation, such as the ADDR sub-module (for generating $WOTS+$ private key sk), WOTS sub-module (for computing $WOTS+$ public key pk or signature sig), L-Tree sub-module (for computing L-tree root node ϕ_{index} from pk), Merkle sub-module (for computing Merkle tree root node PK), BDS sub-module (for computing authentication path $auth$), and Control sub-module.
   A *terminal device* consists of all the sub-modules required for signature verification, such as the WOTS sub-module (for computing $WOTS+$ public key pk from signature sig), L-Tree sub-module (for computing L-tree root node ϕ_{index} from pk), AUTH sub-module (for computing *XMSS* public key PK from ϕ_{index} and $auth$), and Control sub-module.
   One or more Hash modules will be provided as Hash accelerator which will be common to various sub-modules.

3. **EP3910872A1 : Parallel friendly variant of *SPHINCS*$^+$ signature algorithm** [12].
   This work aims to provide a method and a system to implement *SPHINCS*$^+$ that can execute the operations faster. The system uses a novel method for computing the signature on the message M; here, the signature computing engine is configured to execute two functions in parallel. In the first step, both of these functions execute on the message M simultaneously, where the first function generates a randomization parameter R, and the second function computes the hash of the message. Then, in the second step, only the second function is used to generate the signature on the message M by using the randomization parameter R and the hash of the message–that was already computed in the first step.

4. **US2019319782: Combined *SHA2* and *SHA3* based *XMSS* hardware accelerator** [10].
   This work presents an *XMSS* hardware accelerator, consisting of: a computer-readable memory, to store *XMSS* inputs and intermediate results; one or more processors, to implement *XMSS* operations; a chain function logic module; and a set of registers, shared between a unified *SHA-256/SHA-512* computational logic module and a unified *SHAKE128/SHAKE256* computational

logic module. This *SHA-256/SHA-512* computational logic module has a novel unified configurable architecture that enables single 64-bit *SHA-512* operation or two parallel 32-bit *SHA-256* operations, which enables a significant speedup for *XMSS* operations in *SHA-256* mode when compared to the traditional *SHA2* hardware-based *XMSS* accelerator.

5. **US2021306155: Robust state synchronization for stateful Hash-based signatures** [27].
HBS schemes, such as *XMSS* and *LMS*, are stateful, which means that some state (e.g., a monotonic counter) needs to be securely stored between the successive signature generations. If the signer reuses the same counter (which means reusing the same *OTS* key), it exposes the system to *forgery* attacks. To facilitate the robust state synchronization, the signing facility may comprise: a computer-readable memory block (or register tile), which may be used to store signature operation inputs and intermediate results for the signature operations; a state synchronization manager; a load balancer; and multiple instances of *Hardware Security Module (HSM)*, which are configured to compute signatures using a common *XMSS/LMS* key pair. The state synchronization counter sequences are assigned to each of the *HSM*s; the simplest method being the use of modulo operation ($a \bmod n$), where a is the counter and m is the total number of *HSM*s. Another approach is using a shared duplicate state. The state synchronization manager retains the database of encrypted state in the memory; only the *HSM*s can decrypt the state when a signing operation is requested. The load balancer selects one of the available *HSM*s, the state synchronization manager retrieves the encrypted state from memory (based on the requested signing key), and sends the encrypted state to the selected *HSM*.

6. **US2022109558: *XMSS* management to address randomized hashing and federal information processing standards** [31].
The classical signature schemes commonly utilize a 256-bit hash digest for verification of the signature, whereas *XMSS* requires the entire object image, which demands a large local memory for storage. Also, the former requires only a few hundred bytes of memory for a fixed signature, while, *FIPS* requires a 2 KB fixed signature to be stored in the hardware, which consumes significant memory resources. Furthermore, *FIPS* mandates the verification process to be atomic (or uninterrupted), because interrupts may cause corruption of operation/memory.
This work provides a solution for all the above-mentioned issues. To address the first issue of large local storage, the *XMSS* verification circuitry contains a small local memory. The data from large external memory is streamed into the small local memory one block at a time, and then processed and stored back into the local memory. Hence, the local memory stores only the intermediate results during hash computation and the final hash value required for signature generation and verification.

The second problem is solved by storing the required 2 KB fixed signature or test vector in the *ROM*. During the *Power-On Self-Test (POST)* operation, the test vector is loaded into the local memory. If the verification of the test vector is successful, the verification circuitry is accessible to the User, otherwise not. This reduces the hardware gates in the *XMSS* verification circuity by 48K gates (2 K Bytes $= 2 \times 8$ K bits $= 2 \times 8 \times 3$ K gates).

To address the third issue of uninterrupted execution, the *XMSS* verification circuitry disables all the interrupts during the core *XMSS* verification process, and restores them back, once the verification operation is complete.

7. **US2022131708: Efficient hybridization of classical and post-quantum signatures** [9].

The conventional *XMSS* signature generation and verification operations consume significant memory resources while storing the intermediate results computed during the recursive function calls to calculate the *WOTS+* public key *pk* and signature *sig*, respectively, employed in *XMSS*. This may lead to memory allocation issues in the resource-constrained devices.

This work introduces a novel architecture to implement an efficient hybridization of classical and post-quantum signature schemes. The proposed architecture comprises of: an *RSA/ECDSA* processor, and an *XMSS/LMS* processor, coupled to an *RSA/ECDSA* and *XMSS/LMS* signature verification circuit by a suitable communication bus. The *RSA/ECDSA* processor consists of a *SHA-384* processor and an *RSA/ECDSA* verify circuitry. Similarly, the *XMSS/LMS* processor consists of a *SHA-256* processor and an *XMSS/LMS* verify circuitry. The *SHA-384* processor and the *SHA-256* processor may be coupled by a suitable communication bus and may operate cooperatively or independently and in parallel.

The *SHA-384* processor comprises a circuitry to accept the input message *M* from a remote device, and generate a hash digest *d* for the *RSA/ECDSA* verify circuitry, that verifies the *RSA/ECDSA* signature on *d*. The *SHA-256* processor receives a hash digest *d* from *SHA-384* and generates a randomized digest *d'* for the *XMSS/LMS* verify circuitry, that verifies the *XMSS/LMS* signature on *d'*.

4.2   GPU-Based Optimization

A *Graphics Processing Unit (GPU)*, originally designed as a hardware accelerator for the image and video processing, is capable of performing several 'hard' computations (such as multiplication and division) with extremely high efficiency and throughput. With improved technology, modern *GPUs* are capable of performing extreme computations in parallel, and find applications in neural networks and *cryptocurrency mining*, working at the speed of 100 s of Tera-*FLOPS (Floating-point operations per second)*.

GPUs–technicalized for extremely fast hash operations–can be utilized for key generation and quick signature generation (especially for the servers producing thousands of signatures every second). Below we present an overview of patents, that make use of *GPU*s for improving the performance.

1. **CN116015635: A method and system for parallel implementation of *GPU*s using the *XMSS* signature method against quantum attacks** [14].

   This work presents a unique technique to reduce the execution time of the *XMSS* implementation by using parallel *GPU*s. Here, the main idea is to implement a multi-level parallel scheme, where, at the first level, each Merkle node at the same level is computed in parallel, and at the second level, the *WOTS+* signature and L-tree is computed in parallel. The additional operations undertaken are:

   (a) Data transfer between the controller *CPU* and executing *GPU*s,
   (b) The C Language code adapted for *GPU*,
   (c) Hash functions migrated onto *GPU*, and
   (d) Synchronization after completion of computation at each level.

2. **CN116260588: A *GPU* uses a single key to pair data in parallel methods, systems, media and devices** [15].

   This work provides a unique method for *GPU* data parallelism with a single key, by reusing the *XMSS* tree construction process when the system is generating multiple *WOTS+* and *XMSS* signatures, with the help of load balancing of threads for building the *XMSS* tree.

   For multiplexing the *XMSS* tree, it makes multiple copies of the *XMSS* key and each copy is assigned to each parallel hardware, with an updated index. This allows parallel hash computation on the messages, followed by the *WOTS+* signature generation. The load balancing uses the thread count to distribute the load across the parallel hardware units, thus, increasing *GPU* utilization.

   To parallelize the *XMSS* computations, the *XMSS* signature generation is optimized by common calculation, which provides overall parallel execution efficiency. It includes: customizing the data transmission interface between the *CPU* and *GPU*–for improved efficiency; parallelizing and multiplexing the process (with load balancing) of constructing the *XMSS* tree and the (all independent) *WOTS+* signatures; and parallelizing the computation of authentication paths.

   To parallelize the *XMSS* verifications, each thread executes a different signature verification process that does not affect each other, resulting in parallel construction with high throughput.

   Finally, to speed up the communication, the common data for all the signatures is transferred using *synchronous transmission*, and the data specific to each signature is multi-streamed using *asynchronous transmission*, leading to more efficient use of transmission capacity.

4.3 Platform-Dependent Optimization

Several applications are built on a very specific platform (such as *Snapdragon* and *Dimensity*) and would require optimization of all the functions and computations to be executed efficiently on that particular platform. For such applications, it is mandatory to refine and optimize the implementation of the schemes, by utilizing the supported instruction set and resources contained.

In the following discussion, we present the patents that optimize the design of the *HBS* schemes for the target platform(s).

1. **CN114329639: *XMSS* algorithm software and hardware collaborative acceleration computing system based on *ZYNQ*** [18].
   This work provides a solution to the problem of insufficient computing power of general-purpose embedded devices by using a *ZYNQ*-based computation system. The solution comprises three modules: a *hardware acceleration* module; a *software-hardware interaction interface* module; and a *software-end control* module.
   The *hardware acceleration* module consists of 3 acceleration operators: Hash module for computing hash of message; THASH_F module for chaining function in *WOTS+*; and THASH_H module for randomized hash function in Merkle tree and L-tree structures. The *hardware acceleration* module interfaces with the *software-hardware interaction interface* module through *AXI-Stream (AXIS)*. The *software-hardware interaction interface* module has 2 components: *AXIS* interface and *Direct Memory Access (DMA)*. The *DMA* connects with the *software-end control* module through an *AXI* interface. The *software-end control* module includes logical design and input data frame construction for the software-end to send and receive data.

4.4 Hash Function-Based Optimization

The *HBS* schemes involve a substantial number of hash operations. For instance, *XMSS* requires 3 million to 20 million operations during key generation, and up to 10 thousand operations each during signature generation and verification; here, in signature generation, we have not included the cost of computation of the authentication path, which would be roughly around 20 thousand operations per L-tree root node.

For resource-constrained devices and for servers generating thousands of signatures every second, it is practically impossible to execute these operations in a short period. So, it is necessary to optimize the number of hash operations, without altering the design or compromising the security. In this subsection, we elucidate some patent reports that aim to optimize the number of hash function operations, by using various innovative methods.

1. **CN113794558: L-tree calculation method, device and system in *XMSS* algorithm** [36].
   This work focuses on reducing the space required for storing the values during the computation of the L-tree from the *WOTS+* public key *pk*. It uses

a stack data structure $Stack$, that optimally stores the generated nodes along with their heights. A node can be: a public key block $pk[i]$ (that forms the leaf nodes of the L-tree) at height 0; or an intermediate L-tree node (computed using the public key blocks $pk[0]$, $pk[1]$, $\cdots$, $pk[i-1]$) at height $1, 2, \cdots, \lceil \log_2(len) \rceil$. Using $Stack$, the memory requirement reduces from $(len \times n)$ bytes to $((\lceil \log_2(len) \rceil + 1) \times n)$ bytes.

When a node at an even index is generated, it is stored on the top of the $Stack$, and, when a node at an odd index is generated, it is immediately hashed with the previous node (stored in the $Stack$) at the same height to compute the node at successive height. When all the len nodes are processed according to the above steps, the remaining nodes in the $Stack$ are processed by modifying the top node of the $Stack$ by increasing the height by 1 and decreasing the index to half, until the top node and the next node in the $Stack$ is at the same height (to be hashed together to generate their Parent Node).

2. **US2019319797: Accelerators for post-quantum cryptography secure hash-based signing and verification** [32].

This work uses the in-depth knowledge of the design of the SHA-256 function to optimize the operations in the software, hardware and hybrid implementations of $XMSS$ and LMS. It reduces the number of rounds in SHA-256 by: identifying the sections of code that use identical values/carry out identical computations, such as those in the PRF and $Keyed\ Hash$ functions of the $chaining$ and $randomized\ hashing$ functions; pre-computing and re-using the redundant values; and removing the redundant computations from the code. Using this technique, the number of hash operations required for the $XMSS$ signature verification was reduced by 38%. Here, the additional cost of generating the authentication path $auth$ has not been considered during signature generation.

3. **US2019319800: Fast $XMSS$ signature verification and nonce sampling process without signature expansion** [26].

This work presents efficient methods to compute candidate nonce and a desirable message representative, which reduces the overhead of signing operation during nonce sampling. The solution comprises of: identifying the identical 96-byte prefix $OPCODE \parallel SK\_PRF \parallel idx\_sig$; pre-computing the hash on the identical prefix to get the internal $state$; and re-using the $state$ with various values of $nonce$ to compute the final message randomizer r.

This method is useful in applications where the target is to reduce the signature verification cost, even if the signature generation cost is increased. This technique also preserves the size of the signature and cost of communication, by avoiding to send the sampled $nonce$, and improves overall efficiency by reducing the computations. To further improve the efficiency, it implements accelerators for hash-based PQC algorithms, such as a parallel chaining function for each $WOTS+$ public key block $pk[i]$.

4. **US2022123949: Side channel protection for *XMSS* signature function** [8].
   This work provides a method for side-channel protection for an *XMSS* signature function, more particularly for implementing the *SHA-256* operation used in *XMSS*. *SHA-256* operation receives a set of 8 state variables which represent the hash values generated from a message representative. Then, it implements a series of compression operations on the state variables; these operations may be vulnerable to side-channel attacks. To address this, they proposed the implementations of addition operations in a pseudo-random order, thereby inhibiting the ability of side-channel attacks to decipher the signing key.

4.5 Application-Based Optimization

The *HBS* schemes are used for achieving integrity and authentication in various applications. However, due to the diverse specifications and flexible nature of applications, the deployment of signature schemes may not be always identical. Furthermore, in some cases, the functionality of the applications can itself be modified, in an immensely specific manner, to attain maximum benefits of optimization. For this reason, it is worth analyzing the applications and customizing the use of signature schemes, to achieve maximum benefits.

Below we present the novelty of some patents, that optimize the design of the applications and use of *HBS* schemes for reducing the computations.

1. **US2019319796: Low latency post-quantum signature verification for fast secure-boot** [11].
   This work introduces *XMSS* hardware accelerator to implement low-latency post-quantum signature verification for fast secure-boot operations. *XMSS* hardware accelerator may comprise of: a computer-readable memory block (or register file), which may be used to store *XMSS* inputs and intermediate results for the *XMSS* operations; an *XMSS* verification manager; a *WOTS+* (or *WOTS*) public key generation manager logic; an L-tree computation logic; a tree-hash operations logic; and a chain function logic. All the aforementioned components are coupled to a low-latency *SHAKE* hardware engine via an interface. The *SHAKE* hardware engine configures with 256-bit state, 1344-bit input block and 256-bit output.

2. **US2021119799: Post-quantum secure remote attestation for autonomous systems** [7].
   Most of the modern automotive are built upon several heterogeneous computing devices, that may have resource-constrained micro-controllers, with limited memory. In many cases, it is important to check for the up-to-date security footprint of the entire system, and the best way is to verify a single attestation, which would consist of the security footprint of the individual components. However, the individual *ECU*s are, usually, constrained on both - the memory (to store the *XMSS* private key *SK*) and computation power

(for *XMSS* key and signature generation), and therefore, incapable of producing a hash-based Post-Quantum secure signature.

This work presents an approach, where, the gateway device: stores the *XMSS* key SK, having more memory and non-volatile storage; and computes the hash-based Post-Quantum secure signature, possessing a higher capacity for computing millions of hash operations. The resource-constrained *ECU*s generate a symmetric attestation for the gateway device, that requires only a few operations for generation as well as verification. Thus, off-loading the memory-intensive and computationally-intensive tasks to the capable device, the gateway.

3. **US2022100873: Computation of *XMSS* signature with limited runtime storage [35].**

   In this work, the inventors propose to reduce the runtime storage of *XMSS*, by replacing the recursive chaining function with an iterative version of the chaining function, that starts the iteration from the start index i, and goes all the way to the number of steps s, defined in the inputs to the chaining function. For the *XMSS* key generation algorithm, the start index is zero, and the number of steps is $(w-1)$. For the *XMSS* signature generation algorithm, the start index is zero, and the number of steps is the value of $M[i]$ in base w. For the *XMSS* signature verification algorithm, the start index is the value of $M[i]$ in base w, and the number of steps is $(w-1-M[i])$.

   As usual, the chaining function receives as inputs, the input value X, the start index i, the number of steps s, the public seed *seed* and a chaining function-specific address parameter $ADDR$ to store the context variables. Broadly, the operations of a chaining function constitute computing an iteration of the hash function on an n-bit input using a vector of random n-bit bitmask. In each iteration, the bitmask is *XOR*ed with the intermediate result before being processed by the hash function. The final calculation of the hash acts as the output of the method. The *XMSS* signature generation and verification involves 67 executions of the chaining function, each implementing at most 15 recursive function calls.

4. **WO23063957: Stateful hash-based signing with a single public key and multiple independent signers [30].**

   This work proposes a method where each of the multiple independent signers will have a fixed number (or quota) of *OTS* keys, but they will together have a combined *LMS* public key PK. To achieve this, at the first step, the randomization parameter I is either sent to or itself generated by the provisioning server. Then, the provisioning server shares the randomization parameter I, along with the unique starting leaf indices $(0, idx_1, idx_2, \cdots, idx_\ell)$, ranges $(0$ to $idx_1 - 1, idx_1$ to $idx_2 - 1, \cdots, idx_\ell$ to $2^h - 1)$ and heights of sub-trees to each independent $(\ell + 1)$ signers. Each signer generates a random secret $SEED$ to compute *OTS* private keys and then, computes the *OTS* public keys, and using them, computes the root of the sub-tree assigned to them. If a signer is computationally constrained, they supply the masked $SEED$ to the provisioning server, to compute the root of the sub-tree assigned. Once the individual roots are computed by all the signers, the provisioning server collects and orders them, and generates the combined public key PK.

To compute the signature, a signer generates the OTS signature sig, as elaborated by the OTS scheme, and computes a part of the authentication path $auth$. The other signers append their shares to complete the authentication path $auth$, before releasing the full signature Sig.

4.6   Substitution Attack Detection

In a substitution attack, the adversary replaces a genuine cryptographic software with a malicious one, that appears to work identical to the original algorithm. However, the malicious software usually contains backdoors and allows the disclosure of secret information, among other attacks. The substitution attacks pose a great threat, as they leak information and go undetected.

In this subsection, we discuss the summary of patents that aim to detect substitution attacks by various innovative means.

1. **CN116192454 A: State detection method and system for stateful algorithm substitution attack** [13].
   This work describes a method of state detection for replacement attacks on stateful algorithms; the attacks can be easily carried out by storing intermediate states generated during operation, besides other malicious read-write operations. The solution is a 2-step state detection method for a stateful algorithm substitution attack.
   In the first step, for the source codes of the symmetric encryption and digital signature algorithms, the intermediate state detection involves a regular expression matching method, to detect if the library function stores the data on the hard disk or not. In the second step, for the executables of the symmetric encryption and digital signature algorithms, the method involves executing the program in a sandbox virtual environment and recording all the system write-related *API (Application Programming Interface)* calls. If the sensitive calls are detected in the logs, an alert is reported. The sensitive *API*s include, but are not limited to, host user's files, registry files, global hook and device drivers.

2. **CN116318651 A: Time detection method and system for stateless algorithm substitution attack** [34].
   This work describes a method of time detection for replacement attacks on the stateless algorithm. The method is based on receiving a request for a time-based response from a user and consists of three steps. In the first step, the system uses open-source libraries to call symmetric encryption and digital signature algorithms, and then sample its encryption or signature time at different security strengths. These symmetric encryption and digital signature algorithms comprise the standard AES encryption, RSA-PSS signing

and *ECDSA* signing algorithms, and the post-quantum algorithms *Dilithium*, *FALCON* and *SPHINCS+*.

In the second step, the sample data is grouped and the 0,1 conversion is performed to obtain a 0,1 sequence as the next sequence to be detected. Further in the third step, *NIST* SP800-22 specification is used to detect the converted data and determine if the algorithm to be detected is under an algorithm replacement attack. For key detection, out of a total of 15 tests, 4 tests (frequency detection, intra-block frequency number test, overlapping sub-sequence detection, accumulation and detection) are selected from the *NIST* randomness detection test suite. The random detection of any one of the detected processes is considered highly likely to be attacked by an algorithm replacement.

5 Discussion

OTS schemes, such as *Winternitz OTS* [17], and *Lamport-Diffie OTS* [19], are useful in cases when two users wish to exchange a large amount of data only once. When the data needs to be exchanged multiple times, using these schemes becomes infeasible, specifically due to the costs associated with the key generation and certification. This led the path to the invention of the *MTS* schemes, such as the *Merkle signature scheme* [23] in 1979. Although *Merkle signatures* addressed several challenges like combining a few *OTS* keys and effectively computing the authentication path using the Merkle tree, it suffered a few drawbacks such as signature on a limited number of messages and storage of a large private key.

In 2011, Buchmann, Dahmen, and Hülsing, brought a revolution in this area with the proposal of *XMSS*, which was a remarkable improvement over the *Merkle signature scheme* [4], and later standardized as RFC-8391 [16]. *XMSS*, which is now a popular hash-based digital signature scheme, has several advantages like compact implementations, small key and signature sizes, resists side-channel attacks, and resists quantum-computer-aided attacks. Also, it has a minimal security assumption that its security completely relies on the security of the underlying hash function. In case when the *XMSS* private key is completely used or it is compromised much before its *lifetime*, the multi-tree *XMSS*, usually denoted as $XMSS^{MT}$, can prove to be a savior, as it provides several sets of *XMSS* keys which can be used for another *lifetime*(s). However, the substantial number of hash operations makes it less favorable for implementation in resource-constrained devices, which have increased exponentially over the last decade.

Another parallel significant development was the *LMS*, an alternate remarkable improvement over *Merkle signature scheme*, which shares similarities and differences with *XMSS*. Several attempts have been made to optimize the computations required in *XMSS* and *LMS*, the BDS08 algorithm being the most notable one [5], to develop a more efficient and compact implementation. Being a stateful signature scheme, *XMSS* and *LMS* enjoy the advantage of pre-computing the authentication path(s) for the key(s) to be used in the next signature(s), which the stateless signatures cannot (because the index is chosen randomly, and not sequentially).

Many applications prefer the stateful signature scheme over others. However, stateful signatures have their limitations, especially during the synchronization of the state. Several attempts have been made, to convert the stateful signatures into the stateless ones, to inherit as many advantages as possible, *SPHINCS* [2] and *SPHINCS*+ [1,3], being the most popular ones. Furthermore, *XMSS* and *LMS* – being stateful signature schemes – could not qualify to be a candidate for the *NIST PQC* competition. However, *SPHINCS*+ could not only qualify but also become a finalist in the third round of *NIST PQC* competition.

Due to the flexible and simple design, the bunch of advantages it offers, and the standardization and recommendation by *NIST*, the *HBS* schemes have become the hotspot of research and innovation. This claim is further supported by the fact that the keyword "Hash-based signature" has the highest number of patents filed in the survey. Inventors have focused on all the aspects – from high-level structure (such as in the BDS08 algorithm [5]) to the internal rounds of the hash operations (such as in *PQC* accelerators [32]), from platform-dependent implementation (as in the *XMSS* accelerator for *ZYNQ* [18]) to application-based optimizations (as in remote attestation [7]), from *Stack*-based hardware accelerator [37] to hardware with additional dedicated components [27], and from GPU-based accelerators (such as parallel *GPU*s [14]) to resistance against substitution attacks (such as those for stateful signature schemes [13])– to improve the throughput.

6   Conclusion

In this work, we present a landscape of the trend of patents being filed, along with a summary of some of the major innovations in the field of *HBS* schemes. Although *XMSS*, *LMS* and *SPHINCS*+ are computationally expensive, they seem to have emerged as the first choice for quantum-resistant digital signatures. Extrapolating on the current improvements from the earliest design, it is evident that *XMSS*, *LMS* and *SPHINCS*+ would be soon implementable in constrained devices such as *IoT*s and *ECU*s.

References

1. Bernstein, D.J., et al.: SPHINCS+ submission to the NIST post-quantum project (2017)
2. Bernstein, D.J., et al.: SPHINCS: practical stateless hash-based signatures. In: Oswald, E., Fischlin, M. (eds.) EUROCRYPT 2015. LNCS, vol. 9056, pp. 368–397. Springer, Heidelberg (2015). https://doi.org/10.1007/978-3-662-46800-5_15
3. Bernstein, D.J., Hülsing, A., Kölbl, S., Niederhagen, R., Rijneveld, J., Schwabe, P.: The SPHINCS+ signature framework. In: Proceedings of the 2019 ACM SIGSAC Conference on Computer and Communications Security, pp. 2129–2146 (2019)

4. Buchmann, J., Dahmen, E., Hülsing, A.: XMSS - a practical forward secure signature scheme based on minimal security assumptions. In: Yang, B.Y. (ed.) PQCrypto 2011. LNCS, vol. 7071, pp. 117–129. Springer, Heidelberg (2011). https://doi.org/10.1007/978-3-642-25405-5_8

5. Buchmann, J., Dahmen, E., Schneider, M.: Merkle tree traversal revisited. In: Buchmann, J., Ding, J. (eds.) PQCrypto 2008. LNCS, vol. 5299, pp. 63–78. Springer, Heidelberg (2008). https://doi.org/10.1007/978-3-540-88403-3_5

6. Cooper, D.A., Apon, D.C., Dang, Q.H., Davidson, M.S., Dworkin, M.J., Miller, C.A.: Recommendation for stateful hash-based signature schemes. NIST Spec. Publ. **800**, 208 (2020)

7. Ghosh, S., Juliato, M., Sastry, M.: Post-quantum secure remote attestation for autonomous systems. Patent number US20210119799 (2021)

8. Ghosh, S., Sastry, M.: Side channel protection for XMSS signature function. Patent number US20220123949 (2022)

9. Ghosh, S., Sastry, M., Yoon, K.: Efficient hybridization of classical and post-quantum signatures. Patent number US20220131708 (2022)

10. Ghosh, S., et al.: Combined SHA2 and SHA3 based XMSS hardware accelerator. Patent number US20190319782 (2019)

11. Ghosh, S., et al.: Low latency post-quantum signature verification for fast secure-boot. Patent number US20190319796. (2019)

12. Giulia, T., et al.: Parallel friendly variant of SPHINCS+ signature algorithm. Patent number EP3910872 (2021)

13. Jiang, H., Liu, Y., Ma, Z.: State detection method and system for stateful algorithm substitution attack. Patent number CN116192454 A (2023)

14. Chen, H., Dong, X., Kang, Y., Wang, Z.: A method and system for parallel implementation of GPUs using the XMSS signature method against quantum attacks. Patent number CN116015635 (2023)

15. Chen, H., Dong, X., Kang, Y., Wang, Z.: A GPU uses a single key to pair data in parallel methods, systems, media and devices. Patent number CN116260588 (2023)

16. Huelsing, A., Butin, D., Gazdag, S.-L., Rijneveld, J., Mohaisen, A.: XMSS: eXtended Merkle signature scheme. RFC 8391 (2018)

17. Hülsing, A.: W-OTS+ – shorter signatures for hash-based signature schemes. In: Youssef, A., Nitaj, A., Hassanien, A.E. (eds.) AFRICACRYPT 2013. LNCS, vol. 7918, pp. 173–188. Springer, Heidelberg (2013). https://doi.org/10.1007/978-3-642-38553-7_10

18. Chen, J., He, S., Shi, Y., Shi, Z., Wu, J., Zhang, J.: XMSS algorithm software and hardware collaborative acceleration computing system based on ZYNQ. Patent number CN114329639 (2022)

19. Lamport, L.: Constructing digital signatures from a one way function (2016)

20. Leighton, F.T., Micali, S.: Large provably fast and secure digital signature schemes based on secure hash functions. Patent number US5432852 A (1995)

21. McGrew, D., Curcio, M., Fluhrer, S.: Hash-based signatures. Internet-Draft draft-mcgrew-hash-sigs-10, Internet Engineering Task Force. Work in Progress (2018)

22. McGrew, D., Curcio, M., Fluhrer, S.: Leighton-Micali Hash-based signatures. RFC 8554 (2019)

23. Merkle, R.C.: Secrecy, authentication, and public key systems (1976). http://www.ralphmerkle.com/papers/Thesis1979.pdf

24. Merkle, R.C.: A digital signature based on a conventional encryption function. In: Pomerance, C. (ed.) CRYPTO 1987. LNCS, vol. 293, pp. 369–378. Springer, Heidelberg (1988). https://doi.org/10.1007/3-540-48184-2_32

25. Merkle, R.C.: Method of providing digital signatures. Patent number US4309569 A (1982)
26. Misoczki, R., Suresh, V., Wheeler, D., Ghosh, S., Sastry, M.: Fast XMSS signature verification and nonce sampling process without signature expansion. Patent number US20190319800 (2019)
27. Sastry, M., Misoczki, R., Loney, J., Wheeler, D.M.: Robust state synchronization for stateful hash-based signatures. Patent number US2021306155 (2021)
28. Shor, P.W.: Algorithms for quantum computation: discrete logarithms and factoring. In: Proceedings 35th Annual Symposium on Foundations of Computer Science, pp. 124–134 (1994)
29. Shor, P.W.: Polynomial-time algorithms for prime factorization and discrete logarithms on a quantum computer. SIAM J. Comput. **26**(5), 1484–1509 (1997)
30. Stefan, K., Vadimovich, P.A., Jett, R., Vadim, S.: Stateful Hash-based signing with a single public key and multiple independent signers. Patent number WO23063957 A1 (2023)
31. Suresh, V., et al.: XMSS management to address randomized hashing and Federal Information Processing Standards. Patent number US20220109558 (2022)
32. Suresh, V., Mathew, S., Sastry, M., Ghosh, S., Kumar, R., Misoczki, R.: Accelerators for post-quantum cryptography secure hash-based signing and verification. Patent number US2019319797 (2019)
33. Hu, X., Song, Y., Tian, J., Wang, W., Wang, Z.: Key generation hardware acceleration architecture and method based on hash post-quantum signature. Patent number CN113225185 A (2021)
34. Liu, Y., Jiang, H., Ma, Z.: A time detection method and system for stateless algorithm replacement attacks. Patent number CN116318651 A (2023)
35. Yoon, K., Dobles, G.S., Ghosh, S., Sastry, M.: Computation of XMSS signature with limited runtime storage. Patent number US20220100873 (2022)
36. Cao, Y., Chen, S., Zhang, R.: L-tree calculation method, device and system in XMSS algorithm. Patent number CN113794558 (2021)
37. Cao, Y., Chen, S., Zhang, R.: Full-hardware implementation architecture of XMSS algorithm and system thereof. Patent number CN113922955 (2022)

Author Index

Printed in the United States
by Baker & Taylor Publisher Services